Monte Carlo Calculations in Nuclear Medicine
Second Edition
Applications in Diagnostic Imaging

Series in Medical Physics and Biomedical Engineering

Series Editors: John G Webster, Alisa Walz-Flannigan, Slavik Tabakov, and Kwan-Hoong Ng

Series in Medical Physics and Biomedical Engineering

Monte Carlo Calculations in Nuclear Medicine
Second Edition
Applications in Diagnostic Imaging

Michael Ljungberg
Lund University Hospital, Sweden

Sven-Erik Strand
Lund University Hospital, Sweden

Michael A. King
University of Massachusetts Medical School
Worchester, USA

CRC Press is an imprint of the
Taylor & Francis Group, an **informa** business
A TAYLOR & FRANCIS BOOK

CRC Press
Taylor & Francis Group
6000 Broken Sound Parkway NW, Suite 300
Boca Raton, FL 33487-2742

First issued in paperback 2019

ISBN-13: 978-0-4398-4109-9 (hbk)
ISBN-13: 978-0-367-86542-9 (pbk)

Library of Congress Cataloging-in-Publication Data

Monte Carlo calculations in nuclear medicine : applications in diagnostic imaging / [edited by] Michael Ljungberg, Sven-Erik Strand, Michael A. King. -- Second edition.
pages cm. -- (Series in medical physics and biomedical engineering)
Includes bibliographical references and index.
ISBN 978-1-4398-4109-9 (hardback)
1. Monte Carlo method. 2. Diagnostic imaging. I. Ljungberg, Michael, editor of compilation. II. Strand, Sven-Erik, editor of compilation. III. King, Michael A., editor of compilation.

RC78.7.D53M66 2013
616.07'54--dc23

2012028848

Visit the Taylor & Francis Web site at
http://www.taylorandfrancis.com

and the CRC Press Web site at
http://www.crcpress.com

Contents

About the Series

The *Series in Medical Physics and Biomedical Engineering* describes the applications of physical sciences, engineering, and mathematics in medicine and clinical research.

The series seeks (but is not restricted to) publications in the following topics:

- Artificial organs
- Assistive technology
- Bioinformatics
- Bioinstrumentation
- Biomaterials
- Biomechanics
- Biomedical engineering
- Clinical engineering
- Imaging
- Implants
- Medical computing and mathematics
- Medical/surgical devices

- Patient monitoring
- Physiological measurement
- Prosthetics
- Radiation protection, health physics, and dosimetry
- Regulatory issues
- Rehabilitation engineering
- Sports medicine
- Systems physiology
- Telemedicine
- Tissue engineering
- Treatment

The *Series in Medical Physics and Biomedical Engineering* is an international series that meets the need for up-to-date texts in this rapidly developing field. Books in the series range in level from introductory graduate textbooks and practical handbooks to more advanced expositions of current research.

The *Series in Medical Physics and Biomedical Engineering* is the official book series of The International Organization for Medical Physics.

The International Organization for Medical Physics

The International Organization for Medical Physics (IOMP), founded in 1963, is a scientific, educational, and professional organization of 76 national adhering organizations, more than 16,500 individual members, several corporate members, and four international regional organizations.

IOMP is administered by a council, which includes delegates from each of the adhering national organizations. Regular meetings of the council are held electronically as well as every 3 years at the World Congress on Medical Physics and Biomedical Engineering. The president and other officers form the executive committee and there are also committees covering the main areas of activity, including education and training, scientific, professional relations, and publications.

Objectives

- To contribute to the advancement of medical physics in all its aspects
- To organize international cooperation in medical physics, especially in developing countries
- To encourage and advise on the formation of national organizations of medical physics in those countries which lack such organizations

Activities

The official journals of the IOMP are *Physics in Medicine and Biology*, *Medical Physics*, and *Physiological Measurement*. The IOMP publishes a bulletin *Medical Physics World* twice a year, which is distributed to all members.

A World Congress on Medical Physics and Biomedical Engineering is held every 3 years in cooperation with IFMBE through the International Union for Physics and Engineering Sciences in Medicine (IUPESM). A regionally based International Conference on Medical Physics is held between World Congresses. IOMP also sponsors international conferences, workshops, and courses. IOMP representatives contribute to various international committees and working groups.

The IOMP has several programs to assist medical physicists in developing countries. The joint IOMP Library program supports 69 active libraries in 42 developing countries and the Used Equipment Programme coordinates equipment donations. The Travel Assistance Programme provides a limited number of grants to enable physicists to attend the World Congresses.

The IOMP website is being developed to include a scientific database of international standards in medical physics and a virtual education and resource center.

Information on the activities of the IOMP can be found on its website at www.iomp.org.

Preface

Monte Carlo methods have become an important tool for exploring complicated systems and especially for investigation of imaging parameters in nuclear medicine. By using sampling methods based on probability distribution function in combination with methods that simulate the various particle interaction that can occur, a detailed radiation transport can be simulated from the patient into the imaging system. Monte Carlo simulation does not replace experimental measurements but offers a unique possibility to gain understanding in the underlying physics phenomena that form nuclear medicine images. It also provides a substantial help to researchers to develop methods for image improvement. When combining an accurate model of the imaging system and a realistic model of the patient's geometry and activity distribution, the simulated images can be highly clinically realistic and almost undistinguishable from a real patient measurement.

The first edition of this book was published in 1998. It was one of the first books that combined a description of the Monte Carlo methods and principles with relevant Monte Carlo programs and applications in the field of diagnostic nuclear medicine. It is now 14 years since that publication and we therefore felt that it was important to have a second edition since new and very powerful Monte Carlo programs and methods have become available.

This new edition provides the background to, and a summary of, the current Monte Carlo techniques that are in use today. The focus is still on the diagnostic imaging application but several programs that are described in the book also allow for charge-particle simulations applicable to dosimetry-related applications. The physics and technology behind scintillation camera imaging, SPECT/PET systems, and several MC simulation programs are described in detail. We have retained the aim from the first edition of the book, that is, to explain the Monte Carlo method and introduce the reader to some Monte Carlo software packages, developed and used by different research groups, and to give the reader a detailed idea of some possible applications of Monte Carlo in current research in SPECT and PET. Some of the chapters in the first edition of the book have been omitted to allow space for coverage of new programs and topics. Other chapters have been retained but updated. This does not mean that the chapters that we were not been able to include are unimportant; so thus a good suggestion is to have access to both editions.

Some chapters in this book describe in detail the physics and technology behind current simulated imaging detectors and systems in nuclear medicine and molecular imaging. The reason behind this is to let the reader be familiar with the imaging systems and to provide an understanding of the complexity of the systems that Monte Carlo programs are intended to simulate.

The editors are very happy that so many excellent authors were able to contribute to the present book and share their knowledge and experience in the field of Monte Carlo simulation and application. We would therefore like to thank all of them for their hard work and willingness to contribute to this book.

The text is intended to be useful both for education on graduate and undergraduate levels, and as a reference book on the Monte Carlo method in diagnostic nuclear medicine.

Editors

Michael Ljungberg is a professor at the Department of Medical Radiation Physics, Lund University, Sweden. He started his research in the Monte Carlo field in 1983 with a project involving simulation of whole-body counters, but changed focus to more general applications in nuclear medicine imaging and SPECT. As a parallel track to his development of the Monte Carlo code SIMIND, he started working in quantitative SPECT and problems related to attenuation and scatter in 1985. After completing his PhD at Lund University in 1990, he received a research assistant position at the department that allowed him to continue developing SIMIND for quantitative SPECT applications and establish successful collaborations with international research groups. In 1994, he became an associate professor and in 2005 a full professor. His current research also includes an extensive project in oncological nuclear medicine, where he develops dosimetry methods based on quantitative SPECT, Monte Carlo absorbed dose calculations, and registration methods, for accurate 3D dose planning for internal radionuclide therapy. He is also involved in the undergraduate education of medical physicists and supervises several PhD students. He is the deputy director of the Department of Medical Radiation Physics. In 2012, Dr. Ljungberg became a member of the European Association of Nuclear Medicines task group on dosimetry.

Sven-Erik Strand is a professor in medical radiation physics at Lund University. He received his Master of Science degree at the science faculty in 1972, his medical bachelor's degree at the medical faculty in 1981, and his PhD in 1979. At the University of Lund, he was before the director of undergraduate studies of the Department of Medical Radiation and presently he is the director of the Department of Medical Radiation. He has about 200 peer-reviewed publications and book chapters. Strand is a member in the steering group for creating the Lund University BioImaging Center (LBIC) and is now a member in its research board. He has been a member of several task groups in dosimetry and general nuclear medicine for the AAPM, ICRU, and the EANM task group, and recently, he has been a member of the EANM Radionuclide Therapy and Dosimetry Committee 2005–2011. He was awarded the SNM Loevinger–Berman Award for "Excellency in Radiation Dosimetry" in 2002.

Michael A. King received a BA in physics from the State University of New York at Oswego in 1969, and an MS in physics from the State University of New York at Albany in 1972. In 1978, he received his PhD in radiation biology and biophysics from the University of Rochester. From 1977 to 1979, Dr. King was a postdoctoral fellow in medical physics at the University of Alabama Birmingham. In 1979, he joined the Department of Nuclear Medicine at the University of Massachusetts Medical School where he is currently professor of radiology, vice-chairman of radiology for research, and director of medical physics. He is certified by the American Board of Radiology in diagnostic radiological physics and medical nuclear physics. Dr. King is the author of over 160 peer-reviewed publications and the principal investigator of two current National Institutes of Health research grants. In 2006, he received the Ed Hoffman Memorial Award for "Outstanding Scientific Contributions to the Field of Computers and Instrumentation in Nuclear Medicine" from the Society of Nuclear Medicine. His areas of research interest include correction of SPECT studies for attenuation, scatter, and spatial resolution; detection and correction of patient motion; performance of human and numerical observer studies of lesion detection; quantification of activity; image registration; and image segmentation.

Contributors

Didier Benoit received his professional master's degree in optics and laser instrumentation in 2007 from Aix-Marseille University, France. In 2007, he entered the French CNRS (Centre National de la Recherche Scientifique) at CPPM (Centre de Physique de Particule de Marseille) in Marseilles and worked on the modeling, using the GATE Monte Carlo simulation tool, of a CT scanner developed by the imXgam group headed by Christian Morel, and on the optimization of CT system simulations in GATE. In 2009, he was hired as a GATE engineer by the IN2P3/CNRS (National Institute of Nuclear and Particle Physics) in Orsay (Paris Sud University) to assist GATE users, to integrate developments by GATE developers in new release of GATE, to maintain the OpenGATE website, and to manage the mailing list. Since 2010, he is also highly involved in developing iterative reconstruction algorithms for an original preclinical multimodular system of high resolution and high sensitivity dedicated to planar and SPECT imaging. His research is based on using GATE to characterize the imaging performance obtained using different original collimator design, and to optimize the reconstruction strategies.

Irène Buvat received her PhD degree in particle and nuclear physics from Paris Sud University, France in 1992. During her PhD work, she oriented her career toward applications of nuclear physics for medical imaging. She spent 1 year at the University College London, UK, working on single photon emission computed tomography (SPECT) and 2 years at the National Institutes of Health, Bethesda, USA, specializing in positron emission tomography (PET). In 1995, she entered the French Centre National de la Recherche Scientifique (CNRS) and is currently the head of a Quantification in Molecular Imaging team of the Imaging and Modelling in Neurobiology and Cancerology CNRS lab in Orsay, France. Her research activities focus on developing correction and tomographic reconstruction methods in PET and SPECT to improve the accuracy and reduce the variability of measurements made from PET and SPECT images. Her methodological approach is based on the use of Monte Carlo simulations to investigate all details of the forward imaging process so as to identify key aspects to be considered when developing quantification methods. She is currently the spokesperson of the worldwide OpenGATE collaboration developing the GATE Monte Carlo simulation tool dedicated to Monte Carlo simulations in emission and transmission tomography and radiotherapy. Irène Buvat is also largely involved in making quantification in SPECT and PET a clinical reality. She contributed to a number of studies demonstrating the clinical values of quantification to improve image interpretation, and obtained many research contracts with major companies in the field. She has authored and coauthored more than 90 peer-reviewed articles. Irène Buvat teaches medical physics in several French Universities. She is an associate editor of *IEEE Transactions on Medical Imaging* and *IEEE Transactions on Nuclear Science* and serves on the editorial board of the *Journal of Nuclear Medicine* and on the international advisory board of *Physics in Medicine and Biology*.

Claude Comtat graduated in engineering physics at the EPFL in Lausanne, Switzerland in 1989. From 1990 to 1996, he was a teaching assistant at the Institute of High Energy Physics, University of Lausanne, Switzerland, where he earned a PhD in physics in 1996. From 1996 to 1998, he was a post-doctoral fellow at the PET Facility, University of Pittsburgh Medical Center. He is working since 1998 at the Service Hospitalier Frédéric Joliot (SHFJ), French Atomic Energy Commission (CEA). His research interests are in PET instrumentation, in particular statistical image reconstruction algorithms and analytical simulation tools.

Magnus Dahlbom received his BSc degree in physics from the University of Stockholm in 1982 and his PhD in medical physics from the University of California, Los Angeles in 1987. From 1987 to 1989, he was a research scholar at the Karolinska Institute in the Department of Radiation Physics. In 1989, he joined the faculty of the Department of Molecular and Medical Pharmacology at the University of California where he is now a professor. His interests are in nuclear medicine instrumentation, tomographic image reconstruction, and image processing. He has authored and coauthored more than 120 papers that have been published in scientific journals. Dr. Dahlbom is a member of the IEEE, the Society of Nuclear Medicine, and the AAPM.

Yuni K. Dewaraja received her BS in electrical engineering from the University of Western Australia in 1986 after which she worked for 2 years at the Atomic Energy Authority in Colombo, Sri Lanka. She received her MS in nuclear engineering from Kansas State University in 1990 and PhD in nuclear engineering from the University of Michigan in 1994. In 1996, she joined Dr. Kenneth Koral's group in the Division of Nuclear Medicine at the University of Michigan Medical Center where she is now an associate professor (Department of Radiology). Her current research interests include quantitative SPECT and patient-specific dosimetry in internal emitter ther-apy. Dr. Dewaraja is a member of the Society of Nuclear Medicine's Medical Internal Radiation Dose (MIRD) Committee.

Brian F. Elston is a senior computer specialist at the University of Washington. He received a BS degree in computer science from Western Washington University in 1999. After working as a game programmer and software engineer developing electron microscope control systems, he joined the Division of Nuclear Medicine at the University of Washington in 2008. He is currently a member of the physics group in the Division of Nuclear Medicine. His professional interests include research in positron emission tomography (PET), simulation tools and studies, and software coding and algorithmic development. He is currently working on ASIM, an analytic simulator for PET, studies related to delta SUV measurements and assessing their impact on therapy in regard to disease classification as progression or regression, and comparative studies of display characteristics of various common vendor DICOM workstations used in the clinical practice of nuclear medicine.

Kjell Erlandsson was born in Stockholm, Sweden. He received his BSc and PhD (1996) in radiation physics at Lund University, Sweden. Since then, he has worked as a researcher at the Institute of Cancer Research, UK, Columbia University, USA, and the University College London, UK. His main interests are tomographic reconstruction, kinetic modeling, and partial volume correction for PET and SPECT. He is the author and coauthor of more than 60 peer-reviewed articles in the field of medical imaging.

Robert L. Harrison, PhD, is a research scientist at the University of Washington's Imaging Research Laboratory, Seattle. For more than 20 years, he has worked on the development and use of simulation for the study of emission tomography. He leads the development of SimSET (simulation system for emission tomography) and provides development support for the ASIM software (an analytic simulation of PET). With his colleagues at the University of Washington, he has used these simulations to characterize emission tomographs, prototype tomograph design changes, optimize clinical research studies, and design and test new image reconstruction algorithms and data corrections, resulting in important contributions in the areas of simulation efficiency, variance reduction, scatter and randoms correction, time-of-flight tomography, and tomograph optimization. His current research has two main foci: development of improved importance sampling for SimSET and using ASIM to understand how different sources of variability (e.g., coincidence counting noise, biologic variability, calibration error) affect PET measurements of response to therapy.

David R. Haynor, MD, is a neuroradiologist and professor of radiology at the University of Washington. His research interests range from GPU computing and image processing to bioinformatics.

Sébastien Jan graduated in fundamental physics in 1998, and obtained his PhD in nuclear physics from Joseph Fourier University, Grenoble, France in 2002. He joined the French Atomic Commission (CEA) in 2002 in a postdoctoral position, where he worked in the field of preclinical nuclear molecular imaging and Monte Carlo simulation. Since 2004, Dr. Jan has a CEA permanent research position and is leading a group in physics and image processing for molecular imaging. He is the technical coordinator of the GATE Monte Carlo platform since 2003.

Paul Kinahan received his BSc and MSc in engineering physics from the University of British Columbia in 1985 and 1988, and his PhD in bioengineering from the University of Pennsylvania in 1994. He became an assistant professor of radiology and bioengineering at the University of Pittsburgh where he was a member of the group that developed the first PET/CT scanner in 1998. In 1997, he was awarded the IEEE-NPSS Young Investigator Medical Imaging Science Career Award. In 2001, he moved to the University of Washington in Seattle, where he is a professor of radiology, bioengineering, and electrical engineering and also the director of PET/CT physics at the University of Washington Medical Center. His work centers on the development and implementation of reconstruction algorithms for fully 3D PET scanners and the measurement of image quality. His research interests also include multimodality medical imaging, the use of statistical reconstruction methods, scanner optimization, and the use of quantitative analysis in PET oncology imaging. In 2011, he was appointed a fellow of the IEEE. He is a past president of the Society of Nuclear Medicine Computer and Instrumentation Council and the American Board of Science in Nuclear Medicine. He is a member of the Science Council of the American Association of Physicists in Medicine. He was the program chair of the 2002 IEEE NPSS Medical Imaging Conference (MIC), served twice on the Nuclear Medicine Imaging Steering Committee for MIC, and was the chair of the NMISC Awards Committee. He is an associate editor of *IEEE Transactions on Nuclear Science* and a member of the international advisory board for *Physics in Medicine and Biology*.

Kenneth F. Koral, PhD, became research professor emeritus at the University of Michigan, Ann Arbor in 2007 and received the Society of Nuclear Medicine's Loevinger–Berman Award in 2011. He has contributed chapters for three other books (*Nuclear Medical Physics*, Vol. 2, L Williams, ed., CRC Press, Inc., Boca Raton, Florida, 1987; *Therapeutic Applications of Monte Carlo Calculations in Nuclear Medicine*, H Zaidi and G Sgouros, eds., Institute of Physics Publishing, 2003; and *Quantitative Analysis of Nuclear Medicine Imaging*, H Zaidi, ed., Springer Science and Business Media, New York, 2005). He continues to publish with his colleague Dr. Yuni Dewaraja. He receives technical assistance from Charles Schneider and administrative assistance from Linda Brandt. He referees papers in nuclear medicine for various journals. In addition, he has taken up editing similar papers for a private company, TEXT Co. Ltd., which mainly has clients in Japan.

Erik Larsson received his PhD from Lund University in 2011, the topic of his thesis being the development of more realistic dosimetry models for internal dosimetry calculations in nuclear medicine. His research works include small-scale dosimetry models of mouse and rats and small-scale anatomic dosimetry models of tissues with differentiated cell architecture. In these models, the Monte Carlo technique played an extensive part, with the main work performed with the MCNP codes. He now works part-time as a clinical medical physicist with the main objectives being labeling, quality control, administration, and dosimetry of radionuclide therapies, and part-time as a researcher at the Department of Medical Radiation Physics, Lund University, Sweden, where he is involved in the development of new dosimetry models.

Tom K. Lewellen is a professor of radiology and electrical engineering at the University of Washington. He received a BA in physics from Occidental College in 1967 and a PhD in experimental nuclear physics in 1972. After a postdoctoral fellowship designing beam optics for neutron therapy applications, he joined the Division of Nuclear Medicine at the University of Washington (UW) in 1974. Dr. Lewellen is currently the director of the physics group in the Division of Nuclear Medicine. His major research interests are positron emission tomography (PET) system development and improving methods for quantitative imaging (both in PET and in single photon emission tomography). The UW group is currently working on design and construction of new high-resolution animal PET scanners and MRI inserts, improved quantitative data corrections for PET/CT and SPECT/CT systems, faster Monte Carlo simulation software for emission tomographs, and new data analysis techniques for a wide variety of nuclear medicine studies.

Robert S. Miyaoka received a BS degree in general engineering from Harvey Mudd College in 1983. After briefly working for Hughes Aircraft Company, he went on to obtain his MS and PhD degrees in electrical engineering from the University of Washington in 1987 and 1992, respectively. He is currently a research associate professor in the Department of Radiology and an adjunct associate professor in the Department of Electrical Engineering at the University of Washington. He serves as the director of the Small Animal PET Imaging Resource and as the director of SPECT/CT Physics at the University of Washington. He has more than 20 years of experience in nuclear medicine instrumentation and physics research. His research has included time-of-flight PET and dual-head coincidence imaging. He also has developed a series of microcrystal element (MiCE) detectors for high-resolution PET imaging. His recent efforts have focused on PET detector designs that provide depth of interaction positioning and support multimodality imaging. His research interests also include preclinical PET imaging; multimodality PET/MR instrumentation development; and quantitative nuclear medicine and SPECT/CT imaging. Dr. Miyaoka has authored/coauthored more than 100 journal articles and conference proceedings.

Per Munck af Rosenschöld is the head of medical physics research at the Department of Radiation Oncology, Rigshospitalet, Copenhagen, Denmark. He received his MSc and PhD in radiation physics at Lund University,

Sweden in 1999 and 2003, respectively. During this work, he made extensive use of the Monte Carlo simulation code MCNP. He was a certified medical physics expert in Sweden and Denmark in 2009 and 2010, respectively. His research interest lay in radiation therapy with photon, proton/ions, and neutrons, focusing on methods for improving the precision of delivery, and optimizing dose distribution using radiobiological models and based on advanced CT, PET, and MR imaging. Since 2009, he is leading the medical physics and technology research at the Department of Radiation Oncology with a research group consisting of PhD students and postdocs in medical physics, radiation oncology, nuclear medicine radiology, biomedical engineering, and veterinary medicine.

Yoshihito Namito received a BS degree in nuclear engineering from Nagoya University in 1983 and an MS degree in nuclear engineering from the Tokyo Institute of Technology in 1985. He briefly worked for Ship Research Institute and stayed at the Radiation Physics Group of Stanford Linear Accelerator Center as a visiting scientist. He continued research while working at the National Laboratory for High Energy Physics to receive a PhD in engineering from the Tokyo Institute of Technology in 1995. He currently is an associate professor at Radiation Science Center in High Energy Accelerator Research Organization (KEK). He serves as a deputy manager of radiation control room in KEK. He has participated in the improvement of EGS4 code and the development of the EGS5 code for more than 20 years. His research interest also includes photon and electron transport in matter at an energy range of 1 keV to 1 TeV. He has also a long experience on experiment using synchrotron radiation. Dr. Namito has authored/coauthored more than 100 journal articles and conference proceedings.

Tomas Ohlsson received his radiation physics degree in 1985 and his PhD degree in 1996 (on positron emission tomography), both from the University of Lund, Sweden. In 1996, he became a research assistant at the Department of Radiation Physics, Lund. Since 1985, he has been working part time as a medical physicist in nuclear medicine. His interests are positron emission tomography, modeling in nuclear medicine, and dose planning for radionuclide therapy.

Mikael Peterson received his MSc in radiation physics in 2007 from the University of Lund. Since 2010, he has been a part-time PhD student at the Department of Medical Radiation Physics at Lund University, focusing on small-animal SPECT imaging. Since 2008, he has also been a part-time medical physicist at the Department of Radiation Physics, focusing on nuclear medicine—primarily in the area of radionuclide therapy.

W. Paul Segars is an associate professor of radiology and biomedical engineering and a member of the Carl E. Ravin Advanced Imaging Laboratories

(RAILabs) at Duke University, Durham, North Carolina. He received his PhD in biomedical engineering from the University of North Carolina in 2001. Dr. Segars is among the leaders in the development of simulation tools for medical imaging research where he has applied state-of-the-art computer graphics techniques to develop realistic anatomical and physiological models. Foremost among these are the extended 4D NURBS-based Cardiac-Torso (XCAT) phantom, a computational model for the human body, and the 4D Mouse Whole-Body (MOBY) and Rat Whole-Body (ROBY) phantoms, models for the laboratory mouse and rat, respectively. These phantoms are widely used to evaluate and improve imaging devices and techniques.

Scott J. Wilderman is a computations research specialist in the Department of Nuclear Engineering and Radiological Sciences at the University of Michigan. Since receiving his PhD in nuclear engineering from the University of Michigan in 1990, he has worked extensively on Monte Carlo modeling of radiation transport processes, with special focus in the modeling of electron and photon transport at energies below 1 MeV and more recently on dosimetry in radioimmunotherapy treatment. He has also been involved in the development of statistical image reconstruction methods for electronically collimated nuclear imaging cameras. Dr. Wilderman assisted in the development of the Monte Carlo radiotherapy dosimetry program DPM, which has since been modified extensively to permit patient-specific computation of dose to tumor and bone marrow in internal emitter radionuclide therapy. Dr. Wilderman was also one of the primary developers in the collaboration between the University of Michigan, the Stanford Linear Accelerator Center, and the High Energy Accelerator Research Organization (KEK) in Japan, which produced EGS5, a recent update of the extensively used EGS4 computer code system. In addition, Dr. Wilderman is the founder or cofounder of several companies involved in the U.S. fantasy baseball industry, and possesses unique experience and expertise in algorithms for evaluating and forecasting the performance of baseball players in the context of fantasy sports.

I. George Zubal received a bachelor degree in physics from the Ohio State University (OSU) (1972). He began his graduate work at OSU and focused on medical applications for his master's thesis (1974) in which he developed position-sensitive semiconductor detectors for nuclear medicine applications. He went abroad for his further studies and earned his PhD degree (1981) from the Universitaet des Saarlandes, Homburg, while developing Monte Carlo-based dosimetry maps overlaid onto CT images for therapy planning with fast neutrons at the German Cancer Research Center in Heidelberg, Germany. He returned to the United States to work as a postdoc in the medical department of the Brookhaven National Laboratories, Upton, New York. In 1984, he was recruited into industry (Picker International, Northford, CT) and worked as a senior scientist for nuclear medicine camera and computer systems. He reentered the academic world in 1986 and became the technical director of the Section of Nuclear Medicine at Yale/New Haven Hospital and currently holds the position of associate professor in the Department of Diagnostic Imaging, Yale University School of Medicine. He has continued his work in Monte Carlo simulations of nuclear medicine patient geometries and has extended his image-processing techniques to evaluate functional disorders—notably for localizing seizures in the brains of epilepsy patients. While at Yale, he collaborated with Molecular NeuroImaging LLC, New Haven, on automated analyses for Parkinson's and Alzheimer's brain images since 2003, and transitioned to a full-time position at MNI in 2006.

1

Introduction to the Monte Carlo Method

Michael Ljungberg

CONTENTS

In the literature, we see today an increasing number of scientific papers which use Monte Carlo as the method of choice for the evaluation of a range of nuclear medicine topics such as the determination of scatter distributions, collimator design, and the effects of various parameters upon image quality. So what is the Monte Carlo method and why is it so commonly used as a tool for research and development?

A Monte Carlo method can be described as a statistical method that uses random numbers as a base to perform a simulation of specified situation. The name was chosen during the World War II Manhattan Project because of the close connection to games based on chance and because of the location of a very famous casino in Monte Carlo.

In most Monte Carlo applications, the physical process can be simulated directly. It only requires that the system and the physical processes can be modeled from known probability density functions (pdfs). If these pdfs can be defined accurately, the simulation can be made by random sampling from the pdfs. To obtain reasonable statistical errors, a large number of simulations of histories (e.g., photon or electron tracks) are necessary to get an accurate estimate of the parameters to be calculated.

Generally, simulation studies have several advantages over experimental studies. For any given model, it is very easy to change different parameters and investigate the effect of these changes on the performance of the system under investigation. Thus the optimization of an imaging system can be aided greatly by the use of simulations. A very early Monte Carlo study of the spectral distribution was made by Anger and Davis [1] that calculated the intrinsic efficiency and the intrinsic spatial resolution for NaI(Tl) crystals of different thicknesses and for various photon energies. Also, one can study the effects of parameters that cannot be measured experimentally. For example, it is impossible to measure the scatter component of radiation emitted from a distributed source independently of the unscattered component. By using a Monte Carlo technique incorporating the known physics of the scattering process, it is possible to simulate scatter events from the object and determine their effect on the final image. These studies have included measurements of the scatter to primary ratios, the shape of the scatter response function, the shape of the energy spectrum, proportion of photons undergoing various number of scattering events, and the effect attenuator shape and composition in addition to camera parameters such as energy resolution and window size. Hence, a simulation program can help the understanding of the underlying processes since all details of the simulation are accessible.

Overview papers of the Monte Carlo method and its applications in different fields of radiation physics have been given elsewhere by, for example, Raeside [2], Turner et al. [3], Andreo [4], and recently, Zaidi [5,6]. Here, we will outline only the basic methodology and how this may be applied to nuclear medicine problems.

Random Number Generator

A fundamental part of any Monte Carlo calculation is the random number generator. Basing the number on the detection of true random events, such as radioactive decay, the random number can be calculated but is generally very cumbersome and time consuming. On the other hand, true random numbers cannot be calculated since they, by definition, are randomly distributed and, as a consequence of this, are unpredictable. However, for practical considerations, a computer algorithm can be used to generate uniformly distributed random numbers from calculated seed numbers. An example of such an algorithm is the linear congruential algorithm where series of random numbers I_n are calculated from a first seed value I_0, according to the relationship

$$I_{n+1} = (aI_n + b)\mod(2^k) \qquad (1.1)$$

a and b are constants and k is the integer word size of the computer. If b is equal to zero, then the random number generator is called a multiplicative congruential random number generator. The following FORTRAN statement describes the random number generator in Equation 1.1. SEED is the initial value and RAN is the real random number in the range [0,1].

```
REAL FUNCTION RAN(SEED)
PARAMETER (IA=7141, IC=54773, IM=259200)
SEED = MOD(INT(SEED)*IA+IC, IM)
RAN = SEED/IM
END
```

It is important to realize that using the same value of SEED will give the same sequence of random numbers. Thus, when comparing different simulations, one needs to randomly change the initial value of SEED. This can be done, for example, by triggering a SEED value from a call to the system clock or by storing the value of the previous SEED immediately before exit of the previous simulation and then using this value as an initial value in the next simulation. This approach avoids obtaining the same results if a previous simulation is repeated. Repetition can in some cases be advantageous, for example, in a debugging procedure where small systematic errors can be difficult to spot if errors occur between simulations for statistical reasons.

An effect of this form of digital data representation in a computer is that there is a risk that the initial seed number can appear later in the random number sequence. If this occurs, then it is said that the random number generator has "looped." Although the following numbers are still randomly distributed, they are copies of the values generated earlier in the sequence. The severity of this effect depends on the application. The length of the sequence for the linear congruential generator is 2^k if b is odd. For the multiplicative congruential generator, the sequence length is $2^k - 2$. Other popular and high-quality random number generators are the RANMAR algorithm [7] and the RANLUX algorithm [8]. In the RANLUX algorithm, a user can define the level of accuracy ("luxury level") depending on the need. The lowest level provides a fast algorithm but the numbers may not pass some tests of uniformity. Higher levels move toward a complete randomness of the sequence.

Sampling Techniques

In all Monte Carlo calculations, some *a priori* information about the process to be simulated is needed. This information is usually expressed as probability distribution functions, pdfs, for different processes. For example, when simulating photon interactions, the differential cross-section data represent such information used to calculate the path length and interaction type. From this information, a random choice can be made on which type of interaction will occur or how far a photon will go before the next interaction. A probability distribution function is defined over the range of [a,b]. The function is ideally possible to integrate so that the function can be normalized by integration over its entire range. To obtain a stochastic variable that follows a particular probability distribution function, two different methods can be used.

Distribution Function Method

A cumulated cpdf(x) is constructed from the integral of pdf(x) over the interval [a,x] according to

$$cpdf(x) = \int_a^x pdf(x')dx' \tag{1.2}$$

A random sample x is then sampled by replacing cpdf(x) in Equation (1.2) with a uniform distributed random number in the range of [0,1] and solved for x. Two examples of pdf(x)s and corresponding cpdf(x)s are shown in Figure 1.1.

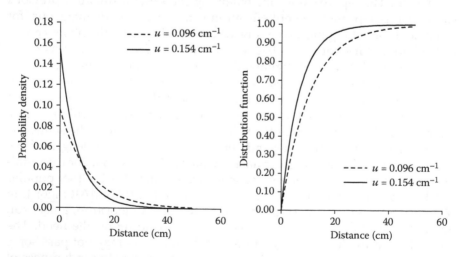

FIGURE 1.1
Two exponential probability distribution functions and their related calculated cumulated distribution function.

"Rejection" Method

Occasionally, the distribution function method is cumbersome to use due to mathematical difficulties in the calculation of the inverse of the cpdf. In these cases, one can use the rejection method that basically can be described by three steps.

Step 1: Let the probability distribution function, pdf(x), be bounded in the range [a,b]. Calculate a normalized function pdf*(x) = pdf(x)/max [pdf(x)] so the maximum value of pdf* is equal to unity.

Step 2: Sample a uniform distributed value of x within the range [a,b] from the relation $x = a + R_1 \cdot (b - a)$ and where R_1 is a random number.

Step 3: Let a second random number R_2 decide whether the sampled x should be accepted. This choice is made by calculating the function value of pdf*(x) from the sampled x value and then checked if $R_2 <$ pdf*(x). If this relation is fulfilled, then x is accepted as a proper distributed stochastic value. Otherwise, a new x value needs to be sampled, according to the procedure in Step 2.

"Mixed" Methods

A combination between the two methods, described so far, can be used to overcome potential problems in developing algorithms, based on either of the two methods alone. Here, the pdf(x) is the product of two probability distribution functions pdf$_A$(x) · pdf$_B$(x). The different steps in using this method are

Step 1: Let pdf$_A$(x) be normalized so that the integral of pdf$_A$(x) over the range [a,b] is unity.

Step 2: Let pdf$_B$(x) be normalized so that the maximum value of pdf$_B$(x) is equal to unity.

Step 3: Choose an x from pdf$_A$(x) by using the distribution function method.

Step 4: Apply the rejection method on pdf$_B$(x) using the sampled value x from Step 3 and check whether or not a random number **R** is less than pdf$_B$(x). If not, then return to Step 3.

Sampling of Photon Interactions

Since this book will mainly focus on Monte Carlo applications for photon transport, describing the basic parts in simulating a photon path can be educative.

Cross-Section Data

Data on the scattering and absorption of photons are fundamental for all Monte Carlo calculations since the accuracy of the simulation depends on the accuracy in the probability functions, that is, the cross-section tables [9–11]. Photon cross sections for compounds can be obtained rather accurately (except at energies close to absorption edges) as a weighted sum of the cross sections for the different atomic constituents.

A convenient computer program developed to generate cross sections and attenuation coefficients for single elements as well as compounds and mixtures as needed is the XCOM [12]. This program calculates data for any element, compound, or mixture, at energies between 1 keV and 100 GeV. The program includes a database of cross sections for the elements. The total cross sections, attenuation coefficients, partial cross sections for incoherent scattering, coherent scattering, photoelectric absorption, and pair production in the field of the atomic nucleus and in the field of the atomic electrons are calculated. For compounds, the quantities tabulated are partial and total mass interaction coefficients, which are equal to the product of the corresponding cross sections and the number of target molecules per unit mass of the material. The sum of the interaction coefficients for the individual processes is equal to the total attenuation coefficient. A comprehensive database for all elements over a wide range of energies has been constructed by combining incoherent and coherent scattering cross sections from [13] and [14], photoelectric absorption from [15], and pair production cross sections from [16]. Figure 1.2a and 1.2b shows differential and total attenuation coefficients for H_2O and NaI, respectively. Note the discontinuity around 30 keV for NaI.

An aspect which deserves further attention is the fact that there exists a variation in the physical cross-section tables included in available Monte Carlo codes. This is of special importance when comparing results from different codes. The use of different cross-section data and approximations will usually yield different results and the accuracy of such results is not always obvious.

Photon Path Length

The path length of a photon must be calculated to decide if the photon escapes the volume of interest. Generally, this distance depends upon the photon energy and the material density and composition. The distribution function method can be used to sample the distributed photon path length x. If the probability function is given by

$$p(x)dx = \mu\exp(-\mu x)dx \tag{1.3}$$

Then the probability that a photon will travel the distance d or less is given by

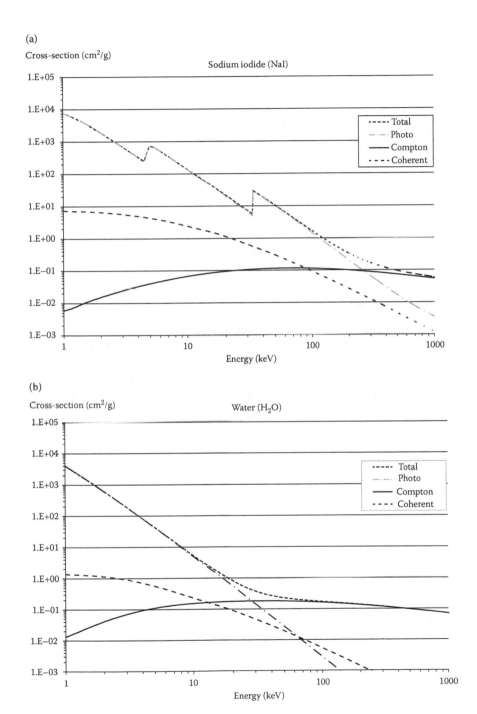

FIGURE 1.2
Distributions of total and differential attenuation coefficients, obtained from the XCOM program, for NaI (a) and H_2O (b).

$$\text{cpdf}(d) = \int_0^d \mu \exp(-\mu x)\,dx = [-\exp(-\mu x)]_0^d = 1 - \exp(-\mu d) \qquad (1.4)$$

To sample the path length, a uniform random number R is substituted for $P(d)$ and the problem solved for d.

$$\left. \begin{aligned} R = P(d) &= 1 - \exp(-\mu d) \\[2mm] d = -\frac{1}{\mu}\ln(1 - R) &= -\frac{1}{\mu}\ln(R) \end{aligned} \right\} \qquad (1.5)$$

Since $(1 - R)$ is also a random number and has the same distribution as R, one can simplify the calculation, according to (1.5).

Coordinate Calculations

After the sample of a new photon path length and direction, the Cartesian coordinate for the end point often is needed to be calculated to check if the photon has escaped the volume of interest. This can be made by a geometrical consideration where the new coordinate (x',y',z') in the Cartesian coordinate system is calculated from the photon path length and direction cosines, according to

$$\left. \begin{aligned} x' &= x + d \cdot u' \\ y' &= y + d \cdot v' \\ z' &= z + d \cdot w' \end{aligned} \right\} \qquad (1.6)$$

where d is the distance between the previous point (x,y,z) and the new point of interest (x',y',z'). Assuming θ and ϕ are the polar and azimuthal angles in the Cartesian coordinate system and Θ and Φ is the polar and azimuthal angle defining the direction change, relative to the initial path of the photon, then the new direction cosines (u',v',w'), necessary to calculate the (x',y',z'), are calculated from

$$\left. \begin{aligned} u' &= \cos\theta \cdot u + \sin\theta[\cos\phi \cdot w \cdot u - \sin\theta\sin\phi \cdot v]/\sqrt{1 - w^2} \\ v' &= \cos\theta \cdot v + [\sin\theta\cos\phi \cdot w \cdot v + \sin\theta\sin\phi \cdot v]/\sqrt{1 - w^2} \\ w' &= \cos\theta \cdot w - \sin\theta\cos\phi \cdot \sqrt{1 - w^2} \end{aligned} \right\} \qquad (1.7)$$

Selecting Type of Photon Interaction

The probability for a certain interaction type to occur is given by the differential attenuation coefficients. These are tabulated for different energies and materials. The sum of the differential attenuation coefficients for photoelectric effect (τ), Compton interaction (σ_{inc}), coherent interaction (σ_{coh}), and pair production (κ) is called the linear attenuation coefficient $\mu = \tau + \sigma_{inc} + \sigma_{coh} + \kappa$, or mass-attenuation coefficient if normalized by the density. To select a particular interaction type during the simulation, the distribution function method can be used. A uniform random number R is sampled and if the condition $R < \tau/\mu$ is true, then a photoelectric interaction will be simulated. If this condition is false, then the same value of R is used to test whether $R < (\tau + \sigma_{inc})/\mu$. If this is true, then one continues with a Compton interaction. If not, then the test $R < (\tau + \sigma_{inc} + \sigma_{coh})/\mu$ will determine if a coherent interaction has taken place. If all conditions are false, then a pair production is to be simulated. Obviously, this will only occur if the photon energy is >1.022 MeV.

Photo Absorption

In this process, the photon energy is completely absorbed by an orbital electron. In the simplest way, the photon history is terminated and the energy (and other parameters) is scored. However, it is possible that secondary characteristic x-rays and Auger electrons can be emitted. The relative probability for the two emissions is given by the fluorescence yield. If a characteristic x-ray is selected, then a new photon energy and an isotropic direction are sampled. The new photon is then followed until absorption or escape. Note the discontinuity in Figure 1.2. There are relatively large differences in attenuation coefficients close to these energies. Therefore, care should be taken so that the characteristic x-ray energy is properly set when sampling subsequent cross-section data. The deposit energy will be the incoming photon energy minus the binding energy for the rejected electron.

Incoherent Photon Scattering

Incoherent scattering, commonly denoted Compton scattering, means an interaction between an incoming photon and an atomic electron where the photon loses energy and changes direction. The energy of the scattered photon depends upon the initial photon energy, $h\nu$, and the scattering angle θ (relative to the incident path), according to

$$h\nu' = \frac{h\nu}{1 + (h\nu/m_0 c^2)(1 - \cos\theta)} \tag{1.8}$$

where hv' is the energy of the Compton scattered photon, m_o is the electron mass, and c is the speed of light in vacuum. One very commonly used method to sample the energy and direction of a Compton-scattered photon uses the algorithm developed by Kahn [17]. This algorithm is based on the Klein–Nishina cross-section equation:

$$d\sigma^e_{\gamma,\gamma e} = \frac{r_e^2}{2}\left(\frac{hv'}{hv}\right)^2\left(\frac{hv}{hv'} + \frac{hv'}{hv} - \sin^2\theta\right)d\Omega \tag{1.9}$$

The sampling method is based on a mixed method and is shown in the following Fortran statement:

```
ALPHA = HV / 511
TEST = (2*ALPHA+1)/(2*ALPHA+9)
        RANDOM = 2*RAN(SEED)
IF(RAN(SEED).LT.TEST) THEN
        UU = 1+ALPHA*RANDOM
        IF(RAN(SEED).GT.4*(UU-1)/(UU*UU))GOTO 1
        COSTET = 1-RANDOM
ELSE
        UU = (2*ALPHA+1)/(ALPHA*RANDOM+1)
        COSTET = 1-(UU-1)/ALPHA
        IF(RAN(SEED).GT.0.5*(COSTET*COSTET+(1/UU))) GOTO 1
ENDIF
```

The mathematical proof for the algorithm has been described by Raeside [2]. The rejection method, described earlier, is derived assuming scattering from a free electron at rest using the Klein–Nishina cross section. For situations where the incoming photon energy is the same order as the binding energy of the electron, the assumption of a free electron at rest becomes less justified. The cross section for this occurrence is given by

$$\frac{d\sigma_{incoh}}{d\Omega} = \frac{d\sigma_{KN}}{d\Omega} \cdot S(x,Z) \tag{1.10}$$

where $S(x,Z)$ is the incoherent scattering function [13], Z is the atomic number, and $x = (\sin(\theta)/2)/\lambda$ is the momentum transfer parameter that varies with the photon energy and scatter angle. It can be shown [18] that

$$\frac{d\sigma_{incoh}}{d\Omega} = \frac{d\sigma_{KN}}{d\Omega} \cdot \frac{S(x,Z)}{S_{max}(x,Z)} \cdot K(hv,Z) \tag{1.11}$$

where $K(hv,Z)$ is constant for a fixed Z and energy. A scattering angle is sampled from Equation 1.8 using, for example, the Kahn's method. A momentum transfer parameter, x, is then calculated and θ (obtained from the sampled x) is accepted only if a random number $R < [S(x,Z)/S_{max}(x,Z)]$. Otherwise, a new scattering angle is sampled.

Coherent Photon Scattering

Coherent scattering is an interaction between an incoming photon and an electron where the direction of the photon is changed but without energy loss. This type of interaction leads to photons that are scattering mostly in the forward direction. The sampling technique for a coherent scattering is based on the Thomson cross section multiplied by the atomic form factor [13] $F(x, Z)$:

$$\frac{d\sigma}{d\Omega} = \frac{r_o^2}{2}(1 + \cos^2\theta)[F^2(x,Z)]d\theta d\varphi \tag{1.12}$$

It can be shown [18] that the probability of a photon being scattered into the interval $d\theta$ around θ is given by

$$P(\theta)d\theta = K(h\nu,Z) \cdot G(\theta) \cdot f(x^2,Z) \tag{1.13}$$

where $K(h\nu,Z)$ is constant for a fixed energy and atomic number, $G(\theta)$ has a fixed range and

$$f(x^2,Z) = \frac{F^2(x,Z)}{\displaystyle\int_0^{x_{max}^2} F^2(x,Z)dx^2} \tag{1.14}$$

A value of x^2 is sampled from a precalculated distribution function of $f(x^2,Z)$. From this value, a scattering angle, θ, can be calculated provided that the relation $R < G(\theta)$ is fulfilled.

Pair Production

Simulating pair production is mainly a book-keeping procedure. An initial photon is assigned 511 keV and emitted in an isotropic direction. The location (x,y,z) and direction cosines (u,v,w) are stored and the photon is followed until absorption or escape. The current position is set to the annihilation location and a second 511 keV photon is emitted but in a direction opposite to the first and followed until absorption or escape. In some cases, there might be a need to simulate the effect of annihilation in flight—an effect that results in a non-180° emission between the two photons—and also account for the path length of the positrons.

Example of a Calculation Scheme

Figure 1.3 shows a flowchart of a photon simulation in a volume including photo absorption, incoherent, coherent scattering, pair production, and simulation of characteristic x-ray emission at the site of photo absorption.

Sampling of Electron Interactions

In many cases in Monte Carlo simulation of nuclear medicine applications and especially for imaging, the energy released by secondary electrons can be regarded as locally absorbed at the interaction site. However, there are some applications that this assumption does not hold. One example is absorbed dose calculations in small regions such as preclinical dosimetry [19]. Another example is the simulation of bremsstrahlung imaging where interacting electrons can produce photons required to be used for imaging [20–22]. In these applications, access to Monte Carlo codes that include a detailed charged particle simulation can be necessary.

A Monte Carlo simulation of charged particles, such as electrons and positrons, differs from a photon simulation in that most electrons interact by the weak Coulomb force which means that for each electron, history typically in the order of millions of interactions may occur before termination as compared to photons that undergo relatively few interactions (order of up to 10) before absorption. Most of the electron interactions will be inelastic scatterings with atomic electrons, which results in small angular deflections at each interaction site with a very small energy loss. Only a few of the electron interactions occur by elastic scattering with the atomic nuclei, production of secondary large-energy electrons, and bremsstrahlung photon generation. Inelastic electron interactions that result in a large change in kinetic energies and directions are sometimes called "catastrophic" events, whereas interactions resulting in only small changes in direction and energy are categorized as "noncatastrophic" events. Since noncatastrophic events will be in a vast majority, it becomes very time consuming to simulate the radiation transport in a detailed mode, that is, simulating each particle interaction explicit. Therefore, in order to reduce the calculation time, it is common to implement the so-called multiple-scattering methods where many electron interactions are condensed into larger steps (Figure 1.4).

The condensed history of electron transport was first suggested by Berger in 1963 [23] where he proposed simulation of the diffusion of electrons by a number of "snapshots" taken at different time or range intervals. The interactions between these snapshots were thus combined into "large"

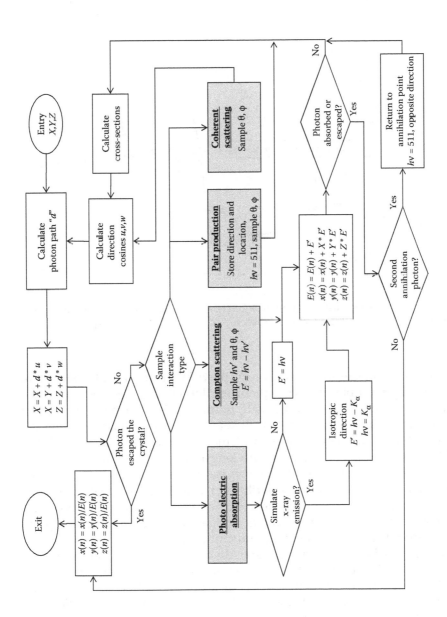

FIGURE 1.3
A flowchart describing the basic steps for simulation of photon transport in a defined volume.

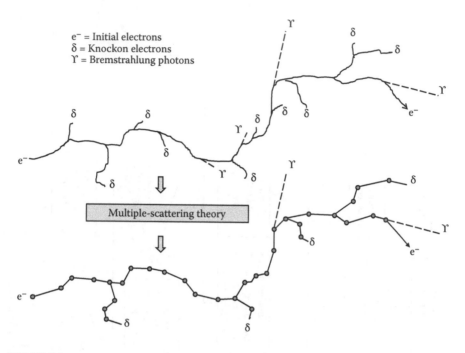

FIGURE 1.4
Schematic figure of a hypothetic electron track and how this is discretized in small step-length by the application of a multiple-scattering theory.

events where changes in kinetic energy, particle direction, and position are described by multiple-scattering theories. Berger suggested two different strategies, named Class-I and Class-II, to make this condensed simulation feasible. The Class-I method relies on sets of predetermined path lengths (or predetermined energy losses) and includes the effect of both elastic and inelastic interactions along a path length using approximate multiple-scattering theories. The Class-II schemes individually simulate the inelastic interactions from probability distribution functions and elastic interactions from multiple-scattering theories. Today, many Monte Carlo codes that use multiple-scattering methods for electrons are based on sampling the angular deflection upon the Goudsmit and Saunderson theory [24], a theory that includes an exact angular distribution after n interactions. It does not include the energy loss but this can be sampled by the continuous slowing-down approximation (CSDA) or more accurately, by considering the statistic fluctuations of the energy loss [25] and the spatial deflections [26]. More details concerning electron sampling are described in Chapters 9 and 10.

Variance Reduction Methods

In many cases, Monte Carlo simulations can be very time consuming. This is particularly true where most of the photon histories generated by direct Monte Carlo simulation are likely to be rejected, for example, when simulating a point source at a large distance from a small detector so that the probability that a photon will hit the detector is small. Variance reduction techniques can then be applied to improve the efficiency of the simulation and hence the statistical properties of the images produced. These techniques are based on calculating a photon history weight, W, that is applied to a photon. This weight represents the probability that the photon passes through a particular history of events. In a direct Monte Carlo simulation, W is always unity but when applying variance reduction methods, W will be less than unity. A detailed description of these various methods is given in Chapter 2.

References

1. Anger HO, Davis DH. Gamma-ray detection efficiency and image resolution in sodium iodide. *Rev. Sci. Instrum.* 1964;35:693–697.
2. Raeside DE. Monte Carlo principles and applications. *Phys. Med. Biol.* 1976;21:181–197.
3. Turner JE, Wright HA, Hamm RN. A Monte Carlo primer for health physicists. *Health Phys.* 1985;48:717–733.
4. Andreo P. Monte Carlo techniques in medical radiation physics. *Phys. Med. Biol.* 1991;36:861–920.
5. Zaidi H, Sgouros G. *Therapeutic Applications of Monte Carlo Calculations in Nuclear Medicine.* Bristol: IOP Publishing; 2002.
6. Zaidi H. Relevance of accurate Monte Carlo modeling in nuclear medical imaging. *Med. Phys.* 1999;26(4):574–608.
7. James F. *A Review of Pseudorandom Number Generators.* CERN-Data Handling Division, Report DD/88/22, Geneva, Switzerland, 1988.
8. James F. RANLUX: A FORTRAN implementation of the high-quality pseudorandom number generator of Lüscher. *Comput. Phys. Commun.* 1994;79:111–114.
9. Hubbell JH. Photon cross sections, attenuation coefficients and energy absorption coefficients from 10 keV to 100 GeV. *Natl. Stand. Ref. Data Ser.* 1969; 29:1–80.
10. McMaster WH, Del Grande NK, Mallett JH, Hubbell JH. Compilation of X-ray cross sections. Lawrence Livermore Lab UCRL-50174 1969.
11. Storm E, Israel HI. Photon cross sections from 1 keV to 100 MeV for elements $Z = 1$ to $Z = 100$. *Nucl. Data Tables.* 1970;A7:565–681.
12. Berger MJ, Hubbell JR. *XCOM: Photon Cross-Sections on a Personal Computer.* Washington DC: National Bureau of Standards; 1987.

13. Hubbell JH, Veigle JW, Briggs EA, Brown RT, Cramer DT, Howerton RJ. Atomic form factors, incoherent scattering functions and photon scattering cross sections. *J. Phys. Chem. Ref. Data.* 1975;4:471–616.
14. Hubbell JH, Øverbø I. Relativistic atomic form factors and photon coherent scattering cross section. *J. Phys. Chem. Ref. Data.* 1979;8:69–105.
15. Scofield JH. Photoionization cross sections from 1 to 1500 keV. Lawrence Livermore National Laboratory UCRL-51326 1973.
16. Hubbell JH, Gimm HA, Øverbø I. Pair, triplet and total atomic cross sections (and mass attenuation coefficients) for 1 MeV-100 GeV photons in elements Z = 1 to 100. *J. Phys. Chem. Ref. Data.* 1980;9:1023–1147.
17. Kahn H. *Application of Monte Carlo.* Rand Corp. Santa Monica, CA RM-1237-AEC 1956.
18. Persliden J. A Monte Carlo program for photon transport using analogue sampling of scattering angle in coherent and incoherent scattering processes. *Computer Programs Biomed.* 1983;17:115–128.
19. Larsson E, Jonsson BA, Jonsson L, Ljungberg M, Strand SE. Dosimetry calculations on a tissue level by using the MCNP4c2 Monte Carlo code. *Cancer Biother. Radiopharm.* 2005;20(1):85–91.
20. Minarik D, Sjogreen Gleisner K, Ljungberg M. Evaluation of quantitative ^{90}Y SPECT based on experimental phantom studies. *Phys. Med. Biol.* 2008;53(20):5689–5703.
21. Minarik D, Ljungberg M, Segars P, Gleisner KS. Evaluation of quantitative planar ^{90}Y bremsstrahlung whole-body imaging. *Phys. Med. Biol.* 2009;54(19):5873–5883.
22. Minarik D, Sjogreen-Gleisner K, Linden O, et al. 90Y Bremsstrahlung imaging for absorbed-dose assessment in high-dose radioimmunotherapy. *J. Nucl. Med.* 2010;51(12):1974–1978.
23. Berger MJ. Monte Carlo calculation of the penetration and diffusion of fast charged particles. In: Alder B, Fernbach S, Rotenberg M, eds. *Methods in Computational Physics,* Vol. 1. New York: Academic Press; 1963:135.
24. Goudsmit S, Saunderson JL. Multiple scattering of electrons. *Phys. Rev.* 1940;57:24–29.
25. Landau L. On the energy loss of fast particles by ionization. *J. Phys. (USSR).* 1944;8:201.
26. Lewis HW. Multiple scattering in an infnite medium. *Phys. Rev.* 1950;78:526–529.

2

Variance Reduction Techniques

David R. Haynor and Robert L. Harrison

CONTENTS

The popularity of the Monte Carlo method in nuclear medicine research arises from the possibility of accurately simulating the physical processes undergone by particles originating in and passing through heterogeneous media and then interacting with collimator and detector systems. In a typical nuclear medicine acquisition, nearly all the photons which arise within a patient are absorbed within the patient, miss the camera entirely, are absorbed in the collimator, or pass through the detector without interacting. Consider, for example, an acquisition in which 25 mCi of activity is administered to a patient and 500,000 counts are accumulated in a planar image over a period of 1 min. In that period of time, $0.025 \times 2.2 \times 10^{12} = 5.5 \times 10^{10}$ disintegrations take place within the whole body; only 5×10^5, or approximately one disintegration in 110,000, gives rise to a detected event. A so-called analog Monte Carlo simulation, in which particle histories are generated and followed according to the underlying physics, would follow the same statistics: only one complete photon history out of every 110,000 generated by the Monte Carlo code would give rise to a detected event. Assuming our interest lies primarily in the detected events, this brute force computation would be extremely inefficient. Put another way, the fluxes of detected particles would be extremely small, and consequently the variance of the flux estimates would be high. The Monte Carlo methods known collectively as *variance reduction techniques* have been developed to deal with this problem.

For variance reduction techniques to be effective, a *weight* must be attached to each photon history that is generated by the Monte Carlo code. This is discussed in more detail later, but roughly speaking the weight of a history may be thought of as the number of "real-world" particles represented by that history. In the case of an analog simulation, all history weights are the same. When variance reduction techniques are used, this is no longer the case. Variance reduction techniques are effective because they increase the ratio of detected events to histories over that obtained with analog Monte Carlo. They achieve this by simulating a different random process, one which is obtained from the analog process by modifying it so that the histories which arise are enriched in those that give rise to detectable events. In order that the particle fluxes estimated from the new process be accurate estimates of the true (analog) fluxes, the weight of each history is adjusted. Generally, this is done each time a simulated physical event (particle emission, scattering, interaction with the collimator, etc.) occurs during a history. As a consequence, each history winds up with a unique weight. Because the statistical quality of the simulated detected events is adversely affected by widely varying history weights, variance reduction methods must also include techniques that attempt to equalize the weights attached to individual histories as much as possible.

In summary, variance reduction techniques must do three things: (1) enrich the generated histories with those that give rise to detected events; (2) adjust the history weights correctly, so that flux estimates remain unbiased; and (3) ensure that the final history weights do not vary too widely. The remainder of this chapter will be devoted to making these three notions more precise and to illustrating the principles of variance reduction with typical examples that are used in contemporary nuclear medicine simulations. It should be emphasized that the basic principles of variance reduction techniques are not new; they have been well understood since the first appearance of the Monte Carlo technique. The reader is referred to [1–4] for a fuller discussion of the basic principles of variance reduction methods.

Weights and Their Management

Weights and the Monte Carlo Evaluation of Definite Integrals

It is easier to understand the basis of variance reduction techniques if we start with the problem of estimating the expected value of a random variable by the Monte Carlo method. The extension to estimating the solutions of transport equations, such as those that describe the birth, scattering, and absorption or detection of photons, is then straightforward (see the following section). For the purposes of exposition, let $\mathbf{D}$ (for "detector") be a random variable whose expectation we wish to estimate. $\mathbf{D}$ is a function on a probability space X, which we take to be a set equipped with an integration method and a probability density $p(x)$ with the properties that $p(x) \geq 0$ for all x in X and $\int_X p(x)\,dx = 1$. The expectation of $\mathbf{D}$ is then defined by the equation $E(\mathbf{D}) = \int_x \mathbf{D}(x)p(x)\,dx$. The analog Monte Carlo approach to the calculation of $E(\mathbf{D})$ is to draw $N \geq 1$ values of x from the space $\mathbf{X}$, according to the distribution $p(x)$, and then to estimate $E(\mathbf{D})$ from the average of the sampled values of $\mathbf{D}$

$$E(\mathbf{D}) \approx D_1 = \frac{1}{N}\sum_{i=1}^{N}\mathbf{D}(x_i) \tag{2.1}$$

D_1 is an unbiased estimator of $E(\mathbf{D})$, that is, $E(\mathbf{D}_1) = F(\mathbf{D})$. The variance of D_1 may be calculated from Equation 2.1:

$$\begin{aligned}\operatorname{var}(D_1) &= \frac{\operatorname{var}(\mathbf{D})}{N} = \frac{E(\mathbf{D}^2) - (E(\mathbf{D}))^2}{N} \\ &= \frac{\int_x \mathbf{D}^2(x)p(x)\,dx - \left(\int_x \mathbf{D}(x)p(x)\,dx\right)^2}{N}\end{aligned} \tag{2.2}$$

Now, suppose that we sample from $\mathbf{X}$ using a different probability density, but that we still wish to estimate $E(\mathbf{D})$ using the same number of particles in the second Monte Carlo simulation as in the first. If our new probability density function is given by $q(x)$, with $q(x) \geq 0$ for all x in $\mathbf{X}$ and $\int_X q(x)\,dx = 1$, then we have

$$E(\mathbf{D}) = \int_x \mathbf{D}(x)p(x)\,dx = \left(\int_x \mathbf{D}(x)[p(x)/q(x)]q(x)\,dx\right) \tag{2.3}$$

This gives us a new way of estimating $E(\mathbf{D})$:

$$E(\mathbf{D}) \approx D_2 = \frac{1}{N}\sum_{i=1}^{N}\mathbf{D}(x_i')\frac{p(x_i')}{q(x_i')} = \frac{1}{N}\sum_{i=1}^{N}\mathbf{D}(x_i')w(x_i')x_i \tag{2.4}$$

where the primes serve to remind us that the x_is are sampled using the new probability density $q(x)$. The factor $w(x_i') = p(x_i')/q(x_i')$ is just the *weight* that must be attached to the ith particle history as a result of our sampling from $q(x)$ instead of $p(x)$. Note that the weight factor depends only on the history; it does not depend on the random variable $\mathbf{D}$, except for the requirement that, if $\mathbf{D}(x)$ is nonzero, $w(x)$ must be finite; that is, $q(x)$ must be nonzero. In other words, if we have a whole family of "detectors" $\mathbf{D}$, as would be the case in a nuclear medicine simulation, our sampling density $q(x)$ must be positive for all "detectable" histories. However, $q(x)$ can vanish whenever x is not detectable. This concentration of the sampling density on the particle histories of potential interest is the key to the success of variance reduction techniques.

A possible advantage of using the weighted estimator D_2 instead of D_1 (assuming that q is chosen appropriately) may be seen when we compare the variance of the two estimators. In most nuclear medicine simulations, the random variable $\mathbf{D}$ that we are interested in represents the flux of a small specified subset of all the particles generated, such as the flux into a particular crystal in a block detector system or into a particular xy-position in an Anger camera. In this case, $\mathbf{D}(x)$ is either 0 or 1 for all histories x. The second term in the numerator of Equation 2.2 will then be the square of the first term. Since the first term will be small, the second term will be substantially smaller, and we may approximate var(D_1) and var(D_2) by using the first term in the numerator of Equation 2.2 only. We obtain

$$\text{var}(D_1) \approx \frac{1}{N}\int_x \mathbf{D}^2(x)p(x)\,\mathrm{d}x \quad \text{and} \quad \text{var}(D_2) \approx \frac{1}{N}\int_x \mathbf{D}^2(x)w^2(x)q(x)\,\mathrm{d}x \quad (2.5)$$

We may approximate the integrals in Equation 2.5 with their Monte Carlo estimates:

$$\text{var}(D_1) = \frac{1}{N^2}\sum_{i=1}^{N}\mathbf{D}^2(x_i) \quad \text{var}(D_2) = \frac{1}{N^2}\sum_{i=1}^{N}\mathbf{D}^2(x_i')w^2(x_i') \quad (2.6)$$

To better understand (2.6), consider again a random variable $\mathbf{D}$ which takes on only the values 0 and 1. Without loss of generality, assume that it is precisely the first M terms of the summation for var(D_1) and the first M' terms of the summation for var(D_2) that are nonzero. Then we have, after some manipulation

$$\text{var}(D_1) \approx M/N^2 \quad (2.7)$$

$$\text{var}(D_2) \approx \frac{M}{N^2}\left[\left(M'\sum_{i=1}^{M'}w^2(x_i')\right)\middle/\left(\sum_{i=1}^{M'}w(x_i')\right)^2\right]\left[\frac{1}{M'M}\left(\sum_{i=1}^{M'}w(x_i')\right)^2\right] \quad (2.8)$$

The second and third terms on the right-hand side of Equation 2.8 can be simplified. The Cauchy–Schwartz inequality states that, for any real numbers $x_1, x_2, \ldots, x_n$, we have $\left(\sum_{i=1}^{n} x_i\right)^2 \leq n \sum_{i=1}^{n} x_i^2$; it follows that the middle right-hand bracketed term is at least 1, and we write it as $(QF)^{-1}$, where QF is the *quality factor* for the simulation. The QF is always ≤ 1, and is equal to 1 if and only if all the histories have equal weights (i.e., we are doing an analog simulation). The simulation is assumed to be unbiased, and so $\left(\sum_{i=1}^{n} w(x_i')\right)/N$ is precisely the D_2 estimate for $E(\mathbf{D})$ (see Equation 2.3); therefore, it must be approximately equal to M/N, the D_1 estimate for $E(\mathbf{D})$ (Equation 2.2). Making these simplifications, we get

$$\frac{\text{var}(D_1)}{\text{var}(D_2)} \approx \frac{QF\,M'}{M}, \quad \text{with QF} = \frac{\left(\sum_{i=1}^{M'} w(x_i')^2\right)^2}{M' \sum_{i=1}^{M'} w^2(x_i')} \tag{2.9}$$

our sought-after result. While several approximations were made in deriving Equation 2.9, it is nonetheless quite useful. It demonstrates that it is possible to reduce the variance of our estimate for $E(\mathbf{D})$ by substituting D_2 for D_1, if two conditions can be met: (1) the number of histories that end in detections that are derived by sampling from $q(x)$, M', must be substantially larger than M, the number of such histories derived from sampling from $p(x)$; and (2) the resulting dispersion of weights must not become too great, which would result in a small quality factor. The reduction in variance is measured not by the ratio M/M', as would be expected with pure Poisson statistics (analog Monte Carlo). Instead, the effective number of particles in the D_2 simulation is reduced by the factor QF, which never exceeds 1. In principle, the QF for a simulation is different for each random variable $\mathbf{D}$ that is to be estimated; in our experience with nuclear medicine problems, a single QF, gotten by evaluating Equation 2.9 for *all* histories, works well and is a useful yardstick for monitoring the overall performance of a set of variance reduction strategies.

Finally, we note in passing that to estimate the *variability* of a quantity computed by Monte Carlo methods, one may use resampling techniques such as the bootstrap [5], being careful to sample from the original (p) rather than the weighted distribution (q). Alternately, if an estimate of λ, the mean "real-world" rate of events arriving at a detector, is known from a variance-reducing simulation, the variance of that rate, which is a Poisson random variable, will also be λ.

Weights in the Simulation of Random Walk Processes

In the previous section, the concept of weights was introduced for sampling from a single probability distribution. In a Monte Carlo simulation in nuclear medicine, there typically are a series of random events which, collectively, constitute a single-simulated photon history. The particle must be *born*, that is, assigned an initial energy and direction. It then travels a random distance in the assigned direction and comes to the next point of interaction. Assuming the particle has not escaped the object entirely, it then undergoes photoelectric absorption, coherent scattering, or Compton scattering. In the first case, it is absorbed at the point of interaction. In the second or third case, a new direction and energy is generated and the process repeated. Finally, if the photon escapes the object, it may miss the collimator/detector entirely, or it may collide with the front face of the detector. Interaction with the collimator and detector may be treated either deterministically or stochastically.

In applying variance reduction techniques to the simulation of random walk processes, various techniques will be used in the simulation of different events (particle birth, absorption, collimator interaction, etc.). An initial weight is attached to each particle history as it is generated. The history weight is then updated after each event by multiplying it by the correct weight factor for the variance reduction technique, if any, used at that step of the simulation. At each step, the principles outlined earlier are used to calculate the correct weights. The final value of the weight is assigned to the history as a whole, and is used for the calculation of the overall QF of the simulation (Equation 2.8). Thus, good control of weight variation (QF close to 1) ideally requires good control of the weight variations produced at each step of the simulation. The use of Russian roulette, particle splitting, and weight windows, which are described later, makes it possible to correct to some extent for the effects of excessive weight variations introduced by sampling techniques, thereby preventing the QF for the simulation from becoming too small.

Applications of Variance Reduction Techniques in Nuclear Medicine Simulations

We will now illustrate the application of variance reduction techniques for random walks to nuclear medicine simulations through a series of examples, arranged roughly according to complexity. While these examples are described in the context of single-photon imaging, they are easily adapted to positron imaging as well.

Photoelectric Absorption

Our first, and simplest example, relates to the handling of photoelectric absorption. For imaging simulations, photoelectric absorption within the object is an undesired outcome, since it cannot lead to a detected event. To calculate the weight factor, suppose that, given the photon energy E and its current location in space (which determines the material property at that point and, therefore, the probability of absorption), the particle will be absorbed with probability α and scattered with probability $1 - \alpha$. Consider an event space X consisting of two events: $X = [absorption, scatter]$. The physics assigns probabilities $p(x) = [\alpha, 1 - \alpha]$ to the points of X. Because none of our detectors will detect histories which end in the *absorption* state, we take for our sampling probability $q(x) = [0,1]$. According to the principles described previously, then, each history must be multiplied by the weight $p(scatter)/q(scatter) = 1 - \alpha$, and only histories which do not end in absorption are generated. This step is performed each time an interaction within the object being simulated (the patient or phantom) occurs. Typically, α is small, and the variation in weights that occurs as a result of this modification is also small. By the same token, the effective enrichment in particles (M'/M) is also small, although it becomes more significant for particles undergoing larger numbers of interactions.

Sampling from Complex Angular Distributions: Coherent Scattering and Use of Form Factors in Compton Scattering

For higher photon energies $(E \geq 100 \text{ keV})$, coherent (Rayleigh) scattering can be neglected, and the classical Klein–Nishina formula for the outgoing angular distribution of a photon scattering off a free electron may be used. At lower photon energies, these approximations are less valid (see (2) for a more detailed discussion). For example, the coherent cross section amounts to about 10% of the total cross section at 50 keV. Moreover, the distribution of coherently scattered photons is very different from those that scatter via a Compton process: it is strongly peaked in the forward direction, although there is no perfectly forward coherent scattering. The exact calculation of the coherent distribution is difficult (it is a many-electron effect) and so semiempirical form factors $C(Z,E,\mu)$ (μ = cosine of scattering angle), which are multiplied by the classical Thomson cross section $T(E,\mu)$ to yield the final cross section p, are used. Similarly, for low photon energies, the Klein–Nishina cross section for scattering $K(E,\mu)$ is modified by a form factor that depends on the Z of the scattering material, E, and μ: $p(E,\mu) = I(Z,E,\mu)K(E,\mu)$.

Sampling may be performed from these empirical angular distributions in either of two related ways. The first method is simpler, but produces more variation in particle weights than the second method. In the first method, we sample first from $q(x) = K(E,\mu)$ or $T(E,\mu)$ to obtain μ, using classical algorithms such as the Kahn method for sampling from $K(E,\mu)$ [6–8]. The weight of the particle history is then simply multiplied by $I(Z,E,\mu)$ for incoherent scattering or by $C(Z,E,\mu)$ for coherent scattering. Note that the method of weight calculation described earlier requires that q be a true density; thus, $I(x)$ must be normalized so that $\int I(Z,E,\mu)K(E,\mu)d\mu = 1$ for all E, Z. The second method sets $q()$ equal to a discretized approximation to the true semiempirical scattering distribution $p()$. For a given Z and E, $q(\mu)$ is defined to be proportional to $p(E,Z,\mu_i)$ for $\mu_i \leq \mu \leq \mu_{i+1}$, where $0 = \mu_0 < \mu_1 < \cdots < \mu_n = 1$ forms a set of grid points, and is normalized so that $\int q(E,Z,\mu)d\mu = \sum p(E,Z,\mu_i)(\mu_{i+1} - \mu_i) = 1$. It is then easy to sample from q to obtain μ, and the particle history is multiplied by $p(E,Z,\mu)/q(E,Z,\mu)$ as before. By choosing a sufficiently fine discretization, the weight variation produced by this method can be made arbitrarily small. The second method is also useful in designing an effective variance reduction technique for forced detection, as described later.

Detector Interactions

Simplified nuclear medicine simulations often treat the interaction of an escaped photon with the detector crystal only approximately. The deposition of energy within the crystal is assumed to occur at a single point along the photon path, the exact location being chosen according to the usual techniques for exponential sampling along a ray [9]. Variance reduction in this case is similar to the handling of photoelectric absorption described before: the weight of the history is simply multiplied by the probability of interaction within the crystal, and all photons whose paths intersect the crystal and which pass through the collimator will be detected. The variability in weights introduced at this step is related to the dispersion of the energy and inclination of the escaped photons, but is typically small. Thus, all the simulated photons that reach the detector will result in detected events, albeit with varying weights.

More sophisticated simulations, which are available in most modern programs, allow multiple interactions within the crystal, taking the deposition-energy-weighted centroid of the interactions as the detector output. Techniques similar to those already described may be used to force at least one interaction to occur within some detector crystal or, more generally, to force enough interactions to occur so that the total deposited energy within a single crystal exceeds a preset threshold. In such cases, the weight of the final event must be further adjusted to reflect the fact that, in the real world, not all photons entering the detector crystals will deposit enough energy to be detected.

Preventing Particle Escape

In an analog simulation, particles which escape the object, but whose path does not intersect the detection system, are similar to particles that undergo photoelectric absorption: computational effort on them is wasted. If forced detection (below) is not in use, these particles may be handled in a similar way to those undergoing absorption. Specifically, after each interaction, the probability π of photon escape is calculated by integrating the total attenuation along the outgoing direction

$$\pi = \exp\left(-\int_0^{\text{escape}} A(E, x + t\mathbf{n})\mathrm{d}t\right) \qquad (2.10)$$

where $A(E,r)$ = total cross section for a photon with energy E at position r, x = current position, and $\mathbf{n}$ is the unit vector in the outgoing direction. Two copies of the history are now created. The first, with weight = $W\pi$ (W is the weight prior to the scattering event), is allowed to escape and possibly interact with the detector. The second is assigned weight $W(1 - \pi)$ and is forced to interact at some point in the object along the direction $\mathbf{n}$ before escaping. Particle histories are terminated in this scheme only when the particle energy drops below a set threshold. Preventing particle escape has two potential disadvantages. The fluctuation in weights resulting from the repeated multiplications can be substantial, but this can be offset through the use of weight windows (see later). Second, multiple detected particles can arise from a single particle birth and consequently subsequent detected events may be highly correlated. In practical simulations, however, the existence of these correlated events is rarely a problem.

Collimator Penetration/Scatter

Most of the inefficiency of conventional gamma cameras arises from the effects of collimation. Accordingly, considerable care must be taken with the collimator simulation, or large numbers of photons will be lost. In single-photon imaging, collimators typically collimate both in the axial plane (φ-direction) and in the azimuthal direction (θ-direction). In positron imaging, collimators (if present at all) collimate primarily in the θ-direction. In either case, however, it may be of interest to produce a stream of escaping photons all pointing primarily in the same φ-direction. This corresponds to obtaining a single projection of a SPECT or PET image. This may be done by using techniques similar to those discussed later under forced detection. Otherwise, it is both more efficient and computationally much simpler to collect a full set of projections, or to study rotationally symmetric activity/ attenuation distributions.

For low-energy photons, standard collimators are effectively "black," that is, there is little penetration of the collimator material itself. In this case, an excellent approximation to the geometric response of parallel hole collimators was given by Metz [10] and extended to cone and fan beam collimators in [11]. An effective aperture function for these collimators is given in these articles which may be used to multiply the weight of the photon as it intersects the front face of the collimator. Accurate, efficient modeling of collimator scatter in the general case in which the collimator is not treated as perfectly absorbing remains problematic. In [12], an approach is described in which a detailed Monte Carlo simulation of a detector–collimator system is done, and the results summarized in empirical parametric form; the simulation of the detector–collimator system is then table-driven, rather than based on detailed calculations for each particle.

Whether or not scatter is modeled, the potential for serious weight dispersion exists at the collimation stage. This is because, even in the geometric model in which collimator scatter is neglected, particle weights may get multiplied by 0 or extremely small positive numbers if the incident particle is sufficiently far from being perpendicular to the collimator face (in the case of a parallel hole collimator) or from the local hole direction in the case of fan beam and cone beam collimators. The best solution to this problem is to produce escaped photons that are enriched in photons whose direction of travel is favorable for passing through the collimator. This may be done using the technique of forced detection that we now describe.

Forced Detection

Forced detection is a technique for increasing the number of photons that escape the object with a direction of travel that makes it likely that they will pass through the collimator [13]. At each interaction point in a photon history, a "descendant" of the photon is created which undergoes one additional scatter into a "productive" direction of travel. Weighting of the photon descendant is performed using a combination of the techniques described earlier, including corrections for the additional travel within the attenuator and appropriate sampling from the Klein–Nishina distribution. Modifications for coherent scattering and bound-electron form factors may be performed as described earlier. Tracking of the original photon, now with a reduced weight, will then resume. Properly done, forced detection allows an average number of particles per computational history on the order of 1 or more.

Stratification

All the techniques described so far increase M' in Equation 2.8, at the expense of increasing weight dispersion. This latter effect decreases the QF, undercutting the apparent increase in M' and making the reduction in the variance of flux estimates lesser than expected (see Equation 2.8). This section and the next describe techniques for improving the QF of a simulation. The technique of stratification, described here, is applied at the time of particle birth, while the weight window technique (next section) may be applied at any point during a particle history [13].

The number of scatters undergone inside a patient by a photon emitted by a nuclear medicine isotope is small (typically 0–4). Consequently, the ultimate fate of a photon (detection vs. no detection) is strongly influenced by the photon's position and direction at its birth (or at any subsequent step in its history). Suppose that we crudely divide the object into a set of spatially compact regions $R_1, \ldots, R_n$ and divide the set of photon directions into $T_1, \ldots, T_m$ (in the case of 2π detection geometry, it is usually sufficient to consider only the photon's inclination, or θ value). We define each of the combinations (R_i, T_j) as a *stratification cell*. If we do a preliminary run with a small number of histories, we can calculate a *productivity* π_{ij} for each stratification cell:

$$\pi_{ij} = \frac{\text{RMS detected weight for particles born in } (R_i, T_j)}{\text{RMS original weight for particles born in } (R_i, T_j)} \qquad (2.11)$$

π_{ij} may be thought of as an estimate of the probability that a photon starting from a point in R_i in a direction lying in the set T_j will be detected. If the photon is born into the (ij)-th stratification cell with weight W, the predicted total detected weight of the photon's descendants is approximately $W\pi_{ij}$. This suggests that, in order to achieve approximate equality of the weights of the detected particles, the *initial* weight of the particles in the (ij)-th cell should be inversely proportional to π_{ij}. This weight can be adjusted by adjusting the initial sampling density. Specifically, if the activity in the (ij)-th cell is A_{ij}, then the number of particles born in the (ij)-th cell should be proportional to $A_{ij} \pi_{ij}$; each such particle will have an initial weight of C/π_{ij}, but an average detected weight of approximately C, where C is a constant whose value is adjusted according to the desired value of M' in Equation 2.8. The more productive stratification cells will be sampled more frequently than less productive cells, but the final weight of each history at detection will be approximately equal, that is, independent of the starting cell.

In practice, the fineness of the stratification can be rather coarse, and the estimates of the π_{ij} need only to be rough, in order to achieve a significant improvement in QF. Therefore, the initial run required to estimate their values can be short. The dependence of the productivity on the energy may also be neglected. As mentioned earlier, the productivities may be defined by Equation 2.9 separately for unscattered and scattered photons, and are generally rather different. Stratification is, therefore, performed separately for scattered and unscattered photons, using the appropriate productivities.

Use of Weight Windows, Splitting, and Russian Roulette

The use of *weight windows* [13] is similar in spirit to stratification. Using the productivities calculated via Equation 2.11 for scattered photons, the expected value of the detected weight may be approximately calculated for a photon at each point in its history (again, neglecting the dependence of the productivity on photon energy). If the predicted detected weight is too large, the particle may be *split* into several particles, each carrying a fraction of the weight of the original particle. The daughter particles are then tracked independently. If the expected detected weight is too small, the weight of the particle may be increased by a factor κ (with probability 1/κ) or the particle history may be terminated (with probability 1–1/κ). This technique (known in the literature as *Russian roulette*) preserves expected values while preventing the code from spending too much time tracking particles that have small weights or that have wandered into unproductive stratification cells.

Modifications for Positron Imaging

The modifications of the techniques described earlier for positron imaging are straightforward. To simplify the description, let us call the two photons that arise from a positron annihilation the "blue" and "pink" photons. The decay event itself is assigned a certain weight (which is similar to the initial weight in a single photon's simulation). The blue and pink photons now each start with a weight of 1, which is then further decreased as they are tracked through the object and tomography. The final detected weight of a coincidence pair is then the product of the decay weight times the product of the blue and pink weights. In order to calculate expected values correctly, the possibility that several blue and several pink detection events might arise from a single annihilation must be considered. In this case, the coincidence events that arise from all possible blue–pink pairings arising from a single decay are all valid (simulated) coincidences, although only one at most will be an unscattered coincidence.

Conclusions

A framework for understanding the goals of variance reduction techniques—to increase the number of detected events, while maintaining rough equality of weights—has been presented, and the major techniques that have been used in practice have been described. These techniques are used in currently available public domain Monte Carlo codes (see Chapter 7) and typically result in effective speedups by factors of 3–100 over analog methods after allowing for the reduction in quality factors produced by variance reduction methods and the increased computation time per history required. The exact speedup depends on the geometry of the detection system, and is produced by a large increase in the average number of detected particles produced per computational history. Since this average figure is already on the order of 1 or more per particle birth, further improvement in computational efficiency will come primarily through increases in quality factors and improvements in collimator simulation.

References

1. Haynor DR, Harrison RL, Lewellen TK. The use of importance sampling techniques to improve the efficiency of photon tracking in emission tomography simulations. *Med. Phys.* 1991;18(5):990–1001.
2. Spanier J, Gelbard EM. *Monte Carlo Principles and Neutron Transport Problems.* Reading, Massachusetts: Addison-Wesley Pub. Co.; 1969.
3. Bielajew AF, Rogers DWO. Variance reduction techniques. In: Jenkins TM, Nelson WR, Rindi A, eds. *Monte Carlo Transport of Electrons and Photons.* New York: Plenum; 1998: pp. 407–420.
4. Bielajew AF. Fundamentals of the Monte Carlo method for neutral and charged particle transport. http://www.personal.umich.edu/~bielajew/MCBook/book.pdf.
5. Haynor DR, Woods SD. Resampling estimates of precision in emission tomography. *IEEE Trans. Med. Imag.* 1989;8(4):337–343.
6. MCNP—A general Monte Carlo code for neutron and photon transport. Los Alamos 1979. LA-7396-M.
7. Williamson JF, Morin RL. Concerning an efficient method of sampling the coherent angular scatter distribution. *Phys. Med. Biol.* 1983;28(1):57.
8. Zerby CD. A Monte Carlo calculation of the response of gamma-ray scintillation counters. In: Alder B, Fernbach S, Rotenberg M, eds. *Methods in Computational Physics, Volume I.* New York: Academic Press; 1963: pp. 89–134.
9. Chen CS, Doi K, Vyborny C, Chan HP, Holje G. Monte Carlo simulation studies of detectors used in the measurement of diagnostic x-ray spectra. *Med. Phys.* 1980;7(6):627–635.
10. Metz CE, Doi K. Transfer function analysis of radiographic imaging systems. *Phys. Med. Biol.* 1979;24(6):1079–1106.
11. Tsui BM, Gullberg GT. The geometric transfer function for cone and fan beam collimators. *Phys. Med. Biol.* 1990;35(1):81–93.
12. Song X, Segars WP, Du Y, Tsui BM, Frey EC. Fast modelling of the collimator-detector response in Monte Carlo simulation of SPECT imaging using the angular response function. *Phys. Med. Biol.* 2005;50(8):1791–1804.
13. Haynor DR, Harrison RL, Lewellen TK. The use of importance sampling techniques to improve the efficiency of photon tracking in emission tomography simulations. *Med. Phys.* 1991;18:990–1001.

3

Anthropomorphic Phantoms: Early Developments and Current Advances

I. George Zubal and W. Paul Segars

CONTENTS

Monte Carlo modeling of the transmission and attenuation of internal radiation sources has led to a better understanding of the image formation process in diagnostic radiology. Since much higher statistics are necessary to model imaging simulations (compared to dosimetry simulations), the speed of computing individual gamma ray histories is of paramount importance for imaging physics calculations. As a result, the software phantoms modeled in these imaging simulations have historically been highly simplified, limited to simple point, rod, and slab shapes of sources and attenuating media, in order to achieve reasonable turnaround times for Monte Carlo simulations. Such simple geometries are useful in studying more fundamental issues of scatter and attenuation, but clinically realistic distributions cannot be adequately evaluated. The modeling of the intricate protuberances and convolutions of human internal structures are necessary for such evaluations. It is, therefore, essential in imaging research to enhance our computer models in order to insure that our simulations are representative of the reality we try to emulate. Due to the current availability of inexpensive memory and disk storage as well as ever-increasing execution speeds, software models are currently being made more complex in order to model the human anatomy more realistically and still permit statistical Monte Carlo computations to be completed within acceptable time limits. In this chapter, we discuss some of the historical developments in computer models as well as highlight some of the more recent anthropomorphic phantoms.

Early Anthropomorphic Phantoms

The essence of a phantom used for radiological calculations is the ability to mathematically capture two physical distributions; the geometrical characteristics of a radiation source (internal or external) and the spatial distribution of attenuating material.

In order to make 3D anatomical data suitable for use in any such patient-oriented radiologic calculations, we must be able to delineate the surfaces and internal volumes which define the various structures of the body. From most applications what we mean by various structures is those internal organs or physiological substructures which can be imaged and diagnosed through radiological procedures. More specifically in nuclear medicine, we refer to structures which are targeted by any of the numerous radiopharmaceuticals available to the clinician. Clear and obvious examples of such structures are: cardiac ventricular volume or myocardium, liver, lungs, bones, kidneys, and virtually any organ found within the human body. As new radiopharmaceuticals are developed, new structures need to be delineated and modeled in our simulations. For example, the advent of new neuroreceptor agents has created an interest in understanding the imaging characteristics of deep-seated structures in the brain (e.g., the striatum and caudate nucleus). A realistic anthropomorphic phantom for research in this area will need to contain these structures through some method of segmenting their internal volumes. These segmented volumes can then be indexed to activity distributions or other physical characteristics (density or elemental composition).

In order to yield faster computation times, early phantoms were constrained to model a single slice extracted from a 3D volume. The advantage of single slice models is that the mathematical computation may be limited to two dimensions. Radiation leaving the 2D slice may either be ignored; or more efficiently, radiation may be forced to remain within the slice of interest within the simulation. This has obvious advantages in speed of computation but falls short in modeling the radiation (in particular scattered radiation) from slices above and below the slice of interest. A full 3D modeling of radiation transport is essential for a complete understanding of the detection process. Although innovative ideas can quickly be tested by modeling within two dimensions, it is essential to make a more accurate, 3D evaluation of any proposed method. Concomitant with the increased resolution and realistic representation of internal structures, anthropomorphic phantoms have evolved into full 3D descriptions. Such anthropomorphic phantoms typically fall into three general classes: mathematical, voxelized, and the recently introduced hybrid phantoms.

Mathematical Phantoms

Mathematical phantoms define the organs and structures in the body using analytical equations or simple geometric primitives [1]. Since they are analytically defined, exact ray-tracing calculations can be performed on these phantoms. Mathematical phantoms can also be easily manipulated through the equations that define them to model variations in the organs or organ motion. However, the simplicity of the equations upon which they are based does limit their ability to realistically model the human anatomy. One of the earliest mathematical phantoms used for developing improved SPECT reconstruction techniques is referred to as the Shepp–Logan phantom [2]. We see a representation of this phantom in Figure 3.1. It is meant to represent a "head-like" object with some internal structures. These are captured through the use of simple ellipses, which, when carefully placed on a 2D plane, work well to represent the essence of simple structures in the human head. The varying thickness of the skull as well as simple internal structures (striatum), including a small tumor, are easily recognized in Figure 3.1.

One of the earliest three-dimensional mathematical anthropomorphic phantoms is the MIRD phantom, which was developed for estimating doses to various human organs from internal or external sources of radioactivity. The MIRD phantom served to calculate the S-factors for internal dose calculations in nuclear medicine [3]. This phantom models internal structures as

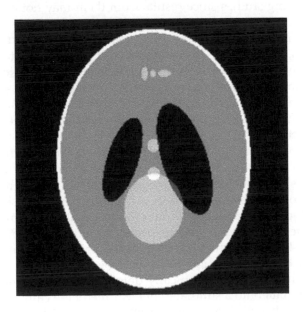

FIGURE 3.1
Shepp–Logan head phantom illustrating an early single slice anthropomorphic representation of the head of a human by using simple ellipses.

(a)   (b)   (c)

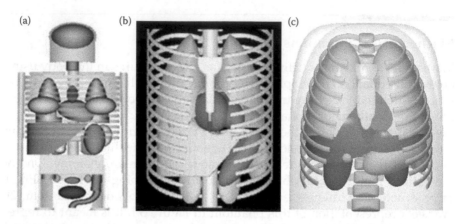

FIGURE 3.2
Anterior views of the (a) MIRD, (b) MCAT, and (c) superquadric mathematical phantoms.

ellipsoids, cylinders, or rectangular volumes (Figure 3.2a). As we can see, the representation of internal organs is quite crude since the simple equations can only coarsely capture the most general description of the organ's position and geometry. For internal dosimetry purposes, such simple human model approximations serve quite sufficiently and have the advantage of allowing very fast calculation of the intersection of ray lines with the analytical surfaces which delineate the organs. However, such simple models are not as useful in imaging studies since results from them may not be indicative of what would occur in actual patients.

In order for mathematical phantoms to represent the structures with more realistic detail, the equations defining them need to incorporate higher-order complicated terms. However, this renders them more cumbersome and computationally intensive for most applications. Much research has been undertaken to make more realistic mathematical phantoms for imaging research while balancing the need to keep them computationally efficient.

The 4D Mathematical Cardiac-Torso (MCAT) phantom, as seen in Figure 3.2b, was developed along these lines for use in nuclear medicine imaging research, specifically for SPECT and PET. The anatomy of the 4D MCAT phantom was constructed using simple geometric primitives based on the MIRD model. However, overlap, cut planes, and intersections of the geometric objects were used to form a more realistic anatomy. In order to study the effects of cardiac and respiratory motions on SPECT and PET imaging, the 4D MCAT phantom included a beating heart model [4] and a model for the breathing motion based on known respiratory mechanics [5]. The MCAT phantom was specifically detailed in order to quickly generate projection data onto a simulated nuclear medicine camera for research. The MCAT has been widely used for evaluating and improving imaging instrumentation, data acquisition techniques, and image processing and reconstruction methods in nuclear medicine.

The use of higher-level surfaces has also been investigated as a means to create more realistic computerized models of the human anatomy. A phantom based on superquadric surfaces was developed by J. Peter to be used in Monte Carlo simulations for SPECT [6] (Figure 3.2c). Whereas the most complex geometric solid in the 4D MCAT phantom is defined as a quadric surface, the phantom by J. Peter is defined by superquadric surfaces which are more mathematically complex. Consequently, the phantom is more realistic than the 4D MCAT.

Other current mathematical phantoms of note include a series of models developed by the Oak Ridge National Laboratory (ORNL) that represent adults and pediatric patients at various ages [7], updated versions of the original MIRD model [8], and the FORBILD phantom series developed for x-ray CT imaging research [9]. Over the years, mathematical phantoms have improved in their modeling of anatomy, but they still do not come close to the level of voxelized phantoms.

Voxel-Based Phantoms

Voxelized phantoms [10–17] are based on segmented patient data and therefore provide a very realistic model of patient anatomy. Transmission computerized x-ray tomography (CT) or magnetic resonance images (MRI) can supply us the required high-resolution three-dimensional human anatomy necessary to construct a volume segmented phantom. By selecting "normal" male or female data, typical voxel-based phantoms can be created by segmenting the structures of interest and assigning unique index values to them. Figure 3.3 shows two examples of voxelized phantoms, a whole-body phantom based on CT data and a brain phantom based on MRI data, both

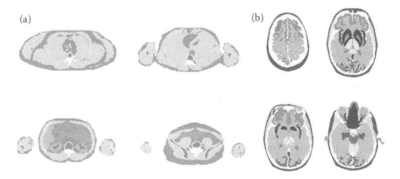

FIGURE 3.3
Transaxial slices from (a) a male whole-body voxelized phantom and (b) a male brain phantom developed by Zubal et al. [12].

TABLE 3.1

Organs for the Whole-Body Phantom

Void outside of phantom	Liver	Gas volume (bowel)
Skin/body fat	Gallbladder	Bone marrow
Brain	Left and right kidney	Lymph nodes
Spinal chord	Bladder	Thyroid
Skull	Esophagus	Trachea
Spine	Stomach	Diaphragm
Rib cage and sternum	Small intestine	Spleen
Pelvis	Colon/large intestine	Urine
Long bones	Pancreas	Feces (colon contents)
Skeletal muscle	Adrenals	Testes
Lungs	Fat	Prostate
Heart	Blood pool (all vessels)	Liver lesion

of male subjects. Both phantoms were defined using manual outlining and designation of anatomical structures. The different colors in the image slices indicate the segmented structures. Tables 3.1 and 3.2 show the organs and structures defined in each of these phantoms.

TABLE 3.2

Organs for the Brain Phantom

Skin	Medulla oblongata	Septum pellucidium
Brain	Fat	Optic nerve
Spinal cord	Blood pool	Internal capsule
Skull	Artificial lesion	Thalamus
Spine	Frontal lobes	Eyeball
Dens of axis	Bone marrow	Corpus callosum
Jaw bone	Pons	Special region frontal lobes
Parotid gland	Third ventricle	Cerebral falx
Skeletal muscle	Trachea	Temporal lobes
Lacrimal glands	Cartilage	Fourth ventricle
Spinal canal	Occipital lobes	Frontal portion eyes
Hard palate	Hippocampus	Parietal lobes
Cerebellum	Pituitary gland	Amygdala
Tongue	Cerebral fluid	Eye
Pharynx	Uncus (ear bones)	Globus pallidus
Esophagus	Turbinates	Lens
Horn of mandible	Caudate nucleus	Cerebral aquaduct
Nasal septum	Zygoma	Lateral ventricles
White matter	Insula cortex	Prefrontal lobes
Superior sagittal sinus	Sinuses/mouth cavity	Teeth
	Putamen	Lesion

One of the most difficult and interesting image processing problems deals with this task of delineating organ outlines within a diagnostic image. Even high-contrast, high-resolution images (like MRI or CT) are difficult to automatically segment into independent structures or organs. At the time of this writing, such computer vision algorithms, which intelligently locate surfaces of internal structures, are being investigated for very specific applications and have not been developed to the level where they can generally locate internal structures within three-dimensional volumetric human diagnostic images with high accuracy. All too often, still, we must rely on human operators to manually draw and connect edges in order to designate internal human structures. Although this would be intolerable if applied diagnostically to individual patients, it is well worth the effort for creating a very realistic typical representation of the human anatomy. As such, many different voxelized models have been developed for imaging research. A detailed review of what is currently available can be found in [18–20].

Despite being ultrarealistic, voxelized phantoms do have their disadvantages as compared to mathematical phantoms. Since they are fixed to a particular patient anatomy, it is difficult to model anatomical variations or motion with voxelized models. To model these things, one would have to assemble models based on many different patient datasets. This would take a great amount of work and a long time to achieve since every structure in the body would have to be segmented for every phantom, most manually. This can take many months to a year per phantom. Other disadvantages of a voxelized model are that generation of the phantom at other resolutions requires interpolation, which induces error. Also, exact ray-tracing calculations are difficult and time consuming since the organ shapes are not analytically defined. The calculations can only be approximated through a tedious search voxel per voxel along each ray in order to find the points of intersection with the structures defined in the voxelized phantom.

Hybrid Phantoms

Hybrid phantoms are the latest development in anthropomorphic phantoms. Such models seek to combine the advantages of voxelized and mathematical phantoms [21,22]. Hybrid phantoms are initially defined in a similar manner to that of their voxelized counterparts. Patient data, CT or MRI, is first segmented creating an initial voxelized model. The segmented structures are then fit with continuous, smooth surfaces. Nonuniform rational b-splines (NURBS) [23,24] or polygon meshes have been typically used to define each anatomical object. Based on patient data, the NURBS surfaces or polygon meshes can accurately model each structure in the body providing the realism of a voxelized model. They also have the flexibility of a mathematical phantom to model anatomical variations and motion. NURBS surfaces can be easily modified through the control points which define their shape while the polygon meshes can be easily altered through the vertex points that define the polygons.

Hybrid phantoms are typically rendered or revoxelized into a 3D image format, with the organs set to index values or user-defined property values, in order to input them into existing analytical or Monte-Carlo-based simulation packages. Since the organs and structures are defined by continuous, smooth surfaces, the phantoms can be generated at any resolution without using interpolation. In addition to voxelization, direct ray-tracing calculations are possible from hybrid phantoms. Ray-tracing calculations of polygon objects are straightforward and can be done analytically. NURBS surfaces, on the other hand, involve more complicated, iterative procedures which require more computational time. However, efficient ray-tracing techniques from computer graphics have been investigated to speed up this process [25].

One of the first hybrid phantoms was the 4D NURBS-based Cardiac-Torso (NCAT) phantom [5,21,26] (Figure 3.4), which was developed as the next-generation MCAT phantom. NURBS surfaces were used to define the organ shapes in the NCAT using the Visible Human Male CT data from the National Library of Medicine (NLM) [27] as their basis. The NLM CT data slices were manually segmented and 3D NURBS surfaces were fit to each segmented structure using the Rhinoceros NURBS modeling software [28]. The NCAT phantom was designed for nuclear medicine research and consisted of just the chest region. A separate female phantom was not created; female NCAT anatomies were modeled simply by adding breast surfaces to the base male torso. Like its MCAT predecessor, the NCAT was extended to four dimensions to model common patient motions such as the cardiac and respiratory motions. The motions were based on 4D tagged magnetic resonance imaging (MRI) data and high-resolution respiratory-gated CT data, respectively.

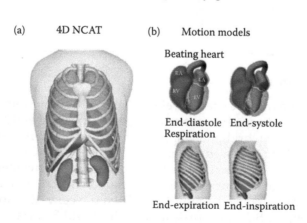

(a)   4D NCAT   (b)   Motion models

Beating heart

End-diastole   End-systole
Respiration

End-expiration   End-inspiration

FIGURE 3.4
(a) Anterior view of the 4D NCAT phantom. (b) Cardiac and respiratory motions.

To model motion, time curves were defined for the control points of the anatomical structures. Each control point's location in 3D space was determined by the time curves. At different points in time, the control points change locations altering the surfaces.

The 4D NCAT phantom greatly improved upon the realism of the geometry-based MCAT. However, the phantom was still limited in that it only modeled the torso region, did not include a separate model for the female, and was a highly smoothed representation of the anatomy having been designed for low-resolution nuclear medicine research. The NCAT phantom was recently revamped to include highly detailed whole-body male and female anatomies based on the high-resolution Visible Male and Female anatomical datasets from the NLM. The anatomical data was taken at a much higher resolution than the corresponding CT data. The anatomical data was segmented and the structures converted into 3D NURBS surfaces resulting in the next-generation NCAT phantom, the 4D extended Cardiac-Torso (XCAT) phantom [29] (Figure 3.5). The cardiac and respiratory models were also updated based on more state-of-the-art gated CT data. With its vast improvements, the 4D XCAT is applicable to nuclear medicine as well as more high-resolution imaging modalities such as CT or MRI. Combined with accurate models of the imaging process, the phantom can simulate realistic imaging data (Figure 3.6).

With their ability to maximize the advantages of both voxelized and mathematical phantoms, many 3D and 4D hybrid phantoms [14,18,22,30–35] have been developed to model various anatomical variations within different age groups from pediatric to adult. Figure 3.7 shows 3D models from the University of Florida created for different ages. In addition to human subjects, many animal models are also being developed [36,37] for using in small animal imaging research.

Discussion

Much progress has been made in phantom development in recent years, from the simple geometrically based phantoms to the more detailed hybrid models. As computational phantoms get more and more sophisticated, they allow users to simulate diagnostically realistic images close to that from actual patients. With such advances, computational phantoms give us the ability to perform experiments entirely on the computer with results that are indicative of what would occur in live patients. Phantoms provide a known truth from which to optimize, evaluate, and improve imaging devices and techniques. Since we are able to model a known source distribution and known attenuator distribution, Monte Carlo simulations give us projection data which not only closely resemble clinical data, but also include additional information not determinable in patient studies. Such data sets can

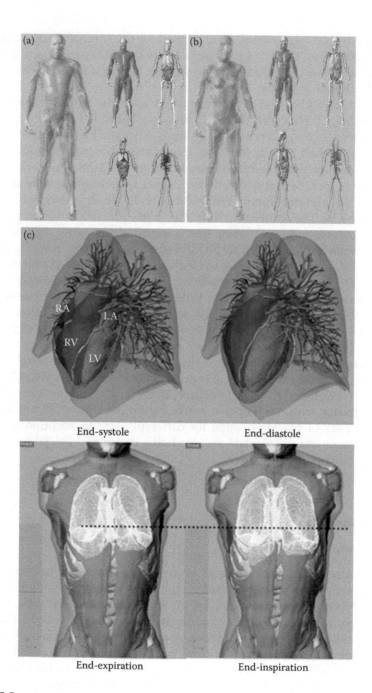

FIGURE 3.5

(a,b) Anterior views of the male and female 4D XCAT phantoms. Different levels of detail are shown with transparency. (c) Improved cardiac and respiratory motions of the XCAT phantom.

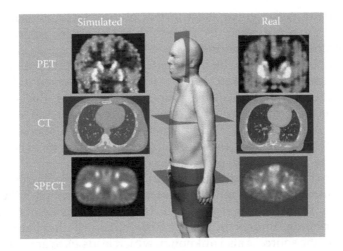

FIGURE 3.6
Simulated images of the XCAT phantom compared to actual patient images for different modalities.

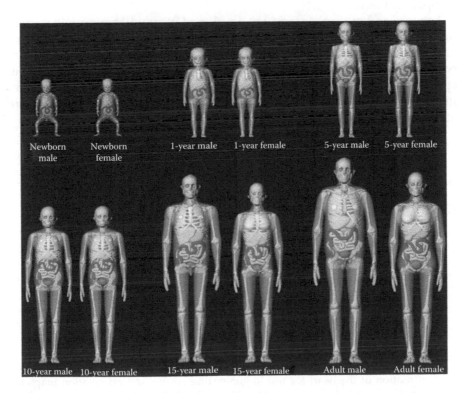

FIGURE 3.7
Anterior views of the series of UF hybrid pediatric and adult phantoms.

help to better understand the image formation process for clinically realistic models, and can prove especially interesting in testing and improving tomographic reconstruction algorithms.

New imaging devices can be investigated using *in vivo* simulations. Early design changes can be realized before studies are conducted in living models. One of the advantages of developing realistic human is that such simulations can decrease the necessity of conducting experimental studies using live animal subjects—particularly primates. Dose calculations for internal and external radiation sources using computational phantoms can give new insights in the field of health physics and therapy.

The original impetus for developing some of the phantoms described here was to simulate SPECT imaging geometries. Reconstructing SPECT source distributions from projection data is a difficult task since not only is the source distribution unknown, but the distribution of attenuating material surrounding the source is also unknown, which leads us to apply substantial approximations during the reconstruction process. Simulating projection data from software phantoms allows us to compare our reconstructed image to a known "gold standard" where the source and attenuating material are completely known.

As more and more phantoms are developed, they will have a widespread use in imaging research to develop, evaluate, and improve imaging devices and techniques and to investigate the effects of anatomy and motion. They will also provide vital tools in radiation dosimetry to estimate patient-specific dose and radiation risk and optimize dose reduction strategies, an important area of research given the high amounts of radiation exposure attributed to medical imaging procedures.

References

1. Poston J, Bolch W, Bouchet L. Mathematical models of human anatomy. In: Zaidi H, Sgouros G, eds. *Monte Carlo Calculations in Nuclear Medicine: Therapeutic Applications.* Bristol, UK: Institute of Physics Publishing; 2002: pp. 108–132.
2. Shepp LA, Logan BF. The Fourier reconstruction of a head phantom. *IEEE Trans. Nucl. Sci.* 1974;NS-21:21–43.
3. Snyder WS, Ford MR, Warner GG, Fisher HL. Estimates of absorbed dose fractions for monoenergetic photon sources uniformly distributed in various organs of a heterogeneous phantom. *J. Nucl. Med.* 1969;10(Supplement Number 3, Pamphlet #5).
4. Pretorius PH, King MA, Tsui BMW, LaCroix KJ, Xia W. A mathematical model of motion of the heart for use in generating source and attenuation maps for simulating emission imaging. *Med. Phys.* 1999;26:2323–2332.
5. Segars WP, Lalush DS, Tsui BMW. Modeling respiratory mechanics in the MCAT and spline-based MCAT phantoms. *IEEE Trans. Nucl. Sci.* 2001;48:89–97.

6. Peter J, Jaszczak R, Coleman R. Composite quadric-based object model for SPECT Monte-Carlo simulation. *J. Nucl. Med.* 1998;39:121P.
7. Cristy M, Eckerman KF. *Specific Absorbed Fractions of Energy at Various Ages from Internal Photon Sources.* ORNL TM-8381 Oak Ridge, TN: Oak Ridge National Laboratory. 1987.
8. Kramer R, Zankl M, Williams G, Jones DG. The calculation of dose from external photon exposures using reference phantoms and Monte-Carlo methods, Part 1: The male (ADAM) and female (EVA) adult mathematical phantoms. *GSF Bericht.* 1982;S-885.
9. FORBILD Phantom Database. http://www.imp.uni-erlangen.de/phantoms.
10. Shi CY, Xu XG. Development of a 30-week-pregnant female tomographic model from computed tomography (CT) images for Monte Carlo organ dose calculations. *Med. Phys.* 2004;31(9):2491–2497.
11. Xu XG, Chao TC, Bozkurt A. VIP-man: An image-based whole-body adult male model constructed from color photographs of the visible human project for multi-particle Monte Carlo calculations. *Health Phys.* 2000;78(5):476–486.
12. Zubal IG, Harrell CR, Smith EO, Rattner Z, Gindi GR, Hoffer PB. Computerized three-dimensional segmented human anatomy. *Med. Phys.* 1994;21:299–302.
13. Kramer R, Khoury HJ, Vieira JW, Lima VJM. MAX06 and FAX06: Update of two adult human phantoms for radiation protection dosimetry. *Phys. Med. Biol.* 2006;51(14):3331–3346.
14. Lee C, Williams J, Bolch W. The UF series of tomographic anatomic models of pediatric patients. *Med. Phys.* 2005;32(6):3537–3548.
15. Kramer R, Khoury HJ, Vieira JW, et al. All about FAX: A female adult voxel phantom for Monte Carlo calculation in radiation protection dosimetry. *Phys. Med. Biol.* 2004;49(23):5203–5216.
16. Kramer R, Vieira JW, Khoury HJ, Lima FRA, Fuelle D. All about MAX: A male adult voxel phantom for Monte Carlo calculations in radiation protection dosimetry. *Phys. Med. Biol.* 2003;48(10):1239–1262.
17. Petoussi-Henss N, Zankl M, Fill U, Regulla D. The GSF family of voxel phantoms. *Phys. Med. Biol.* 2002;47(1):89–106.
18. Zaidi H, Xu G. Computational anthropomorphic models of the human anatomy: The path to realistic Monte Carlo modeling in radiological sciences. *Annu. Rev. Biomed. Eng.* 2007;9:471–500.
19. Caon M. Voxel-based computational models of real human anatomy: A review. *Radiat. Environ. Biophys.* 2004;42(4):229–235.
20. Xu G, Eckerman KF. *Handbook of Anatomical Models for Radiation Dosimetry.* Boca Raton, FL: CRC Press/Taylor & Francis; 2009.
21. Segars WP. *Development of a New Dynamic NURBS-Based Cardiac-Torso (NCAT) Phantom.* PhD dissertation, The University of North Carolina; May 2001.
22. Lee C, Lodwick D, Hurtado J, Pafundi D, Lwilliams J, Bolch WE. The UF family of reference hybrid phantoms for computational radiation dosimetry. *Phys. Med. Biol.* 2010; 55(2):339–363.
23. Piegl L. On NURBS: A survey. *IEEE Comput. Graph. Appl.* 1991;11:55–71.
24. Piegl L, Tiller W. *The Nurbs Book.* New York: Springer-Verlag; 1997.
25. Segars WP, Mahesh M, Beck TJ, Frey EC, Tsui BMW. Realistic CT simulation using the 4D XCAT phantom. *Med. Phys.* 2008;35(8):3800–3808.
26. Segars WP, Lalush DS, Tsui BMW. A realistic spline-based dynamic heart phantom. *IEEE Trans. Nucl. Sci.* 1999;46:503–506.

27. Visible Human Male and Female Datasets. http://www.nlm.nih.gov/research/visible/visible_human.html.
28. Rhinoceros (computer program). Version 1.0. Seattle, WA, 1998.
29. Segars WP, Mendonca S, Grimes J, Sturgeon G, Tsui BMW. 4D XCAT phantom for multimodality imaging research. *Med. Phys.* 2010;37:4902–4915.
30. Caon M. Voxel-based computational models of real human anatomy: A review. *Radiat. Environ. Biophys.* 2004;42:229–235.
31. Dimbylow P. Development of pregnant female, hybrid voxel-mathematical models and their application to the dosimetry of applied magnetic and electric fields at 50 Hz. *Phys. Med. Biol.* 2006;51(10):2383–2394.
32. Lee C, Lodwick D, Hasenauer D, Williams JL, Lee C, Bolch WE. Hybrid computational phantoms of the male and female newborn patient: NURBS-based whole-body models. *Phys. Med. Biol.* 2007;52(12):3309–3333.
33. Lee C, Lodwick D, Williams JL, Bolch WE. Hybrid computational phantoms of the 15-year male and female adolescent: Applications to CT organ dosimetry for patients of variable morphometry. *Med. Phys.* 2008;35(6):2366–2382.
34. Lee C, Lee C, Lodwick D, Bolch WE. NURBS-based 3-D anthropomorphic computational phantoms for radiation dosimetry applications. *Radiat. Prot. Dosim.* 2007;127(1–4):227–232.
35. Segars WP, Sturgeon G, Li X, et al. Patient specific computerized phantoms to estimate dose in pediatric CT. *Proceedings of the SPIE Medical Imaging*, Lake Buena Vista, FL, USA. 2009.
36. Padilla L, Lee C, Milner R, Shahlaeel A, Bolch WE. Canine anatomic phantom for preclinical dosimetry in internal emitter therapy. *J. Nucl. Med.* 2008;49(3):446–452.
37. Segars WP, Tsui BMW, Frey EC, Johnson GA, Berr SS. Development of a 4-D digital mouse phantom for molecular imaging research. *Mol. Imaging Biol.* 2004;6(3):149–159.

4

The Gamma Camera: Basic Principles

Sven-Erik Strand, Mikael Peterson, and Magnus Dahlbom

CONTENTS

The gamma camera is an imaging detector where photons are converted to electrical signals carrying information on energy and positioning. Nowadays, gamma cameras can be divided into scintillation crystal-based and solid-state detector-based cameras.

The scintillation camera, also called the Anger camera, was invented by Hal O. Anger in 1956/57 [1,2]. In 1962, the first commercial camera was introduced by Nuclear Chicago Inc., with a thallium-doped sodium iodide crystal, NaI(Tl), 25 cm in diameter and thickness of 1.25 cm, using 19 photomultiplier (PM) tubes. With the development of techniques for growing larger NaI(Tl) crystals, cameras with wider fields were developed from the mid-1970s. Although ^{99}Tc^m was discovered already in 1938 [3], it was not until the middle of the 1960s, ^{99}Tc^m-labeled radiopharmaceuticals became widely available. ^{99}Tc^m has a photon energy of 140.5 keV, which is optimal for scintillation camera imaging and has favorable dosimetric properties. It promoted the rapid growing interest in the scintillation camera technique, with the camera's possibility of high sensitivity, large field of view, and enabling dynamic studies. Since its invention, the scintillation camera has been developed and the performance has improved significantly; however, the basic operational principles remain the same. Today, about a dozen vendors are manufacturing cameras with many different models. Figure 4.1 shows an example of a dual-headed scintillation camera.

The scintillation camera is the most used registration system in nuclear medicine today for both planar and tomographic studies [4]. The photon energy range covered is from 40–50 keV up to 400 keV normally. Some cameras are shielded and can be used up to 511 keV, making it possible to study positron-emitting radionuclides/pharmaceuticals. Today's system can be characterized as composed of scintillation camera head(s), gantry, and computer. The computer is often separated in acquisition and processing stations.

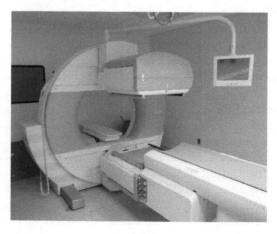

FIGURE 4.1
Modern dual-headed scintillation camera with SPECT capability and with CT.

The components together make up an integrated system. To increase the sensitivity, multihead cameras with two, three, or four heads are available, where camera with two heads is the most commonly used configuration in clinical routine. The improvements in scintillation camera performance in recent years are in sensitivity, uniformity, intrinsic spatial resolution, energy resolution, and dead time. Details about camera construction and function can be found in several textbooks and review articles [5–7].

The NaI(Tl) crystal-based gamma camera is by far the most commonly used device in conventional nuclear medicine imaging. However, over the last decade or so, there have been a number of other imaging devices developed for gamma imaging. Many of these are devices, dedicated to specific imaging tasks, such as cardiac imaging, breast imaging, and preclinical imaging. In addition to new system geometries, some systems are also utilizing new detector materials, including scintillation materials other than NaI(Tl) and solid-state detectors both as gamma ray detectors as well as photodetectors. A brief overview of these specialized systems will also be discussed in this chapter.

Although many parameters for the camera can be numerically calculated, the Monte Carlo technique is an important tool for optimizing those imaging parameters, difficult to calculate or measure, as shown throughout this book.

Principles for the Scintillation Camera

The scintillation camera can be considered as a photon registering and an image-forming detector. Photons emitted from the activity distribution in the patient are projected through the collimator onto the NaI(Tl) crystal, where they interact and create scintillation light. The scintillation light is detected by an array of PM tubes, and transformed into current pulses, proportional to the flux of the incoming light. These pulses are run through and processed by analog and/or digital electronic circuits, and computed with suitable algorithms resulting in output signals (data) representing the centroid of the scintillation light distribution created in the crystal and the energy depositions in the crystal.

Detector

The detector part of the scintillation camera consists of a NaI(Tl) crystal, optically coupled to the PM tubes. The crystal can be of different sizes and shapes, and the number of PM tubes varies between system designs. Circular camera heads have been designed with diameters up to 50 cm. The number of PM tubes coupled to the crystal varies greatly between systems and can be between 37 up to over 100 tubes, circular, square, or hexagonal in shape. Newer cameras are typically designed with rectangular-shaped crystals, with dimensions of up to 66×45 cm and a thickness of 9.5 mm (3/8 in.). In these systems, up to 115 hexagonal or circular PMTs are used with diameters from 1.5 to 7.5 cm.

The primary decay time at room temperature for the scintillation light in NaI(Tl) is 0.23 µs for 60% of the light, with 1.15 µs for the rest. However, there exists a 0.15-s long-lived afterglow component that contributes approximately 9% of the total light yield [8,9]. In the energy range used in nuclear medicine (i.e., 40–511 keV), there is a linear relationship between scintillation light intensity (i.e., light output) and photon energy. The detection efficiency of scintillation light output varies with many factors such as crystal size and thickness, location of interaction in the crystal, and variations in photocathode sensitivity, giving some variations in energy response. During the crystal growing process, the crystals are drawn from the molten mixture of NaI and small amounts of Tl; some variations in thallium concentration can be expected along the primary and radial axes. Typically, scintillation light intensity varies generally <5%.

After the NaI(Tl) crystal is grown, it is cut, etched, and polished, and the surface is treated to maximize the light output. The NaI(Tl) crystal is hygroscopic and is therefore hermetically sealed in a thin aluminum cover (<1 mm thick), except on the side facing the PM tubes where a transparent light guide is applied. The other surfaces are surrounded with a diffuse reflector (aluminum oxide). The whole detector assembly is mounted in a lead housing with walls thick enough (about 1–2 cm) for shielding of photon penetration up to at least 500 keV. Thus, the transmission of low-energy photons through the cover is small. In front of the detector, the lead cover can be even thicker.

Although NaI(Tl) is by far the most commonly used scintillator material, there are other materials that are used in different detector designs. Pixelated arrays of CsI(Tl) is one such example that has been used in detector heads for human imaging [10]. Thin films of the same material have also been grown specifically for the use as a detector material of low-energy gammas such as the 27 keV x-ray emission from I-125, a radionuclide often used in small animal imaging [11].

Optical Coupling: Light Guide

The crystal and the PM tubes are optically coupled via a light guide consisting of a 5–25-mm-thick sheet of glass or poly(methyl methacrylate) (PMMA). To minimize light losses, the components are optically coupled with optical glue, typically silicon oil or grease. The light guide keeps the PM tubes further away from the crystal, thus enabling more PM tubes to collect scintillation light from each light flash. The light guide helps to scatter light into the PM tubes as the angle for total reflection between glass and NaI is 53.1°. The geometrical efficiency of the PM tube is dependent on the total thickness of crystal and light guide and the area of the PM tube.

The PM tubes should be mounted such that the difference between the collected light of neighboring tubes varies linearly with the distance to the scintillation point in the crystal. The larger the difference between these signals, the better the ability for the camera to localize the scintillation event. One important factor is the packing density of the PM tubes. By using hexagonal PM tubes, better packing with less loss of scintillation light is achieved.

To improve the spatial resolution, the thickness of the light guide is reduced. However, then the nonlinearity in positioning is increased. Corrections must be made either mechanically (light guide geometric design, PM tube geometry) or electronically. Both the energy and the spatial resolutions in the camera are dependent on good scintillation light to photoelectron conversion efficiency. Improvements have been achieved by matching the spectral sensitivity of the PM tube to the emission spectrum of the scintillation crystal (using PM tubes with bialkali photocathodes).

Electronics

When a scintillation event occurs in the crystal, scintillation light is collected in the PM tubes in a certain proportion to their solid angle from the light emission point. The current created in the tubes is thus independent of the distance between scintillation light emission point and PM tube. The signals produced in the array of PM tubes is then used to calculate the position of the event. The current in the PM tubes is typically integrated for about 1 μs, to create a pulse, which includes the charge from the entire light signal. The total charge registered by all PM tubes is summed and is proportional to the energy deposition in the crystal. This signal is designated as the "energy signal" or the "Z-signal" and is used for energy discrimination.

Scintillation camera systems are sometimes called analog or digital cameras, depending on the degree of digital electronics in the detector. The general meaning is that the circuits doing the position and the energy calculations can be either analog or digital. In newer cameras, the signals are digitized at an early stage in the processing chain and most pulse processing and corrections are done digitally.

Analog Circuits (Anger Logics)

The pulse shaping occurs in the preamplifiers and the signals are fed through a resistor matrix to summing amplifiers, one for each of four position directions, X_+, X_-, Y_+, and Y_-. All summing, integration, calculation, and normalization take place in analog circuits and position and total energy are calculated from simple algorithms as

$$X = (X_+ + X_-)/Z$$
$$Y = (Y_+ + Y_-)/Z \qquad (4.1)$$
$$Z = X_+ + X_- + Y_+ + Y_-$$

In this simple algorithm, the different PM tube signals are weighed with a factor dependent on the distance from the center of the crystal, the origin of the coordinate system. The division of the position signal with the energy or Z signal makes the image coordinates independent of the statistical variation in pulse amplitude within the energy window.

Some electronic operations are slow (e.g., signal integration and pulse shaping), and the light decay with the electronics creates a system with a large time constant. To prevent large "dead times," the pulses are integrated for a limited time and sometimes also truncated. A typical pulse length in the detector electronics is 0.5–3 μs.

Digital Circuits

Pulse shaping and, sometimes, pulse clipping take place in the preamplifiers. An analog event detector works in real time, detects a valid event including coarse position, and triggers off the digital sampling. The outputs from the preamplifiers are simultaneously digitized in a set of A/D-converters and all signals or a subset of all signals is processed further. Calculations and normalization for position and energy similar to the analog circuit are now done digitally and are processor controlled. In camera systems with digital circuits, the processor is often also used for baseline stabilization and automatic PM-tube/preamplifier gain control.

Pile Up and Mispositioning

Dead time and pulse pile up can make absolute quantification difficult due to nonlinearity between activity (i.e., photon fluence rate impinging on the crystal) and the observed count rate. Mispositioning could further decrease quantification ability and degrade the image quality [12,13]. When the crystal is exposed to a high photon fluence rate, the long decay time of the scintillation light creates such a long afterglow at a preceding event site that the PM tubes, while registering the scintillation light from an event, still collect light from the former event. Thus, the signal from each PM tube is the sum of the total light emission in the crystal and thus the events will be mispositioned. The effects have been described in detail [14]. Remember that it is the photon fluence for the whole energy spectra that is important. All photons interacting in the crystal cause scintillation light and despite their energy they contribute to the signal from the PM tubes.

Correction Circuits

The performance of the scintillation camera detector has improved since the 1950s. However, there are uniformity and image quality problems that are intrinsic to the scintillation camera and are caused by how the gamma events are collected and processed. Nonlinearity is one major problem in the detector, but other causes of image degradation are: (a) defects in the NaI(Tl) crystal, (b) nonperfect light coupling between crystal and PM tubes, (c) limited number of PM tubes, (d) electronic dead time, and (e) collimator defects. A number of built-in corrections can be found in a scintillation camera today to reduce degradation in spatial linearity, energy resolution, time resolution, and sensitivity. As the camera ages, it is not uncommon that these corrections have to be recalculated and/or recalibrated.

Corrections in Light Guide and Preamplifier

Stationary mechanical corrections are specially designed light guides with varying light transmission, which is prefabricated with the camera. Preamplifiers with nonlinear amplification can also be used, where an event under the PM tube is relatively less amplified compared with an event away from the tube. In a digitally controlled detector, the signal from any PM tube directly under an event is attenuated.

Correction for Spatial Nonlinearity

Position errors occur because a discrete array of tubes is used to detect events over a continuous crystal surface. "Coordinate bunching" causes events that occur between PM tubes to be positioned toward the center of the tubes. A matrix of correction factors can be created with the aid of a lead hole, line, or grid pattern mask placed on the camera detector. These correction factors are then used to position the events in the correct position in the field of view (FOV).

Correction for Energy Nonlinearity

Since events occurring directly under a PM tube tend to generate a stronger signal, for the events occurring between the PM tubes, it is necessary to map out this variation in signal intensities which otherwise would produce image nonuniformities. The solution is the same principle of a correction matrix as for spatial nonlinearity. A mask over the camera surface defines the matrix points of corrections.

Correction for Remaining Nonuniformity

In spite of the corrections mentioned before, there is still typically a residual nonuniformity. This is due to imperfections in these corrections and to varying response in the collimator. Correction for this is done with a high-count flood image with homogenous photon fluence density over the field of view.

Correction for Drift

In order for the corrections described earlier to work correctly, the system has to be stable in terms of signal output. However, the PM tubes and the high voltage can suffer from changes in gain, which will influence the signal amplitudes. These drifts have to be monitored and adjusted for. This can be accomplished by exposing each PM tube regularly with stable light source (e.g., LED), or by monitoring the pulse height distribution in a "tracking" window and then automatically adjusting the gain or high voltage.

Correction for Dead Time and Pulse Pile Up

The "dead time" in modern cameras is about 0.5–3 μs due to processing of signals and corrections. The dead time can be reduced by variable integration time and double buffering. Another cause of dead time is pulse pile up at high photon fluence rates. Pile up in the crystal and in the electronic amplifiers can be corrected for by "pile up rejection" circuits. The condition for rejection is set up after analyzing the event pulses. Two examples of pile up rejection are (1) the integrated value of the pulse is compared at a certain predefined time with the level of the original pulse. A reject condition for nontolerable pile up can be set; and (2) by analyzing the leading edge of the event pulse and matching to a predefined criterion for pile up. Pile up pulses can also, after detection, be restored. In a double event pulse, the envelope and value of a single pulse are roughly calculated and subtracted from the pile up pulse. Thus, both pulses are restored giving less system count losses [15].

In recent years, the issues on dead time and pile up have grown in interest because of the increased interest in radionuclide therapy and especially dose planning [16]. Recent studies on the effect of high activities in the FOV have been evaluated in modern cameras showing that this effect still can cause problems in some gamma cameras [17].

Photon Collimation

The collimation is "mechanical" sorting of the photons impinging on the crystal, preventing those not projecting the activity distribution on the crystal from entering the crystal. The type of collimator used is what ultimately determines the final spatial resolution in the image and the overall system sensitivity. The dimension of the collimator holes determines the sensitivity and the spatial resolution. The essential factor is the solid angle from the object through the holes that the crystal can be seen.

Parallel Hole Collimators

The parallel hole collimator is the most widely used collimator, with field of view matched to the useful field of view of the crystal. Specially designed collimators have been made to overcome some collimator limitations [18].

Two parameters characterize collimators: geometric spatial resolution, R_c and collimator efficiency, ε_c. Overall, both R_c and ε_c are dependent on the ratio of the hole diameter and hole length (d/l). R_c varies with d/l and improves when the ratio decreases. ε_c is almost proportional to $(d/l)^2$ and decreases when d/l decreases. Good spatial resolution at large distances needs long and small holes whereas high sensitivity requires short and wide holes [7].

For clinical studies, a good compromise must be arrived at due to the opposed dependence of these parameters. One limiting factor is the activity administered keeping the absorbed dose low to the patient. This will not be the case for applications in radionuclide therapy, where the high photon fluence will make it possible to have specially designed collimators reducing the count rate from the camera below pile up and dead time effects, hence enabling a better spatial resolution.

Other important parameters are collimator scatter and septum penetration that may broaden the point spread function (PSF), by decreasing the spatial resolution and contrast [19,20]. A thumb of rule is that septum penetration should be below 5% thus resulting in a septum thickness t, larger than

$$t > \frac{6d\mu^{-1}}{1 - 3\mu^{-1}} \tag{4.2}$$

where μ is the linear attenuation coefficient of lead. In the choice of collimator, one needs to consider the whole energy spectra for the radionuclide. In studies for ^{123}I, it was shown that the best imaging characteristics were not the collimator optimal for ^{123}I's main photon energy of 159 keV, but a collimator designed for higher energies, due to the septum penetration of the low abundant photons for 440 and 529 keV (7,14,28). The same reasoning will be for ^{131}I where the high-energy photons will cause problems with septum penetration to be corrected for in quantification [21].

Pinhole Collimators

The pinhole collimator, also applicable to newer SPECT cameras (i.e., cardiac or small animal imaging), can be used when small objects should be imaged [22,23]. Pinhole SPECT offers the best combination of spatial resolution and sensitivity for imaging small objects positioned close to the collimator. Although a single pinhole collimator can achieve spatial resolution better than the intrinsic resolution of the detector, limited by pinhole aperture size, the sensitivity is not adequate for large organ clinical diagnostic imaging.

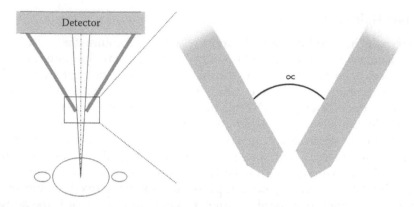

FIGURE 4.2
A general pinhole setting. A infinitesimal small source in an object will emit photons that pass through the pinhole and then hit the detector. The width of the projection on the detector is primarily determined by the pinhole size (true for large pinholes) and the pinhole to object/detector distances

Pinhole collimators for use in small animal SPECT was first thoroughly investigated in the beginning of the 1990s [24]. When using pinhole collimation, the intrinsic resolution of the detector is of great importance [25]. The system resolution in camera systems equipped with parallel hole collimation is primarily determined by the collimator resolution and there is no extra benefit of using detectors with higher intrinsic resolution for these systems. When using pinhole collimation, the image is projected through a small hole onto the detector (see Figure 4.2). The image is magnified by a factor proportional to the ratio of the pinhole-to-detector distance and pinhole-to-image plane distance. If the detector resolution is increased, it is possible to decrease the magnification by moving the detector closer to the pinhole since the image can be projected on a smaller portion of the detector with preserved detail information. This leaves room for more projections onto the detector, that is, more pinholes and as a consequence a better sensitivity. The increase in sensitivity can also be traded for better resolution by decreasing the size of each individual pinhole.

The situation described earlier, where several pinhole projections are projected onto a single detector, is referred to as multipinhole systems. But detector resolution is seldom increased with maintained detector size. But the foregoing conclusions are still valid if one increases the number of detectors since smaller detectors allow a higher packing fraction close to the pinholes. Following this reasoning, it is concluded that the detector performance for pinhole imaging depends on both the detector area and resolution and that one of these quantities alone is not a suitable measure of detector performance. Therefore, Barrett [26] introduced a new imaging quantity for detector performance called space-bandwidth product (SBP).

$$SBP = \frac{(Area\,of\,detector)}{(Area\,of\,PSF)} \qquad (4.3)$$

SBP is the ratio of the active detector area and the smallest possible resolvable element; consequently, a large detector with a high resolution yields a high SBP. In some systems, large scintillation detectors are used with multipinhole collimators. The pinholes are arranged in such a way that they project multiple images onto the camera surface, allowing a high resolution and high sensitivity to be obtained at the same time. Collimators can consist of a tungsten cylinder in some systems with five rings of 15 pinhole apertures. In SPECT mode, the pinhole positions in each ring are rotated with respect to the pinhole positions in adjacent rings, to increase angular sampling of the object. The collimators are surrounded with tungsten shielding that prevents the projections on the detector from severe overlapping [27].

High-energy imaging with pinholes would allow the imaging of PET and SPECT tracers simultaneously. Positron emitters have been imaged using a single traditional pinhole or using uranium pinhole insert. Severe pinhole edge penetration of 511 keV annihilation photons makes the multipinhole collimators currently used unsuitable for high-resolution imaging of positron emitters. Clustered multipinhole based on several pinholes with small acceptance angles grouped in clusters has been suggested [28] and now included in small animal imaging systems.

An alternative way of performing high-resolution SPECT is by using image intensifiers [29,30] and constructing pinhole collimators of Rose metal [31]. By this approach, a simple and inexpensive system can be achieved with high resolution.

Focusing Collimators

Diverging and converging (focusing) collimators are typically used for specialized imaging situations. The diverging collimator allows imaging of organs greater than the physical FOV of the scintillation camera. This type of collimator is still in use in portable systems for lung imaging in the intensive care units (ICU). Converging collimators have a better sensitivity compared to the other collimator types, but have a limited FOV, which depends on the focal length of the collimator. This collimator has been used in both brain and cardiac imaging, where the organ fits within the FOV of the camera. A variation of the converging collimator is the confocal collimator, which is a variable focus converging collimator. The collimator is designed to increase the number of collected count from a specific region (e.g., the heart) and avoids truncation artifacts. This type of collimator was first introduced in the early 1990s [32] and has recently been reintroduced commercially as the IQspect collimator [33]. Using the confocal collimator, the number of counts from the heart can be at least doubled which either would allow a reduction in acquisition time and/or a reduction in injected dose.

Sensitivity and Resolution

The system sensitivity of the scintillation camera is governed by two factors: (1) geometrical efficiency of the collimator, ε_g, defined as the number of photons emitted from the object that passes through the collimator holes, and (2) the photo peak detection efficiency of the crystal, f, defined as the part of the photons impinging on the crystal that results in an event, registered within the photo peak window.

The system spatial resolution of the scintillation camera depends on (c) geometrical resolution of the collimator, R_c, (d) the ability of the crystal to transfer the "absorbed energy image" to a "scintillation light image," and (e) the ability in optical transfer of a scintillation light image to the PM tube cathodes and of the electronics to transfer this "scintillation light image" to a readable image.

The two points (d) and (e) comprise the intrinsic spatial resolution of the camera, R_i. The position resolution in a scintillation camera is its ability to separate two nearby point sources in the image. The spatial resolution is most accurately described by the point or line spread functions, PSF or LSF, or the modulation transfer function, MTF, which is the absolute value of the Fourier transform of the PSF or LSF. A simplified description of the spatial resolution is the Full-With at Half-Maximum or Tenth-Maximum (FWHM, FWTM) of the PSF or LSF. There are two contributions to the detector PSF that can be mathematically described as the convolution of the camera-intrinsic PSF with the collimator PSF.

If the crystal is exposed for an infinitesimal narrow beam of photons, the image will be a count rate distribution, almost Gaussian in shape. To separate two such line sources, these must be separated by a minimal distance R, the resolution distance. The relationship is $R = 0.87 \times \text{FWHM}$ due to the contribution from the LSF tail. The FWHM of the intrinsic resolution is primarily caused by multiple Compton scattering of photons in the crystal, scintillation light scatter, and statistical variation in the production of electrons in the PM tube cathode.

The position of the photon interaction point in the crystal will be the center of gravity of the total emitted scintillation light. The higher the energy and the thicker the crystal, the higher the probability for multiple Compton processes, causing deteriorated intrinsic spatial resolution.

The number of photoelectrons emitted from the photocathode in the PM tubes following a gamma event in the crystal suffers from statistical variations. The fractional standard deviation in the number of photoelectrons is proportional to $1/\sqrt{E}$, where E is equal to the energy deposition. Thus, the position resolution in the X- and Y-signals will vary with $1/\sqrt{E}$ (i.e., the spatial resolution will improve with photon energy). To overcome some of that uncertainty, the electron efficiency in the photocathode should be as high as possible.

The use of PM tubes with as high quantum efficiency, $Q(E)$, as possible at the photon energy E is used. The intrinsic resolution relationship then for different photon energies will be

$$R_i(E') = R_i(E)\sqrt{\frac{E\ Q(E)}{E'\ Q(E')}} \tag{4.4}$$

The intrinsic spatial resolution, R_i, in scintillation cameras has improved since its introduction 30 years ago. Today, R_i is in the order of 2.5–3 mm for 140 keV compared to about 12 mm in 1970. Computer simulations taking into account statistical phenomena for the scintillation light including reflection–refraction–absorption and transmission have shown that theoretically better than 2 mm will hardly be achievable.

Multiple Compton scatter in the crystal can be decreased using thinner crystals improving the spatial resolution, however, at the cost of loss in camera sensitivity.

The energy resolution $\Delta E/E$ is proportional to $1/\sqrt{E}$ for the same reason as the spatial resolution. Energy resolution is however also dependent on a uniform light transmission to the cathodes. The value today is about 8–10% for 140 keV photon energy. The better the energy resolution, the narrower is the energy window to be used, giving less Compton scattered photons registered in the window. Also, multiple radionuclide imaging of nearby photon energies will be feasible [34].

It should also be mentioned that in cameras with discrete detector elements with individual photodetector readouts, the intrinsic resolution is primarily determined by the size of the individual detector elements rather than the amount of light produced per photon absorption. Energy resolution, on the other hand, follows the same behavior as in conventional gamma cameras.

System Sensitivity and System Resolution

The scintillation camera's system sensitivity, S, can be calculated from the collimator geometric sensitivity, ε_c, and the intrinsic efficiency dominated by the photo peak efficiency of the crystal, f, according to

$$S = n \cdot \varepsilon_c \cdot f \tag{4.5}$$

where n is the number of emitted photons per disintegration. Sensitivities for parallel hole collimators are between 100 and 200 cps/MBq. The total or system spatial resolution for the scintillation camera R_s depends on the collimator geometric resolution R_c and the intrinsic spatial resolution R_i. Then

$$R_s = \sqrt{R_i^2 + R_c^2} \tag{4.6}$$

The system resolution will also be distorted by Compton scattered photons collected in the energy window, R_{sc}, then the total spatial resolution for the system will be

$$R_s = \sqrt{R_i^2 + R_c^2 + R_{sc}^2} \tag{4.7}$$

Intrinsic Linearity

Intrinsic linearity is the ability of the detector (without collimator) to image straight radioactive sources as straight lines in the image. Two types of measures are given, intrinsic spatial differential and absolute linearity: (1) differential linearity that is the standard deviation of the deviation from that line in the image, and (2) the absolute linearity is the maximum displacement. Typical differential linearity values for modern cameras are in the order of 0.2–0.5 mm.

Intrinsic Uniformity

The ability of imaging a homogenous photon fluence rate impinging on the scintillation camera crystal is determined by the nonlinearity in the positioning and the sensitivity variations over the crystal. Position nonlinearity is an effect of the "camera optics," where light guide, PM tubes, and electronics will result in small shifts in the true coordinates of the event. Sensitivity variations over the crystal can be due to crystal defects, PM tube variations, optical coupling, and so on. One direct effect is variation in the pulse-height distribution, where the full energy peak will shift between different channels. Values, calculated before any uniformity correction, are in the order of 2–6% and 2–4% for integral and differential uniformity, respectively.

SPECT

When rotating the scintillation camera for tomographic imaging (SPECT), there are besides the reconstruction algorithm, attenuation, and scatter problems, other factors such as mechanical stability and functional variations with rotation to consider. One important parameter is the center-of-rotation (COR) where the mechanical rotation axis projected to the scintillation crystal face should coincide with the center of the image matrix. Also, it is very important that the camera face is parallel to the rotation axis. The same is the coordinate system for the image matrix that must be parallel to the rotation axis.

During rotation, the functional parameters of the camera can be changed. It is mostly magnified on the uniformity parameter. Causes might be mechanical—the coupling between crystal, light guide, and PM tubes due to gravitation, the gain from the PM tubes can vary with the changes in earth's magnetic field (which mostly can be overcome with magnetic shielding around the tubes) and thermal changes within the camera head. Most of these problems are taken care of well in modern cameras.

511-keV Imaging: PET

Already, Anger, in his earlier work, had proposed the use of coincidence techniques for scintillation cameras for imaging positron emitters [35]. However, the early development was toward dedicated ring systems used for PET imaging nowadays. The subsequent developments of specific PET-radiopharmaceuticals have renewed the interest to image them with more inexpensive scintillation cameras. One such radiopharmaceutical, 18FDG of special interest in neurology, cardiology, and oncology, has promoted the scintillation camera technique to be used at 511 keV imaging [36]. The simplest approach is to equip the camera with a super high-energy collimator. These collimators are very heavy, 100–200 kg, have thick septa, 2–4 mm and long holes, 60–80 mm. Their sensitivity is about 50 cps/MBq for 18F with a spatial resolution of 10–15 mm at 10 cm from collimator face [37]. Thus, both planar imaging and SPECT can be performed at 511 keV.

The common technique with a single energy window over the 511 keV photo peak can, for that high photon energy and relatively low interaction probability for total absorption in these thin crystals, be complemented with an additional energy window around 320 keV, just below the energy for the Compton edge of 341 keV. The method increases the sensitivity and also gives better spatial resolution. Also, scatter correction can be implemented for quantification [38,39].

Another approach is to use two antiparallel uncollimated scintillation camera heads in coincidence mode registering both annihilation quanta. Then, transversal tomography—limited angle tomography—is possible [40,41]. When rotating the cameras, a complete tomographic data set is obtained, and transaxial images can be reconstructed [42,43]. A hybrid camera is the PENN-PET consisting of large-area, position-sensitive NaI(Tl) scintillation crystals coupled to 30 PMTs in a hexagonal array. The spatial resolution for uncollimated cameras is 5–7 mm.

As mentioned earlier, specially designed pinhole collimators can image 511 keV photons, thus enabling even the PET radiopharmaceuticals to be imaged. One interesting feature is that this allows multiple imaging with both SPECT and PET radiotracers.

A gamma camera-like approach has been built for a preclinical PET scanner dedicated to mouse studies consisting of detectors of bismuth germanate crystal arrays with a thickness of 5 mm and cross-sectional size of 2.05×2.05 mm. Two or four such detectors are placed forming a dual/quadro-head geometry for imaging mice. The detectors are kept stationary during the scan, making it a limited angle tomography system. 3D images are reconstructed using a maximum likelihood and expectation maximization method. In-plane image spatial resolution was better than 2 mm full width at half maximum for coronal images and better than 3 mm for the anterior–posterior direction [44].

Bremsstrahlung Imaging

Imaging of pure beta emitters can be obtained by proper settings of the energy window and image processing as shown by Minarik et al. [45]. When beta particles are interacting with matter, Bremsstrahlung photons are emitted and these can be used for imaging. Bremsstrahlung imaging has generally been regarded as having a very poor image quality because of the large fraction of septal penetration.

The radionuclide ^{90}Y used for therapy has a Bremsstrahlung energy spectrum as a continuous spectrum up to the maximum energy for the beta particle equivalent to 2.2 MeV. Correction method can significantly improve image quality and make quantitative Bremsstrahlung imaging feasible (46).

Scatter correction for Bremsstrahlung imaging can be overcome by use of model-based scatter correction obtained from Monte Carlo simulations. Both clinical examples as well as dosimetry applications of Bremsstrahlung imaging have been published [46].

Gamma Cameras Based on CdZnTe: CZT

The use of NaI(Tl) as detector in scintillation cameras will for some applications be replaced with cadmium zinc telluride, CZT, a direct conversion semiconductor with an effective atomic number Z_{eff} of 50 and a density of 5.8 g cm^{-3}. The crystal is grown at very high temperatures (1100°C) and then cut into wafers; polished and metal contacts are attached to its surface to collect the electrical signals. When ionizing radiation interacts with the CZT detector, electron–hole pairs are produced and by applying an electrical field, these are collected at the cathode and anode and a current pulse is generated. Thus in a pixelated detector both position and energy can be registered. The superior energy resolution results from a relatively large number of electron–hole pairs generated per keV of deposited energy producing low statistical variation in the signal. The energy resolution in both NaI(Tl) and CZT are limited by Poisson statistics in the number of signal carriers. However, the number of signal carriers produced in semi-conductors is more than an order of magnitude higher for semiconductors (about 30,000 electron–hole pairs for 140 keV photons compared to about 1000 photocathode electrons in a NaI(Tl) crystal coupled to a PM tube). Theoretically, the energy resolution for NaI(Tl) would be in the order of 5% FWHM and for CZT 1%; however, practically those values are around 10% and 6%, respectively.

The spatial resolution of the CZT detector can be improved by reducing the size of the anode pixels and increasing the number of electronic channels. The practical limit of resolution is then given by the limit of the density of the electronics. Pixels of 0.6 mm can be produced.

Scintillation cameras based on the Anger positioning principle have a position invariant spectral response and thus the pixel size can be determined by the users. However, CZT detectors are pixelated with distinct pixel-to-pixel boundaries. Energy deposition in this boundary will contribute to signal in the low-energy tail whereas events within the pixel contribute to the photo peak. Thus the low-energy tail increases with smaller pixel sizes. This effect can partly be overcome with improved data acquisition.

A typical imaging detector consists of modules typically with 256 pixels arranged in a square pattern of 16 × 16, each with 2.46 mm dimension and a thickness of 5 mm. The pixel location and energy deposition in each pixel is recorded over a preset threshold and the anodes are connected to a readout board ending with an application-specific integrated circuit (ASIC) bonding pads. Thus each pixel has its own energy spectrum and its own energy window. The standard deviation of the histogram of photo peak signal per channel is typically 10% for CZT and does not tend to change with time. Dead pixels can exist in the detector; however, the probability is nowadays <1 dead pixel per 256 pixel module.

Organ-Specific Cameras

Most clinically used gamma cameras are based on large-area detectors often in SPECT or hybrid SPECT/CT configurations. Thus large areas/volumes can be examined in relatively short time. However, there is a demand for more specialized cameras, optimizing the spatial resolution and sensitivity together with ideal detection geometry. Small field of view cameras (30 × 40 cm) are today manufactured based on arrays of CsI coupled to pin diodes. This makes the camera thin and compact, and useful for portable planar imaging. An energy resolution of better than 8% and an intrinsic resolution of 3 mm are achieved. Dedicated NaI(Tl) scintillation camera systems have been developed for thyroid imaging.

For dedicated breast imaging, gamma cameras based on planar, dual-head, fully solid-state digital imaging system utilizing CZT technology with 1.6-mm intrinsic resolution and 4.5% FWHM energy resolution are available.

Cardiac Imaging

Cardiovascular diseases are a large patient population that can be studied and accurately diagnosed with scintillation cameras. Approximately one-half of all nuclear medicine studies performed clinically involve cardiac imaging. There is then need for smaller dedicated systems with high sensitivity for several reasons: to reduce the injected activity and thus minimize the absorbed dose, shorten the scan time, and to improve the image quality.

A cardiac study typically requires 20 min of acquisition time on a conventional gamma camera. Because of the relatively long acquisition time, the images are prone to contain motion artifacts. In order to reduce these artifacts, manufacturers have developed systems with a significantly higher sensitivity which in some cases allows acquisition time <5 min to produce images of equivalent diagnostic imaging quality. The higher sensitivity would also allow the use of lower injected activity, which would allow a reduction in absorbed dose.

Curved scintillation cameras placed side by side (three detectors) for cardiac imaging have been employed where horizontal collimation is obtained with a series of thin lead sheets that are stacked vertically, with the gaps between the sheets defining the hole apertures. Vertical collimation can be obtained by using a curved lead sheet with vertical slits that fan back and forth during acquisition.

One system is using an array of 19 pinhole collimators each with four solid-state CZT pixelated detectors. The pinholes are focused to image the heart. The sensitivity is high. Also, combination with CT as a hybrid system is available. A spatial resolution of 5 mm is achieved compared to conventional cardiac imaging with Na(Tl); the better energy resolution will reduce the Compton scatter and improve image contrast.

In one camera, 768 pixelated, thallium-activated CsI(Tl) crystals are coupled to individual silicon photodiodes utilizing Anger positioning techniques. Three of these detector heads are stationary and are mounted at a fixed angular separation of 67.5°. The data is acquired while the patient is sitting in a chair and is rotated through an arc.

Preclinical Systems

Preclinical studies in different species have mostly preceded the introduction of new radiopharmaceuticals in the clinic. Both proof of principle of targeting usefulness for physiological studies is the goal for these studies but important as well is the estimation of dosimetry based on this information. In the past, such studies were conducted with clinical systems that were usually surplus from clinical use. However, the fast development in biomedicine of new compounds for diagnostics and therapy, and the interest in understanding biochemical processes have sped up the development of molecular imaging with dedicated imaging systems for small animals, mainly mice and rats.

The early attempts to optimize clinical systems for small animal imaging are noticeable. In an early study, pinhole collimators with very small inserts were used for revealing the lymph flow in the brain [47]. In the 1990s, the development of pinhole collimator SPECT was presented for rabbits [48] and rats [49]. The earliest attempt to combine MR and Pinhole SPECT was published in the beginning of the 1990s [23,49] in rats with glioma. In the last few years, the Pinhole SPECT has evolved a lot [24].

Now, several preclinical systems are available especially for preclinical studies. Some are based on NaI(Tl) detector technology and some on CZT technology. All have the capability of high-resolution imaging with multipinhole collimation. There are systems with rotating planar detectors and systems with stationary detector, the latest are also available with moving the object for larger field of view with high spatial resolution.

Multimodality

Future applications of functional and anatomical information in combination with pharmacokinetic information with SPECT tracers, especially for preclinical systems, have boosted a huge interest in the development of SPECT inserts in MR cameras. An advantage of SPECT over PET is that simultaneous imaging of multiple radionuclides is better with higher energy resolution. MR can provide high-resolution anatomical information and also chemical and physical information.

This development will lead to both preclinical and clinical systems. Systems can be based on scintillator or solid-state detectors such as CdZnTe. For scintillators, the development of magnetically insensitive photodiode avalanche light detectors is a prerequisite for a compact detector design.

The CZT detector technology (with ASICs) is most promising with compact geometrical design and being mostly magnetically invariant. However, electrons produced in the CZT element will be deflected by the magnetic field by the so-called Lorentz shift. For high-resolution spatial information in the SPECT, multiple pinhole collimators can be used. The energy resolution has been shown not to be decreased by the magnetic field and still be in the order of 5%; a spatial resolution in the submillimeter range should be achievable.

Monte Carlo Simulations for Gamma Cameras

The gamma camera has become the "work horse" in most nuclear medicine imaging procedures and will be for a long time in the future. Many of the camera's parameters have been improved during the years. Many steps in the process from the emission of the photon at the decay point in the patient till the image has been created still can and need to be optimized.

From the earlier presentation of the characteristics for different parameters of the gamma camera, many parameters are obviously measurable; others cannot be measured or depend on factors difficult to measure. Simulations using the Monte Carlo technique therefore became an important tool, where the photon is followed from the emission point until it is totally absorbed. When the absorption happens in the detector of the camera, the creation of the image information can be investigated in detail. In that way multiple scattering of photons can be evaluated, attenuation properties investigated, detector interaction points determined, and so on.

The earliest MC studies of scintillation camera parameters were done by Anger who calculated intrinsic efficiency and intrinsic spatial resolution for NaI(Tl) crystals. Also, collimator influence on image creation was later studied by MC technique. In one of our group's earlier studies on the pile up effect and mispositioning at high photon fluence rates, the Monte Carlo method was used to reveal that problem, almost impossible to evaluate analytically [50]. Another good example where MC technique is an excellent tool is the optimization of collimators (Chapters 9, 10, and 14), and also for temporal resolution (Chapter 11) and scatter correction in SPECT (Chapter 12).

With the powerful tool of Monte Carlo technique with fast computers, one can expect more individual optimization of each investigation procedure, like the development of patient-related dose planning in radionuclide therapy [51–53].

Recently, Monte Carlo simulations of preclinical systems with optimization of pinhole geometry have been published. For example, it is used in the simulation of optimal number of pinholes in multipinhole SPECT based on preclinical biodistribution data for mouse brain imaging or in designing a dedicated high-resolution SPECT system for small animal imaging [27].

References

1. Anger HO. A new instrument for mapping gamma ray emitters. *Biol. Med. Q Rep. U Cal. Res. Lab-3653*, 1957:38.
2. Anger HO. Scintillation camera. *Rev. Sci. Instrum.* 1958;29:27–33.
3. Segre E, Seaborg GT. Nuclear isomerism in element 43. *Phys. Rev.* 2011;54:772.
4. Ott RJ, Flower MA, Babich JW, Marsden PK. The physics of radioisotope imaging. In: Webb S, ed. *The Physics of Medical Imaging*: Adam Hilger, Bristol and Philadelphia; 1988: pp. 142–318.
5. Rosenthal MS, Cullom J, Hawkins W, Moore SC, Tsui BM, Yester M. Quantitative SPECT imaging: A review and recommendations by the Focus Committee of the Society of Nuclear Medicine Computer and Instrumentation Council. *J. Nucl. Med.* 1995;36(8):1489–1513.
6. Sorensson JA, Phelps ME. *Physics in Nuclear Medicine*: Philiadelphia, WB Saunders Co; 1987.
7. Wernick MN, Aarsvold JN. *Emission Tomography; The Fundamentals of PET and SPECT*: San Diego, Elsevier Academic Press; 2004.
8. Birks JB. *The Theory and Practice of Scintillation Counting*: Oxford, Academic Press; 1964.
9. Hine GJ. *Instrumentation in Nuclear Medicine*: New York, Academic Press; 1967.
10. Patt BE, Iwanczyk JS, Rossington C, Wang NW, Tornai MP, Hoffman EJ. High resolution CsI(Tl)/Si-PIN detector development for breast imaging. *IEEE Trans. Nucl. Sci.* August 1998;45(4):2126–2131.
11. Miller BW, Barber HB, Barrett HH, Wilson DW, Liying C. A low-cost approach to high-resolution, single-photon imaging using columnar scintillators and image intensifiers. *Conference Records of the IEEE Nuclear Science and Medical Imaging Conference*, San Diego, CA. 2006;6:3540–3545.
12. Ceberg C, Larsson I, Strand SE. A new method for quantification of image distortion due to pile- up in scintillation cameras. *Eur. J. Nucl. Med.* 1991;18:959–963.
13. Strand SE, Larsson I. Image artifacts at high photon flurence rates in single NaI(Tl) crystal scintillation cameras. *J. Nucl. Med.* 1978;19:407–413.
14. Strand SE, Lamm IL. Theoretical studies of image artifacts and counting losses for different photon fluence rates and pulse-height distributions in single-crystal NaI(Tl) scintillation cameras. *J. Nucl. Med.* 1980;21:264–275.
15. Knoll GF. *Radiation Detection and Measurements*: New York, Wiley; 2000.
16. Sjogreen K, Ljungberg M, Wingardh K, Minarik D, Strand SE. The LundADose method for planar image activity quantification and absorbed-dose assessment in radionuclide therapy. *Cancer Biother. Radiopharm.* 2005;20(1):92–97.
17. Delpon G, Ferrer L, Lisbona A, Bardies M. Correction of count losses due to deadtime on a DST-XLi (SmVi-GE) camera during dosimetric studies in patients injected with iodine-131. *Phys. Med. Biol.* 2002;47(7):N79–N90.
18. Kimiaei S, Larsson SA, Jacobsson H. Collimator design for improved spatial resolution in SPECT and planar scintigraphy. *J. Nucl. Med.* 1996;37(8):1417–1421.
19. De Vries DJ, Moore SC, Zimmerman RE, Mueller SP, Friedland B, Lanza RC. Development and validation of a Monte Carlo simulation of photon transport in an Anger camera. *IEEE Trans. Med. Imag.* 1990;9(4):430–438.

20. Kibby PM. The design of multichannel collimators for radioisotope cameras. *Br. J. Radiol.* 1969;42(494):91–101.
21. Dewaraja YK, Ljungberg M, Koral KF. Accuracy of 131I tumor quantification in radioimmunotherapy using SPECT imaging with an ultra-high-energy collimator: Monte Carlo study. *J. Nucl. Med.* 2000;41(10):1760–1767.
22. Wanet PM, Sand A, Abramovici J. Physical and clinical evaluation of high-resolution thyroid pinhole tomography. *J. Nucl. Med.* 1996;37(12):2017–2020.
23. Weber DA, Ivanovic M, Franceschi D, Strand SE, Erlandsson K, Franceschi M, Atkins HL, Coderre JA, Susskind H, Button T. Pinhole SPECT: An approach to *in vivo* high resolution SPECT imaging in small laboratory animals. *J. Nucl. Med.* 1994;35:342–348.
24. Jaszczak RJ, Li J, Wang H, Zalutsky MR, Coleman RE. Pinhole collimation for ultra-high-resolution, small-field-of-view SPECT. *Phys. Med. Biol.* 1994;39(3):425–437.
25. Rogulski MM, Barber HB, Barrett HH, Shoemaker RL, Woolfenden JM. Ultra-high-resolution brain SPECT imaging: Simulation results. *Conference Records of the IEEE Nuclear Science and Medical Imaging Conference*, Orlando, FL. 1992:1071–1073.
26. Barrett H. *Foundation of Image Science*, New York, Wiley & Sons; 2003.
27. Beekman FJ, Vastenhouw B. Design and simulation of a high-resolution stationary SPECT system for small animals. *Phys. Med. Biol.* 2004;49(19):4579–4592.
28. Goorden MC, Beekman FJ. High-resolution tomography of positron emitters with clustered pinhole SPECT. *Phys. Med. Biol.* 2010;55(5):1265–1277.
29. Miller BW, Barber HB, Furenlid LR, Moore SK, Barrett HH. Progress in BazookaSPECT. *Proc. SPIE.* 2009;7450(7450C).
30. Peterson M, Ljunggren K, Palmer J, Strand SE, Miller B. Construction of a pre-clinical high resolution tomographic scintillation camera system. *Conference Records of the IEEE Nuclear Science and Medical Imaging Conference*, Orlando, FL. 2009:3670–3671.
31. Strand SE, Peterson M, Ljunggren K, Andersson-Ljus L, Miller B. A method for using high density fusible Rose's metal with high precision machining in small animal imaging applications. *Conference Records of the IEEE Nuclear Science and Medical Imaging Conference*, Knoxville, TN. 2010:3155–3157.
32. Hawman PC, Haines EJ. The cardiofocal collimator: A variable-focus collimator for cardiac SPECT. *Phys. Med. Biol.* 1994;39(3):439–450.
33. Esteves FP, Raggi P, Folks RD, Keidar Z, Askew JW, Rispler S, O'Connor MK, Verdes L, Garcia EV. Novel solid-state-detector dedicated cardiac camera for fast myocardial perfusion imaging: Multicenter comparison with standard dual detector cameras. *J. Nucl. Cardiol.* 2009;16(6):927–934.
34. Ivanovic M, Weber DA, Loncaric S, Franceschi D. Feasibility of dual radionu-clide brain imaging with I-123 and Tc-99m. *Med. Phys.* 1994;21(5):667–674.
35. Anger HO, Rosenthal DJ. Scintillation camera and positron camera. In: *Medical Radioisotope Scanning*, Vienna, International Atomic Energy Agency; 1959: 59–75.
36. Kalff V, Berlangieri SU, Van Every B, Rowe JL, Lambrecht RM, Tochon-Danguy HJ, Egan GF, McKay WJ, Kelly MJ. Is planar thallium-201/fluorine-18 fluorode-oxyglucose imaging a reasonable clinical alternative to positron emission tomo-graphic myocardial viability scanning? *Eur. J. Nucl. Med.* 1995;22(7):625–632.
37. Leichner PK, Morgan HT, Holdeman KP, Harrison KA, Valentino F, Lexa R, Kelly RF, Hawkins WG, Dalrymple GV. SPECT imaging of fluorine-18. *J. Nucl. Med.* 1995;36(8):1472–1475.

38. Ljungberg M, Danfelter M, Strand SE, King MA, Brill BA. Scatter correction in scintillation camera imaging of positron-emitting radionuclides. *Conference Records of the IEEE Nuclear Science and Medical Imaging Conference*, Anaheim, CA. 1996;3:1532–1536.

39. Ljungberg M, Ohlsson T, Sandell A, Strand SE. Scintillation camera imaging of positron-emitting radionuclides in the Compton region. *Conference Records of the IEEE Nuclear Science and Medical Imaging Conference*, Anaheim, CA. 1996;2:977–981.

40. Muehllehner G, Atkins FB, Harper PV. Positron camera with longitudinal and transverse tomographic capability. In: *Medical Radionuclide Imaging*, Volume 1 STI/PUB/44C. Vienna: IAEA; 1977:391–407.

41. Muehllehner G, Buchin MP, Dudec JH. Performance parameters of a positron imaging camera. *IEEE Trans. Nucl. Sci.* 1976;23(1):528–537.

42. Paans AM, Vaalburg W, Woldring MG. A rotating double-headed positron camera. *J. Nucl. Med.* 1985;26(12):1466–1471.

43. Sandell A, Ohlsson T, Erlandsson K, Hellborg R, Strand SE. A PET system based on 2-18FDG Production with a low energy electrostatic proton accelerator and a dual headed PET scanner. *Acta Oncol.* 1992;31(7):771–776.

44. Zhang H, Vu NT, Bao Q, Silverman RW, Berry-Pusey BN, Douraghy A, Williams DA, Rannou FR, Stout DB, Chatziioannou AF. Performance characteristics of BGO detectors for a low cost preclinical PET scanner. *IEEE Trans. Nucl. Sci.* 2010;57(3):1038–1044.

45. Minarik D, Ljungberg M, Segars P, Gleisner KS. Evaluation of quantitative planar 90Y bremsstrahlung whole-body imaging. *Phys. Med. Biol.* 2009;54(19):5873–5883.

46. Minarik D, Sjögreen-Gleisner K, Linden O, Wingårdh K, Tennvall J, Strand SE, Ljungberg M. 90Y Bremsstrahlung imaging for absorbed-dose assessment in high-dose radioimmunotherapy. *J. Nucl. Med.* 2010;51(12):1974–1978.

47. Widner H, Jonsson BA, Hallstadius L, Wingardh K, Strand SE, Johansson BB. Scintigraphic method to quantify the passage from brain parenchyma to the deep cervical lymph-nodes in rats. *Eur. J. Nucl. Med.* 1987;13(9):456–461.

48. Palmer J, Wollmer P. Pinhole emission computed tomography: Method and experimental evaluation. *Phys. Med. Biol.* 1990;35(3):339–350.

49. Strand SE, Ivanovic M, Erlandsson K, Weber DA, Franceschi D, Button T, Sjögreen K. High resolution pinhole SPECT for tumor imaging. *Acta Oncol.* 1993;32:861–867.

50. Strand SE, Lamm IL. Theoretical studies of image artifacts and counting losses for different photon fluence rates and pulse-height distributions in single-crystal NaI(TI) scintillation cameras. *J. Nucl. Med.* 1980;21:264–275.

51. Erdi AK, Erdi YE, Yorke ED, Wessels BW. Treatment planning for radio-immunotherapy. *Phys. Med. Biol.* 1996;41(10):2009–2026.

52. Kolbert KS, Sgouros G, Scott AM, Bronstein JE, Malane RA, Zhang J, Kalaigian H, McNamara S, Schwartz L, Larson SM. Implementation and evaluation of patient-specific three-dimensional internal dosimetry. *J. Nucl. Med.* 1997;38(2):301–308.

53. Tagesson M, Ljungberg M, Strand SE. Transformation of activity distribution in quantitative SPECT to absorbed dose distribution in a radionuclide treatment planning system. *J. Nucl. Med.* 1994;35:123P.

5

Positron Emission Tomography

Kjell Erlandsson and Tomas Ohlsson

CONTENTS

Positron emission tomography (PET) has forwarded the development of nuclear medicine and molecular imaging through the introduction of new radiopharmaceuticals, making it possible to study physiological and bio-chemical processes inaccessible with other techniques. The PET technique is complex, requiring radionuclide production facilities, advanced radiochemistry, PET scanners, mathematical modeling, and trained personnel.

The most commonly used radionuclides in PET are ^{11}C, ^{13}N, ^{15}O, and ^{18}F (Table 5.1). There are several reasons for using these positron emitters. First,

TABLE 5.1

Physical Characteristics of the Most Commonly Used Positron-Emitting Radionuclides

Radionuclide	$T_{1/2}$ (min)	β⁺ Branching Fraction (%)	β⁺ Energy (MeV) Maximum	β⁺ Energy (MeV) Average	β⁺ Range (mm) RMS	β⁺ Range (mm) FWHM	β⁺ Range (mm) FWTM	γ Photons, Energy/Abundance (MeV)	γ Photons, Energy/Abundance (%)
^{11}C	20.4	99.8	0.96	0.385	0.39	0.19	1.86	–	
^{13}N	9.96	100	1.20	0.491	0.59	0.28	2.53	–	
^{15}O	2.04	99.9	1.72	0.735	1.02	0.50	4.14	–	
^{18}F	109.8	96.9	0.64	0.242	0.23	0.10	1.03	–	
^{64}Cu	12.7 h	19	0.65	0.278	0.23	0.10	1.05	–	
^{68}Ga	67.6	88	1.9	0.836	1.19	0.58	4.83	1.08–1.88	3.1
^{76}Br	16.1 h	54	3.7					0.473–3.60	146
^{82}Rb	1.27	95	3.36	1.52	2.59	1.27	10.5	0.777	13
^{89}Zr	3.27 days	23	0.897	0.395				0.909	100
^{124}I	4.18 days	22	1.5					0.603–1.690	23

Source: Adapted from ICRP. Radionuclide Transformations—Energy and Intensity of Emissions. *ICRP Publication. Ann. ICRP.* 1983;38:11–13; Zanzonico P. *Seminars in Nuclear Medicine.* 2004;34:87–111; Lecomte R. *European Journal of Nuclear Medicine and Molecular Imaging.* 2009;36(Suppl. 1):S69–85.

a broad range of tracers can be labeled with only a change of isotope in the case of ^{11}C, ^{13}N, and ^{15}O, while ^{18}F can be used as a hydrogen substitute. Second, the use of isotopes of these biologically ubiquitous elements makes it possible to label molecules that trace biochemical processes precisely, which is rarely possible with radiopharmaceuticals used in scintillation camera imaging. Third, the short half-life of these positron-emitting radionuclides results in a low absorbed dose to the patient, and allows for several PET studies to be made on the same patient the same day. A drawback of the short half-life is that an on-site cyclotron is required for the production of these radionuclides, although ^{18}F can be obtained through regional distribution. Some other positron emitters, such as ^{68}Ga and ^{82}Rb, can be obtained from generator systems.

The interest in PET has increased considerably, from being only a research tool for brain imaging, to being a clinical whole-body imaging modality. PET can be used for measuring blood flow, metabolism, receptor binding, and therapy response. Several specific clinical indications have been defined in cardiology [1], neurology [2], and oncology [3,4]. Some common tracers used in PET are: [^{11}C]-raclopride (receptor binding, psychiatry), [^{11}C]-PIB (amyloid binding, neurology), ^{13}N-labeled NH_3 (blood flow), ^{15}O-labeled H_2O (blood flow), ^{18}F-labeled FDG (glucose metabolism), [^{68}Ga]-Dotatate (oncology), and ^{82}Rb (cardiology). FDG is by far the most widely used PET tracer.

β^+-Decay

Radionuclides with excess protons may decay in two ways to reduce the number of protons and achieve a stabler nuclear configuration. The nucleus may either capture an orbital electron or emit a positron (β^+-particle) and a neutrino. The excess energy from the β^+-decay is divided between the positron and the neutrino. The kinetic energy of the positrons, for a number of disintegrations, therefore exhibits a continuous distribution from zero up to a maximum energy E_{max} dependent on the radionuclide (Table 5.1). After losing almost all of its kinetic energy, the positron interacts with an atomic electron. The two particles form a short-lived hydrogen-like structure called positronium, and then undergo an annihilation process, in which the masses of the two particles are converted into electromagnetic radiation, usually in the form of two γ-photons. These annihilation photons are emitted in nearly opposite directions, each one carrying an energy of 511 keV. The sum of the two photon energies corresponds to the total mass of the two particles. The opposite directions of the photon trajectories are demanded by the law of momentum conservation and are exact only in the center of mass system.

The thermal energy of the positron–electron system at the instant of annihilation results in a slight deviation from 180° in the laboratory system (noncolinearity), with a mean value of ~0.5°. Annihilation can sometimes occur before the positron has lost most of its kinetic energy (annihilation in flight), in which case the total energy of the annihilation photons will be greater than 1022 keV. There is also a small but finite probability for triple-photon annihilation.

PET Scanners

The suggestion of using positron emitters for medical imaging purposes was first made in the early 1950s, the basic principle being that of observing the emission of positrons in tissue through the detection of annihilation radiation. Brownell and Sweet [5] and Anger and Rosenthal [6] described the first PET imaging devices. A PET scanner is composed of at least two radiation detectors operating in time coincidence mode.

When two 511 keV photons are detected in two different detectors simultaneously, an annihilation is assumed to have occurred somewhere on the line connecting the two detection points (Figure 5.1). In this case, "simultaneously" means within a time window of 4–12 ns. If an annihilation occurs outside the region between the detectors, only one annihilation photon can

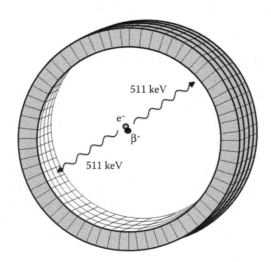

FIGURE 5. 1
PET scanner, ring system, consisting of a number of scintillation crystals operating in the coincidence mode.

be detected and, since this does not satisfy the coincidence condition, the event is rejected. The spatial resolution in PET images is primarily dependent on the physical size and cross section of the detectors, but is also affected by the positron range before annihilation and by the deviation from 180° of the photon trajectories (noncolinearity). The positron range distribution does not have the typical Gaussian shape, normally used for modeling resolution in nuclear medicine. It has a much larger FWTM/FWHM ratio (Table 5.1), that is, a narrow peak and long tails. The influence of the noncolinearity depends on the detector separation.

A PET scanner can in principle be constructed with two opposed large area detectors but the most common design for PET scanners today is a large number of small detectors in a ring (or polygon) surrounding the patient (Figure 5.1). Ring scanners have evolved through several generations; from single-ring to multiring systems, from NaI(Tl) to BGO to LSO crystals, from individual detectors to block detectors, from 2D to 3D scanners. The latest development is the use of time-of-flight (TOF) information.

PET scanners are built with scintillation detectors, which consist of a scintillation crystal coupled to an optical detector. The 511 keV photons interact in the crystal producing a flash of light, which is read out by the optical detector and converted into an electronic pulse. The optical detectors traditionally used are photomultiplier tubes (PMTs), which are vacuum tubes containing a series of dynodes.

Scintillation Crystals

The most important properties of scintillation crystals for PET are: stopping power, light yield, and decay time. High stopping power is needed due to the high energy of the annihilation photons (511 keV). This property depends on the atomic number and density of the material, which should both be as high as possible. High light yield leads to good energy resolution, which is important for discrimination between scattered and unscattered photons. Fast decay time results in high count-rate capability, lower rate of random coincidences, and the possibility for TOF acquisition.

Table 5.2 summarizes the properties of different scintillation materials used in PET scanners. NaI(Tl) has been used but suffers from low efficiency at 511 keV. Bismuth germanate ($Bi_4Ge_3O_{12}$, BGO) has a much higher sensitivity and was for many years the most commonly used material in PET scanners despite some drawbacks. Its relatively low light output leads to poor energy resolution and its long fluorescent decay time to limited count-rate capability. Cerium-doped lutetium oxyorthosilicate (Lu_2SiO_5:Ce, LSO) has a unique combination of high density, atomic number, light emission intensity, and speed that makes it a very suitable scintillation crystal for PET scanners. One drawback is that it contains the radioactive isotope [176]Lu (2.6%). LYSO has similar properties as LSO, and both can be used

TABLE 5.2

Physical Properties of Inorganic Scintillation Crystals Used in PET Scanners

	NaI(Tl)	BGO	GSO	LSO	LYSO	BaF$_2$	LaBr$_3$
Density (g/cm^3)	3.67	7.13	6.71	7.35	7.19	4.89	5.3
Z-effective	50	73	58	65	64	54	46
μ (cm^{-1}) @ 511 keV	0.39	0.89	0.67	0.81	0.79	0.44	0.45
PE (%)	18	44	26	34	33		14
$\Delta E/E$ (%) @ 511 keV	8	12	9	10	10		3 @ 661 keV
$\Delta E/E$ (%), system	10–12	18–25	12–18	12–18	12–18		6–7
Light output (photons/keV)	41	9	8	30	30	2	60
Decay constant (ns)	230	300	60	40	40	0.8	16
λ-Peak (nm)	410	480	440	420	420	220	360
Refraction index	1.85	2.15	1.85	1.82	1.81	1.50	1.9
Hygroscopic	Y	N	N	N	N	Slightly	Y

Source: Adapted from Zanzonico P. *Seminars in Nuclear Medicine.* 2004;34:87–111; Lecomte R. *European Journal of Nuclear Medicine and Molecular Imaging.* 2009;36(Suppl. 1):S69–S85; Muehllehner G, Karp JS. *Physics in Medicine and Biology.* 2006;51:R117–R137; Lewellen TK. *Physics in Medicine and Biology.* 2008;53:R287–R317.

in TOF–PET systems. Historically, BaF$_2$ and CsF were used in early TOF–PET systems due to their rapid decay times [7]. However, the low efficiency as compared to BGO was a clear drawback and this line of development was discontinued. LaBr$_3$ is a new material for TOF–PET systems, which has lower stopping power than LSO/LYSO but better energy and timing resolution.

Block Detectors

The quest for higher spatial resolution without increased complexity of the scanner has led to the development of the block detector. Originally, PET scanners were built with a one-to-one crystal-to-PMT coupling. However, this approach was not practical when improved spatial resolution required reduction in the detector size. While smaller crystals could be easily made, there were limitations in terms of the size and cost of PMTs. Block detectors are essentially small Anger cameras, consisting of a scintillator block that is cut into, for example, 8 × 8 elements and viewed by four PMTs (Figure 5.2). Position logic, similar to that in the scintillation camera, is used for identification of the different elements. The depth of the cuts increases from the center to the edge of the scintillator block in order to achieve a light distribution appropriate for crystal identification. The number of resolution elements per PMT thus increases by a factor of 16 compared to the previously used one-to-one coupling design.

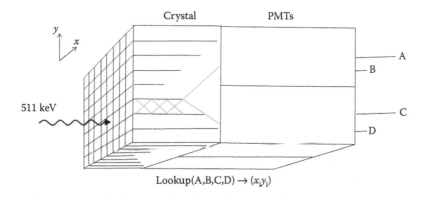

Lookup(A,B,C,D) → (x_i, y_i)

FIGURE 5. 2
Block detector geometry. The scintillation light from the crystal array is shared between four PMTs. The four signals are used to identify the individual crystal in which the scintillation was produced using a lookup table.

Depth of Interaction

Individual crystals in most PET scanners are long and narrow. They therefore have a small cross section when seen from the center of the scanner but, moving away from the center, the cross section becomes more and more elongated. This results in a degradation of the spatial resolution, and is known as the parallax effect. To correct for this effect, it is necessary to know the depth of interaction (DOI) in the crystal, that is, the depth at which the annihilation photon interacted. DOI information can be obtained with the so-called phoswich detectors, which consist of two layers of crystals with different decay times. The two layers can be distinguished using a technique called pulse-shape analysis [8].

New Photodetectors

Recent developments in photodetectors have led to the possibility of replacing the conventional PMTs in PET detectors with solid-state detectors, such as avalanche photodiodes or silicon photomultipliers (SiPMs). This would allow for a return to the on-to-one coupling, result in a more compact system design, and open up the possibility for PET systems operating in magnetic fields [9].

Data Processing

In order to obtain quantitatively correct image values of the activity distribution, it is necessary to correct the acquired PET data for the following effects: efficiency variations, dead-time, random coincidences, scatter, and attenuation.

Normalization and Dead-Time

The intrinsic efficiency for each detector in a PET scanner will not be the same. To correct for this effect, the acquired data is normalized using a measurement with a rotating rod source, providing a uniform exposure of all detectors. At high-count rates, the rate of data acquired is not a linear function of the photon flux due to electronic dead-time in the system, which is the time required to process each detected event. This can be corrected for using a model of the count-rate performance of the system (see Chapter 14).

Random Coincidences

There is no natural background in PET measurements due to the coincidence requirement. However, there are two effects that create background events during a PET scan: random coincidences and Compton scattering. Since the coincidence detection technique is based on a finite time window of 4–12 ns, there is always a chance that two photons originating from different annihilations may cause a coincidence event, as shown in Figure 5.3a. The number of these random coincidences can be calculated with the equation $n_R = n_1 \cdot n_2 \cdot 2\tau$, where n_R is the count rate of random coincidences between detectors 1 and 2, n_1 and n_2 are the single count rates in each detector, and 2τ is the time window (the maximum time difference between the two registrations being τ). It is also possible to measure the randoms count rate using a delayed coincidence window, which means that one signal is delayed so that no true coincidences can be detected.

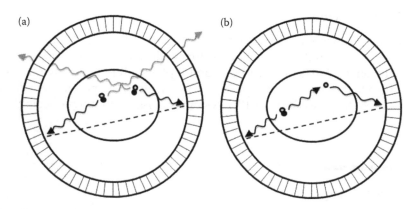

FIGURE 5. 3
Background events. (a) Random coincidences. (b) Compton scattering.

Compton Scatter

The annihilation photons may interact with atomic electrons in the patient either by photoelectric absorption or by Compton (incoherent) scattering. In the latter process (Figure 5.3b), a photon of lower energy and different direction is produced. The energy of the scattered photon is given by

$$hv' = hv\left(1 + hv(1 - \cos\phi)/m_0c^2\right)^{-1} \tag{5.1}$$

where hv and hv' are the energies of the unscattered and scattered photons, respectively, ϕ is the scattering angle, and m_0c^2 is the rest-mass of the electron (511 keV). If detected, the two photons will still be in coincidence. The energy resolution of the detectors used in PET is in general not good enough to allow efficient discrimination of scattered photons. Also, a wide energy window is sometimes used in order to accept events in which a photon does not deposit its entire energy in the detector. Scatter correction is essential for accurate quantification, especially in 3D mode. Several correction methods have been developed both for 2D and 3D PET. One simple method is to utilize the "tails" of the projections outside the contour of the object. After correction for randoms, the data in the tails should correspond only to scattered events. A correction can be done by fitting an analytical function to the tails and subtracting it from the measured data. Scatter correction methods are described further in Chapter 13.

Count-Rate Performance

The count-rate performance of PET scanners is often quantified by the peak value of the noise equivalent count rate (NEC), defined as: $NEC = T^2/(T + S + kR)$, where T, S, and R are the count rates for true, scattered, and random coincidences, respectively, and k is a constant ($k = 1$ or 2, depending on whether variance reduction was or was not used in the estimation of the randoms rate). NEC is essentially the product of the trues-count rate and the trues-fraction.

Attenuation

Attenuation correction is essential for an accurate reconstruction. Some of the emitted photons will never reach the detectors due to absorption or scattering in the patient. The total interaction probability for the two annihilation photons along a line-of-response (LOR) is independent of the annihilation position along the line. The correction factors for each LOR can be calculated if the object is uniform and its contour is known. In practice, the factors are usually derived from a transmission measurement. These can be done using an external radioactive source, but is nowadays usually done with x-ray CT.

Reconstruction

Image reconstruction was in the past mainly done with fast analytical methods such as filtered backprojection. Now there is a trend toward the use of iterative methods, such as ML–EM, OSEM, or MAP, which can take into account the Poisson nature of the noise in the data and also compensate for resolution degradation by incorporating a model for the point spread function (PSF) of the system in the system matrix [10,11].

Time-of-Flight

The basic principle of TOF–PET is to measure the time difference between the detection of the two annihilation photons, and thereby get an estimate of the position of the annihilation along the line joining the two detection points. With the arrival of new faster crystals with high sensitivity (LSO, LYSO, LaBr$_3$), TOF–PET has now become a practical reality. Currently, the timing resolution of commercial TOF–PET scanners is ~0.5 ns, which corresponds to a spatial accuracy of ~7.5 cm. This should be compared, not with the spatial resolution in the reconstructed images (~5 mm), but with the size of the patient. In a conventional PET system, there is no information about the position of the annihilation event along the LOR, except that it was somewhere inside the patient. With a TOF–PET system, an estimation of the position along the LOR is obtained, described by a Gaussian probability distribution. The FWHM of the Gaussian depends on the timing resolution of the system (Figure 5.4). The TOF information can be incorporated into the system matrix and utilized in the image reconstruction process. This leads to a reduction of the noise propagation from raw data to reconstructed image, resulting in an improved SNR. The usefulness of the TOF information depends on the size of the object being imaged. It is more useful in whole-body studies

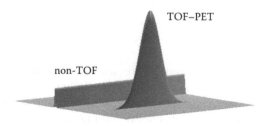

FIGURE 5. 4
Probability distribution for the annihilation position corresponding to a single coincidence event for a time-of-flight (TOF) and a non-TOF PET system (based on an object size of 384 mm and a TOF resolution of 0.5 ns, equivalent to 75 mm).

than in brain studies, and currently not useful at all in small-animal studies. The spatial uncertainty (Δx) on the position of the positron annihilation is related to the timing resolution (Δt) by the equation $\Delta x = c\Delta t/2$, where c is the speed of light, and the SNR improvement can be estimated by the factor $f = \sqrt{D/\Delta x}$, where D is the diameter of the object. Another effect of improved timing resolution in a TOF–PET system is a reduction in the number of random coincidences, further improving the image quality.

PET Systems

2D and 3D PET

PET scanners can operate in either 2D or 3D mode. For 2D mode (Figure 5.5a), scanners are equipped with the so-called septa, which are thin lead shields placed between the detector rings, reducing the number of scattered and random events. Coincidence events are treated as belonging to either direct planes or cross-planes. Coincidences between crystals within the same ring belong to a direct plane. Coincidences between crystals in adjacent rings belong to a cross-plane, parallel to the direct planes, midway between the two rings. Events are also accepted with larger ring differences (typically up to 3 or 4). These are included in either direct or cross-planes, depending on whether the ring difference is even or odd.

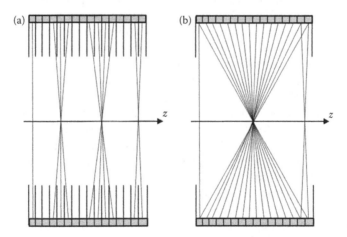

FIGURE 5. 5
Axial ring scanners geometries. (a) 2D mode with septa. (b) 3D mode without septa. In 3D mode, the number of LORs passing through a point in the FOV depends on the axial position (z) of the point. (The axial dimension has been exaggerated for clarity.)

For 3D mode, the septa are removed so that all possible ring combinations can be utilized (Figure 5.5b). The sensitivity increases by a factor of ~5, partly due to a higher number of coincidence lines, partly due to the absence of shielding by the septa. In 3D mode, the sensitivity varies in the scanners axial direction for geometric reasons, with much higher sensitivity in the center than at the edges (as indicated in Figure 5.5b), and a more complex image reconstruction algorithm is required. The reprojection algorithm and the Fourier rebinning algorithm (see [10]) were developed specifically for reconstruction of 3D PET data. The count-rate capability of the system will remain the same in 3D as in 2D mode, but reaches its maximum value with lower activity in the field of view (FOV). 3D PET offers the possibility of lowering the radiation dose to the patient, or to acquire more counts in low count-rate situations. The drawback is that, apart from the sensitivity, the fraction of unwanted scattered and random events also increases. New scintillation crystals and advances in electronic signal processing have led to a reduction in scattered and random events as well as improved count-rate capability. Therefore, more and more scanners are nowadays designed for operating in 3D mode only.

Clinical PET Systems

The first clinical PET scanners were developed in the 1970s [12,13]. The design parameters and performance characteristics of a few modern PET scanners are presented in Table 5.3. The best spatial resolution is achieved by the scanner with the smallest crystals, the ECAT high-resolution research tomograph (ECAT HRRT). This is a system designed exclusively for brain research studies, and it is not used clinically. The highest NEC is obtained with the Discovery RX in 2D mode. This is due to the good timing properties of LYSO and due to the reduction in scatter and randoms fraction by the use of septa. However, a high activity concentration is required. In 3D mode, the peak NEC of this system is slightly lower than that of the Gemini TF, also based on LYSO crystals. The latter system has TOF capability, as does the LaBr$_3$ system. LaBr$_3$ has lower stopping power than LYSO but better timing and energy resolution. This leads to lower scatter fraction, lower randoms rate, and more accurate TOF information, which may partly compensate for lower sensitivity. The reason for the similarity in the scatter fraction for the two systems, despite the difference in energy resolution, is the longer axial FOV of the LaBr$_3$ system.

The latest generation of PET systems usually includes an option for list-mode acquisition, so that the data can be processed on an event-by-event basis. Other common options are cardiac and respiratory gating, in order to reduce motion artifacts. New iterative image reconstruction algorithms may also be available, including PSF modeling in order to improve the spatial resolution.

TABLE 5.3

Comparison of Design Parameters and Performance Characteristics for Some PET Scanners

	Discovery STE/VCT	Discovery RX	Biograph 16 Hi-Rez	ECAT HRRT	Allegro	Gemini TF	(Prototype)
Crystal material	BGO	LYSO	LSO	LSO/LYSO	GSO	LYSO	LaBr$_3$
Operating mode	2D/**3D**	**2D/3D**	3D only	3D (DOI)	3D only	3D, TOF	3D, TOF
Ring diameter (mm)	886	886	830	312	820	903	930
FOV, transaxial (mm)	700	700	585	250	560	576	250
FOV axial (mm)	157	157	162	250	180	180	250
No. of crystals	13,440	15,120	24,336	59,904 × 2	17,864	28,336	38,880
No. of rings/planes	24/47	24/47	39/81	104/207	29/.	44/.	60/.
Crystal size[a] (mm)	4.7 × 6.3 × 30	4.2 × 6.3 × 30	4 × 4 × 20	2.1 × 2.1 × 10 × 2	4 × 6 × 20	4 × 4 × 22	4 × 4 × 30
Crystals/block	8 × 6	9 × 6	13 × 13	8 × 8			
No. of blocks	280	280	144	8 × 9 × 13			
Crystals/module					22 × 29	23 × 44	27 × 60
No. of modules				8	28	28	24
No. of PMTs				8 × 10 × 14	420	420	6 × 72
Axial sampling (mm)	3.27	3.27	2.0				
Coincidence window (ns)	11.7/**9.3**	6.5	4.5	6	7.5	6	5
Timing resolution (ps)	**425**					585	375
Lower energy threshold (keV)	375	425	425	400	410	440	470
Upper energy threshold (keV)	650	650	650	650	665	665	
Sensitivity (s^{-1}/kBq)	2.3/**8.8**	**1.66/7.30**	4.87	25	4.36	6.6	2.2
Scatter fraction (%)	**34**	**32**	34	45	40	27	25

continued

TABLE 5.3 (continued)

Comparison of Design Parameters and Performance Characteristics for Some PET Scanners

	Discovery STE/VCT	Discovery RX	Biograph 16 Hi-Rez	ECAT HRRT	Allegro	Gemini TF	(Prototype)
Peak NEC, $k = 1$ $(k\,s^{-1})$/activity concentration (kBq/mL)	88/47 **68/12**	155/92 **118/22**	85/29	45/12 $(k=2)$	30/9.2	125/17	
FWHM[b] (mm), $R = 1$ cm	4.9/4.4 **5.2/5.2**	5.1/4.8 **5.0/5.8**	4.6/5.1	2.5/2.5	5.5/5.8	4.8/4.8	5.8/.
FWHM[c] (mm), $R = 10$ cm	5.7/5.8/6.0 **5.9/5.5/5.9**	5.9/5.1/6.3 **5.9/5.2/6.5**	5.3/5.3/5.9	2.6/2.8/2.9	5.7/./6.7	5.2/5.2/4.8	6.5/./6.3

Source: Adapted from Teras M et al. *European Journal of Nuclear Medicine and Molecular Imaging.* 2007;34:1683–1692; Kemp BJ et al. *Journal of Nuclear Medicine.* 2006;47:1960–1967; Brambilla M et al. *Journal of Nuclear Medicine.* 2005;46:2083–2091; de Jong HW et al. *Physics in Medicine and Biology.* 2007;52:1505–1526; Surti S, Karp JS. *Journal of Nuclear Medicine.* 2004;45:1040–1049; Surti S et al. *Journal of Nuclear Medicine.* 2007;48:471–480; Daube-Witherspoon ME et al. *Physics in Medicine and Biology.* 2010; 55:45–64.

Note: Bold values represent to identify the values corresponding to the 3D operation mode.

a   Transaxial × axial × radial(× layers).

b   Transaxial/axial.

c   Radial/tangential/axial.

Preclinical PET Systems

A large number of PET systems for imaging small animals, such as mice and rats, have been developed [14]. The basic principles for these systems are the same as for human scanners. Some differences are that preclinical systems have better spatial resolution and operate only in 3D mode, the scatter and attenuation effects are not as important and are often ignored, and TOF capability is not available. Also, the injected activity per unit body weight is much higher than in human studies (often orders of magnitude higher), which results in better statistics in the acquired data.

Multimodality Systems

The combination of PET images with images from other tomographic modalities, such as x-ray CT or MRI, has proven to be very useful for various reasons, mainly for anatomical localization and attenuation correction. Images acquired independently on different systems can be coregistered using software tools [15]. However, while this approach is relatively straightforward in the case of brain studies, it is not as simple in the case of whole-body studies. The brain can be treated as a rigid object, but the motion of the internal organs in the thorax and abdomen are more complex and difficult to model accurately. Therefore, multimodality systems have been developed, combining a PET scanner and a CT or MRI scanner in one single device. In this way, inherently coregistered images from two different modalities can be obtained in one single imaging session [16].

PET/CT is today a well-established clinical tool. The combination of high-resolution anatomical detail from the CT images with the highly specific tracer distribution from the PET data can provide useful clinical information in a range of diseases. In addition, the CT image can be used for attenuation correction (AC) of the PET data. This leads to faster data acquisition and higher patient throughput as compared to previous protocols including transmission measurements with radioactive rod sources. The photon energy used in CT ($\sim$70 keV) is much lower than the one in PET (511 keV), and the measured attenuation coefficient (μ) values need to be scaled before the AC. The scaling is not entirely trivial as it depends on the tissue type (such as soft tissue or bone). A simplified bilinear scaling is normally used, with different slopes above and below the μ value for water. This simple model is not valid for CT contrast agents, which can lead to artifacts in the PET images. Artifacts can also occur due to respiratory motion causing PET/CT mismatch.

While PET/CT has been available for about a decade now, PET/MRI is still in its infancy. The advantages of MRI over CT are higher soft-tissue contrast, resulting in improved clinical information, and the lack of radiation dose to the patient. The main disadvantage is that MRI images cannot be used directly for AC, for example, air and bone cannot be distinguished in a standard MRI image, but they are quite different in terms of attenuation properties. Various software algorithms have been developed to solve this problem, and further work in this field is ongoing.

PET/CT systems consist of two separate scanners placed back to back, and images from the two modalities are obtained sequentially. The patient is transferred between scanners while lying on the couch. PET/MRI systems, on the other hand, are being developed with truly simultaneous data acquisition capability from the two modalities. This opens up the door for a range of new possibilities, such as MRI-based motion correction. As a bonus, the positron range is reduced due to the magnetic field, which can result in improved resolution for radionuclides with high β^+-energy (only in two dimensions, though). The reason for the difference in the design of PET/CT and PET/MRI systems is that a CT scan is much faster than a PET scan, while an MRI scan is of a similar duration. Sequential PET/MRI systems with separate scanners can also be built; however, the two scanners need to be placed quite far apart (≥ 4 m), in order to avoid interference effects.

Monte Carlo Simulations for PET

Monte Carlo (MC) simulation is a very useful tool for studying the performance of a PET system. A number of physical factors, such as source distribution, positron range, noncolinearity of the annihilation photons, scanner configuration, and photon interactions in the object as well as in active and passive detector components, can be included in the simulation. MC can be used for evaluation of sensitivity, resolution, and, in conjunction with other mathematical modeling, also count-rate performance. This is useful both for evaluating existing PET scanners and in designing new ones (Chapters 14 and 15). The possibility of distinguishing between scattered and true events in MC simulations is valuable for development of scatter correction techniques (Chapter 13). MC can also be used for the generation of realistic data for testing new correction or reconstruction methods (Chapter 16).

References

1. Schwaiger M, Ziegler S, Nekolla SG. PET/CT: Challenge for nuclear cardiology. *Journal of Nuclear Medicine*. 2005;46(10):1664–1678.
2. Tai YF, Piccini P. Applications of positron emission tomography (PET) in neurology. *Journal of Neurology, Neurosurgery, and Psychiatry*. 2004; 75(5):669–676.
3. Gambhir SS, Czernin J, Schwimmer J, Silverman DH, Coleman RE, Phelps ME. A tabulated summary of the FDG PET literature. *Journal of Nuclear Medicine*. 2001;42(5 Suppl):1S–93S.
4. Belhocine T, Spaepen K, Dusart M et al. 18FDG PET in oncology: The best and the worst (Review). *International Journal of Oncology*. 2006;28(5):1249–1261.

5. Brownell GL, Sweet WH. Localization of brain tumors with positron emitters. *Nucleonics.* 1953;11:40–45.
6. Anger HO, Rosenthal DJ. Scintillation camera and positron camera. In *Medical Radioisotope Scanning.* Vienna: International Atomic Energy Agency. 1959, pp. 59–82.
7. Mullani NA, Wong W-H, Hartz R et al. Preliminary results with TOFPET. *IEEE Transactions on Nuclear Science.* 1983;30:739–743.
8. Dahlbom M, MacDonald LR, Eriksson L et al. Performance of a YSO/LSO phoswich detector for use in a PET/SPECT system. *IEEE Transactions on Nuclear Science.* 1997;44:1114–1119.
9. Schaart DR, Seifert S, Vinke R et al. LaBr(3):Ce and SiPMs for time-of-flight PET: Achieving 100 ps coincidence resolving time. *Physics in Medicine and Biology.* 2010;55(7):N179–N189.
10. Lewitt RM, Matej S. Overview of methods for image reconstruction from projections in emission computed tomography. *Proceedings of the IEEE* 2003;91(10): 1588–1611.
11. Qi J, Leahy RM. Iterative reconstruction techniques in emission computed tomography. *Physics in Medicine and Biology.* 2006;51(15):R541–R578.
12. Bohm C, Eriksson L, Bergstrom M, Litton J, Sundman R, Singh M. A computer assisted ring-detector positron camera for reconstruction tomography of the brain. *IEEE Transactions on Nuclear Science.* 1978;25:624–637.
13. Hoffmann EJ, Phelps ME, Mullani NA, Higgins CS, Ter-Pogossian MM. Design and performance characteristics of a whole-body positron transaxial tomograph. *Journal of Nuclear Medicine.* 1976;17(6):493–502.
14. de Kemp RA, Epstein FH, Catana C, Tsui BM, Ritman EL. Small-animal molecular imaging methods. *Journal of Nuclear Medicine.* 2010;51(Suppl 1):18S–32S.
15. Hutton BF, Braun M, Thurfjell L, Lau DY. Image registration: An essential tool for nuclear medicine. *European Journal of Nuclear Medicine and Molecular Imaging.* 2002;29(4):559–577.
16. Townsend DW. Multimodality imaging of structure and function. *Physics in Medicine and Biology.* 2008;53(4):R1–R39.
17. ICRP. Radionuclide Transformations—Energy and Intensity of Emissions. *ICRP Publication. Ann. ICRP.* 1983;38:11–13.
18. Zanzonico P. Positron emission tomography: A review of basic principles, scanner design and performance, and current systems. *Seminars in Nuclear Medicine.* 2004;34:87–111.
19. Lecomte R. Novel detector technology for clinical PET. European journal of nuclear medicine and molecular imaging. *European Journal of Nuclear Medicine and Molecular Imaging.* 2009;36(Suppl. 1):S69–85.
20. Muehllehner G, Karp JS. Positron emission tomography. *Physics in Medicine and Biology.* 2006;51:R117–R137.
21. Lewellen TK. Recent developments in PET detector technology. *Physics in Medicine and Biology.* 2008;53:R287–R317.
22. Teras M et al. Performance of the new generation of whole-body PET/CT scanners: Discovery STE and Discovery VCT. *European Journal of Nuclear Medicine and Molecular Imaging.* 2007;34:1683–1692.
23. Kemp BJ et al. NEMA NU 2-2001 performance measurements of an LYSO-based PET/CT system in 2D and 3D acquisition modes. *Journal of Nuclear Medicine.* 2006;47:1960–1967.

24. Brambilla M et al. Performance characteristics obtained for a new 3-dimensional lutetium oxyorthosilicate-based whole-body PET/CT scanner with the National Electrical Manufacturers Association NU 2-2001 standard. *Journal of Nuclear Medicine.* 2005;46:2083–2091.
25. de Jong HW et al. Performance evaluation of the ECAT HRRT: an LSO-LYSO double layer high resolution, high sensitivity scanner. *Physics in Medicine and Biology.* 2007; 52:1505–1526.
26. Surti S, Karp JS. Imaging characteristics of a 3-dimensional GSO whole-body PET camera. *Journal of Nuclear Medicine.* 2004;45:1040–1049.
27. Surti S et al. Performance of Philips Gemini TF PET/CT scanner with special consideration for its time-of-flight imaging capabilities. *Journal of Nuclear Medicine.* 2007;48:471–480.
28. Daube-Witherspoon ME et al. The imaging performance of a LaBr3-based PET scanner. *Physics in Medicine and Biology.* 2010;55:45–64.

6

The SimSET Program

Robert L. Harrison and Tom K. Lewellen

CONTENTS

Since 1990 our laboratory, the Imaging Research Laboratory at the University of Washington, has developed and distributed the Simulation System for Emission Tomography (SimSET), a photon-tracking Monte Carlo simulation of emission tomography. When we began the project, most nuclear medicine simulations either used one of the high-energy physics simulations (e.g., EGS [1], MCNP [2], Géant [3]) or software written for a specific project. The former simulations were too slow and difficult to set up for many simulations of patient scans in tomographs; the latter were often not suitable for distribution and/or not flexible enough to use for research other than the intended project. Thus, our goals were (1) to provide a software system that was highly modular, portable, and efficient; (2) to provide a simulation targeted specifically at emission tomography applications; (3) to optimize the code for simulating heterogeneous objects in three dimensions; and (4) to produce an easy-to-use, well-documented code without unnecessary complexities. To help fulfil these goals, early versions of the software were carefully engineered [4] leading to a clear and consistent modular code structure that has reduced the engineering effort for later revisions.

Before we started on the first version of SimSET, we surveyed potential users on a variety of topics, including asking them for their favored programming language and platform. This led us to choose C as the programming language and to target SimSET for UNIX systems. We have tried to minimize system dependence, and for the most part have been successful: most users are able to install and run SimSET on their UNIX (including on Macintoshes) and Linux systems with little or no help from us. Many users have also installed SimSET on Windows systems, though we do not actively support this. The source code can be downloaded from our website (http://depts. washington.edu/simset/html/simset_main.html). The website also serves as the manual, including installation instructions, a user guide including a detailed description of each module and a tutorial, and news about SimSET. The download package includes all of the source codes, installation scripts, and test data sets for verifying successful installation of the code. The software is designed for both single photon emitters (SPECT and planar systems) and positron emitters (PET systems). As modules in the system are completed, we are placing them in the public domain. The package in release as of August 2012 includes the photon history generator (PHG), the object editor, a collimator module, a detection module, a random coincidence generation module, and a binning module.

The PHG tracks photons through the tomograph field-of-view (FOV), creating a photon history list containing the photons that reach the tomograph surface within user-specified limits. The object editor helps the user define voxelized activity and attenuation objects for the PHG. The current collimator module includes Monte Carlo photon-tracking simulation of PET collimators based on PETSIM [18] and simulation of SPECT collimators based on geometric transfer functions [5]. The detector module includes Monte Carlo photon-tracking simulations of planar detectors [6], cylindrical detectors [7], and block detector modules [8]. This module also allows for Gaussian blurring of the detected energy and, for PET, the detected time (for time-of-flight). The random coincidence generation module, available only for PET simulations, combines single photons from the data stream to make random coincidences and adds them to the data stream; this module also includes calls to user functions that can be used to simulate other aspects of PET detector electronics, like dead time, triples processing, and event mispositioning [9]. The binning module provides binning by number of scatters; by separate binning for randoms (for PET); by photon energy; by spatial position, for example, sinogram or other line-of-response binning; and time-of-flight binning (for PET). Figure 6.1 shows the flow of data between these modules. Many extensions and additions to these modules are planned over the coming years, including photon-tracking simulation of SPECT collimators, more flexibility in the block detector module, improved importance sampling (IS), simulation of isotopes with multiple emissions, and simplified simulation of changes in activity.

SimSET has more than 400 registered users and we often meet unregistered users at meetings (there is no requirement for users to register). Researchers and students on five continents have used SimSET to produce over 250 publications, dissertations, and theses. It is being used in the design and

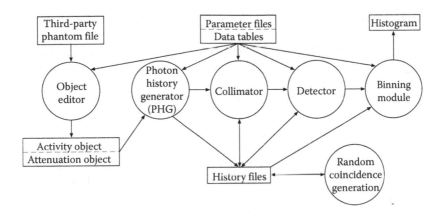

FIGURE 6.1

The main software modules and data flows for the SimSET software. The phantom to simulate is defined using the object editor. The input data for executing SimSET include: the activity and attenuation objects; text parameter files defining the phantom, scan, and tomograph parameters; and data tables defining the physical properties (e.g., attenuation) of the phantom activity and materials and scanner materials. A typical simulation proceeds decay by decay: the photon history generator generates a decay and the resulting photon(s) and tracks the photon(s) through the phantom; the collimator and detector modules track the photon(s) through the tomograph; and the binning module histograms the detected events into a user-formatted output array. A simulation can be broken off between modules and a history file (a list-mode record of the decays and photons) created: the simulation can later be resumed using the history file as input. In PET simulations, the random coincidence generation module can add random coincidences to a history file from the detector module before the binning module processes the history file. (From SimSET, http://depts.washington.edu/simset/html/user_guide/user_guide_index.html. With permission.)

optimization of patient studies, in studies of quantitation and detectability, and in the prototyping and optimization of tomographs and other hardware.

Software Development

To make the original SimSET modules easy to maintain, extend, and transport, we developed them using formal software engineering techniques. We split our development into five stages: analysis, design, implementation, validation, and maintenance. In the analysis stage, we elicited requirements from likely users of the software both within our own group and from groups doing similar work at other institutions. We reduced the users' input to a series of functional requirements and developed data flow diagrams from them. The design process was started by defining functional modules which would be needed for the particular software task. We then developed control flow diagrams of the modules and pseudo-English definitions for each of the functions.

The careful analysis and design focused the effort of the scientific research staff on the functional level, leaving the implementation to be carried out by our programming staff. This division facilitated our goal of producing functionally hierarchical software. The base of the hierarchy is a layer of functionally cohesive routines for both general-purpose and simulation-specific tasks. All operating system dependencies are restricted to these low-level functions.

We have followed a less formal approach with some more recent additions. Often the functionality of the addition is very close to that of a previously engineered module: for instance, when we added the cylindrical detector for PET, we had already implemented the planar detector for SPECT. We were able to use the same design with only minor modifications. Sometimes the similarities go even deeper: often when adding a new binning option we can use the code for another binning option as a template. However, we continue to use more rigorous software engineering methods when developing substantially different extensions to SimSET.

Our validation strategy is composed of functional testing at each layer of the hierarchy; comparison of simulation results with analytic predictions and/or results from previously validated modules; statistical analysis; structured code inspection; and independent beta testing. For example, one step in the photon-transport process is the calculation of a Compton scatter angle. Starting with the Kahn algorithm [10] for sampling from the Klein–Nishina distribution, our programmers wrote a module to compute scatter angle for modeling Compton scatter. The code was inspected line by line with a member of the scientific staff. An independent test shell was created to exercise the scatter module. The resulting distribution of scatter angles was compared to an analytically derived scatter angle distribution. Finally, when enough modules were completed to perform a simple simulation, the results from geometrically simple situations were compared with analytic predictions. Our analytic predictions were computed using Microsoft Excel spreadsheet software. Whenever possible we have used existing third-party applications for validation. This reduces the concern of implementation errors and correlation between code that creates the data and code that analyzes it.

Validation of IS techniques provided a significant challenge. These techniques add complexity to the source code, exacerbating the difficulty of validating both the algorithm and its implementation. We attacked this problem from two angles; first, we developed a statistical test for validating that there was no bias introduced by the use of IS, and then we developed a series of simulations that would highlight potential sources of bias. We performed each test simulation with all combinations of IS techniques turned on and off. The results were then compared, bin by bin, with a simulation with all IS features turned off using Student's *t*-test: the variance for the nonimportance-sampled data can be estimated as it is Poisson, and the data for importance-sampled data can be estimated using the method in the appendix of [11].

The final stage of validation consists of independent beta testing by investigators at other institutions. No amount of in-house testing can substitute for the rigors of this process. In the beta test of our first version of the PHG, 11 bugs were discovered in 11 months of testing. These ranged from innocuous to data corrupting in some circumstances. Not surprisingly, the majority of the bugs occurred in the most recently developed portions of the code, and in areas where new features were added during the testing process. We have had similar results with later releases.

Photon History Generator

SimSET divides space into the object (patient/phantom) space, the collimator space, and the detector space. Photons travel from one to the next, but never in the opposite direction. This division limits SimSET's flexibility (e.g., detectors cannot be placed inside the object) and accuracy (some detected photons do backscatter from the collimator or detector back through the object space), but increases SimSET's tracking efficiency and reduces its functional complexity. The PHG is the module that generates and tracks photons within the object (patient or phantom) space. In the current version, the object space is delimited by two right circular cylinders, the object cylinder and the target cylinder (Figure 6.2). The object cylinder is the largest cylinder contained within the voxelized descriptions of the attenuation and activity objects: no object attenuation or activity is simulated outside this cylinder. The target cylinder lies outside the object cylinder (its radius must be greater than or equal to the object cylinder's radius). When a photon reaches the boundary of the object cylinder, it is projected to the target cylinder (any intervening distance is treated as vacuum). Any photon that will not hit the target cylinder is discarded. The user may also set an axial acceptance angle: any photon with an absolute axial component to its direction vector that is greater than this angle will also be discarded—this is useful for culling undetectable photons when a geometric collimator [5] is being used. All other photons reaching the target cylinder are passed to the collimator/detector/binning modules and/or history file and are considered productive in our current IS algorithms [11].

The PHG module reads the attenuation and activity distributions and overall simulation instructions at run time. The object editor is used to define the attenuation and isotope distributions used by the PHG. To allow for maximum flexibility, the attenuation and activity distributions are modeled as voxelized objects in three dimensions. Arbitrary heterogeneous attenuation and activity

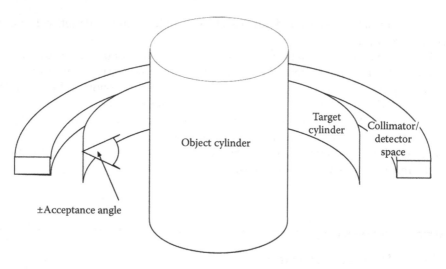

FIGURE 6.2
The PHG generates and tracks photons within the object cylinder. Once a photon reaches the object cylinder, it is projected to a target cylinder between the object cylinder and the collimator/detector. Only photons reaching the target cylinder within a user-specified acceptance angle are passed on to the collimator/detector/binning modules or placed in the PHG history file. (From SimSET, http://depts.washington.edu/simset/html/user_guide/phg.html#target_ cylinder. With permission.)

distributions can be simulated. The object editor provides the choice of initializing each slice with a constant value or from voxel-based physiologically realistic phantoms, for example, the digitized anthropomorphic phantoms of Zubal et al. [12] or Segars [13,14]. After initializing the object with constants or from an anthropomorphic phantom, the user may add geometric elements (e.g., cylinders, ellipsoids) to the object. Tables are used to translate the numbers in the voxelized phantom into actual activity or attenuation values so that the user may change their distributions without changing values in the digitized phantoms (see Figure 6.3). When the PHG processes the object files, it produces three-dimensional integer arrays defining the attenuation and isotope distributions. These may be viewed using most image display programs. For example, Figure 6.4 depicts an attenuation and emission object used for a SPECT simulation in the PHG tutorial (on the SimSET website).

Once SimSET has read in the input instructions and data, it samples decays from the activity distribution until the specified number of decays is reached. (The user can either specify a number of decays or let SimSET compute it from the activity distribution and the scan time.) Each decay produces either a positron or a single photon with a user-specified energy: SimSET does not currently support other options like multiemission isotopes. If the decay is a positron decay, an annihilation position is sampled using the Palmer method [15], and two nearly antiparallel photons are produced—acollinearity is also modeled. The distance the photon should travel before interacting is sampled

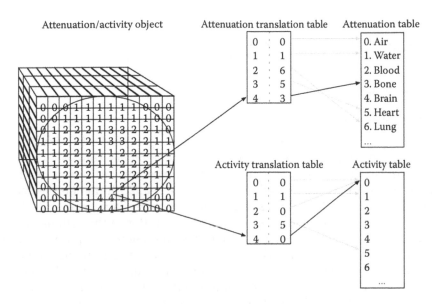

FIGURE 6.3
Translation tables for SimSET attenuation and activity objects. Translation tables link the indexes in the attenuation and activity objects to the SimSET data tables giving the attenuation characteristics and activity values. Here, the same object is used for both the attenuation and activity object. The value 4 in the attenuation object is translated to a 3, the value for bone, in the SimSET attenuation table. The same value in the activity object is translated to the 0 mCi/cc value in the activity table. (From SimSET, http://depts.washington.edu/simset/html/user_guide/object_editor.html#figure2. With permission.)

as described in Chapter 1. SimSET models photo absorption and Compton and coherent scatter. Compton scatter is modeled using the Kahn algorithm [10]; coherent scatter is sampled from tables based on the Livermore Evaluated Photon Data Library [16]. After a scatter, a new travel distance is sampled. This process is repeated until the photon reaches the boundary of

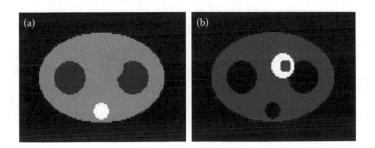

FIGURE 6.4
Examples of an attenuation object (a) and an emission object (b) produced by the object editor. (From SimSET, http://depts.washington.edu/simset/html/user_guide/tutorials.html#creating_object. With permission.)

the object cylinder, the photon is absorbed, or the photon energy falls below a user-specified minimum tracking energy.

Collimator Module

Both SPECT and PET collimator simulations are provided with SimSET. For SPECT, we implemented, in collaboration with researchers from the University of North Carolina, a collimator simulation using geometric transfer functions [17]. There are options for parallel, fan, and cone beam collimation. Geometric transfer functions allow for fast simulations with realistic geometric effects but do not address issues such as penetration, scatter in the collimator, and lead x-rays. A photon-tracking version of the collimator is required to study such effects. We plan to provide such a model in the future; in the meantime, researchers at Johns Hopkins have combined SimSET simulations in the object with MCNP and GATE simulations of the collimator and detector [17].

For PET collimator simulations, we have adapted the model used in Thompson's PETSIM [18]. This model allows for multiple layers of septa, both axially and radially, and for radially tapered septa (Figure 6.5). Different materials can be used for different parts of the collimator. Photons are projected from the target cylinder (the cylindrical surface to which the PHG projects photons) to the innermost surface of the collimator. From there, the photons are tracked until they (1) exit the outermost radial boundary of the collimator and are saved to the photon history list or passed to the detector/binning module for further processing; (2) exit the innermost radial, topmost axial, or bottommost axial boundaries and are discarded; or (3) fall below a user-specified minimum energy and are discarded.

SimSET also provides a slat collimator for use with dual-headed planar cameras for coincidence imaging [7]. These collimators are composed of rectangular septa and can vary radially and axially.

Photons are tracked through the collimator using the same algorithm as described in the section on PHG.

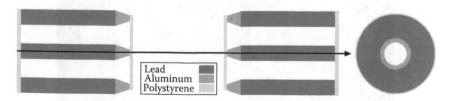

FIGURE 6.5
Axial and transaxial cross sections through a PET collimator. PET collimators are composed of cylinders that can have multiple radial and axial layers. Septa can be tapered, demonstrated with aluminum tips here. (From SimSET, http://depts.washington.edu/simset/html/user_guide/collimator_modelling.html#figure2. With permission.)

Detector Module

SimSET has three photon-tracking detector models: planar (Figure 6.6a), cylindrical (Figure 6.6b), or block detectors (Figure 6.6c).

SimSET's planar detectors are rectangular boxes that can be layered in the radial direction. No material changes are modeled in the axial or transaxial directions (see block detectors below for a more general model). The detectors follow a circular orbit and are always perpendicular to the radius vector. The planar detectors are generally intended for SPECT simulations, but a dual-headed option can also be used for PET.

SimSET's cylindrical detectors are composed of right circular cylinders that can be layered axially and radially. They cannot be split into crystals transaxially.

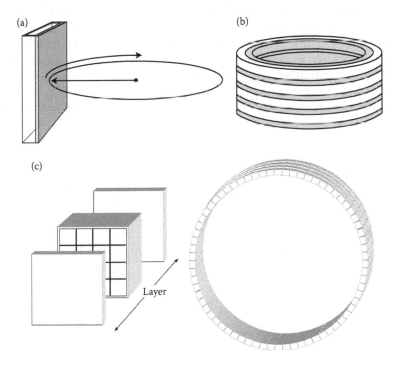

FIGURE 6.6
Examples of SimSET's three detector models. (a) Planar detectors are composed of layers of rectangular boxes. Material changes are allowed only in the radial direction. Only circular detector orbits are modeled. (b) Cylindrical detectors are right circular cylinders. Material changes are allowed in the radial and axial directions, but not in the transaxial direction. (c) In the block detector model, the tomograph is composed of rings axially, with the rings composed of blocks. Each block is a rectangular box that can be further subdivided into smaller rectangular boxes. (From SimSET, http://depts.washington.edu/simset/html/user_guide/detector_modelling.html#supported_models. With permission.)

In the block detector model, the tomograph is composed of rings axially, with the rings composed of blocks. Each block is a rectangular box that can be further subdivided into smaller rectangular boxes. More than one type of block may be used to define a ring, and more than one type of ring to define a tomograph. SimSET does place some limitations on the descriptions: (1) all elements of the detector system must be outside an elliptical cylinder containing the object being imaged and the collimator (if any), (2) the axial faces of the blocks must be parallel to the *x–y* plane, and (3) all blocks within a ring must have the same axial extent as the ring. This model will allow users to simulate typical block-based cylindrical tomographs, pixilated positron emission mammography (PEM) detectors, and many more imaginative tomograph designs.

Tracking a photon through the detectors follows the same logic as given above for the PHG, except that no minimum energy cut-off is used: in the detector photons are tracked until they are absorbed or are on a path that will not intersect the detector(s) again. At that point, SimSET computes detected energy, detected position, and, for time-of-flight PET, detected time. Only interactions in user-specified "active" areas (e.g., scintillation crystal) of the detectors are used in these computations, and in block detectors only the subset of the interactions that occurred in the block in which the photon deposited the most energy. Detected energy and time are computed by adding Gaussian noise with user-specified full-width half maximums to

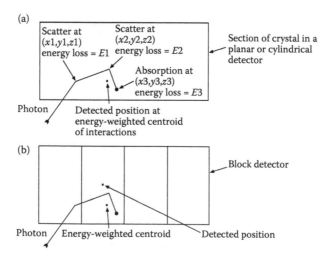

FIGURE 6.7
Detected position. (a) In planar and cylindrical detectors, the detected position is the energy-weighted centroid of the interactions in the crystal. (b) In block detectors, the detected position is the crystal center nearest to the energy-weighted centroid. (From SimSET, http://depts. washington.edu/simset/html/user_guide/detector_modelling.html#figure2 and #figure8. With permission.)

the total deposited energy and last interaction time, respectively. For planar and cylindrical detectors, the detected position, (x_c, y_c, z_c), is calculated as the energy-weighted centroid of the interactions:

$$(x_c, y_c, z_c) = \frac{\sum_{i \in I} E_i * (x_i, y_i, z_i)}{\sum_{i \in I} E_i} \tag{6.1}$$

where the E_i and (x_i, y_i, z_i) are the deposited energy and position of each interaction, respectively (Figure 6.7a). For block detectors, this energy-weighted centroid is then snapped to the nearest center of an active element (crystal) in the block (Figure 6.7b).

Random Coincidence Generation Module

One can estimate random coincidences for PET simulations by running a separate singles simulation using the same detector simulation and then estimate randoms from the detected singles [19]. This is quite accurate in most cases, but does not take into consideration some second-order effects like triple coincidences and dead time. To accommodate these effects, SimSET offers direct simulation of randoms in the coincidence simulation [9]. This option requires several steps and takes considerably more CPU time; we recommend this option only in cases where the computation from singles is potentially inaccurate.

Random coincidence generation occurs between the detector module and the binning module. It reads in a history file from the detector module, re-sorts the history file to be in decay-time order, applies a user-specified time window to the time-sorted history file, and finally writes out a history file with random coincidences inserted (Figure 6.8). If three or more detected photons are seen within a time window, all the decays in the window are discarded (i.e., triple coincidence events are discarded). If two single photons are seen within a time window, a random coincidence is created and written to the output history file. All other coincidences are also written to the output history file.

Random coincidences are just one of the time-related effects that are of interest in simulations. While SimSET does not currently simulate these other effects, it does make calls to user-modifiable functions at critical points in the random coincidence module. These functions could be used to modify the treatment of triple coincidence events or to add other timing-related phenomena. User functions are discussed in greater detail below.

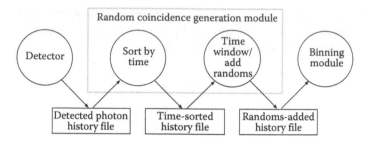

FIGURE 6.8

Random coincidence generation module for PET simulations. The random coincidence generation module takes a detected photon history file, sorts the decays in the history file by time,
and then applies a time window to the time-sorted histories. Any photon that does not fall
in a time window with at least one other photon is discarded (i.e., unmatched single photons
are discarded). Any photon that falls in a time window with two or more other photons is
discarded (i.e., triple coincidences are discarded). When exactly two photons are in a time
window, a decay record and two photon records are written out to the randoms-added history file: when the two photons are from the same decay, the decay and photons are written
out unchanged; when the two photons are from different decays, a new pseudo-decay record
marked as a random coincidence is written out along with the two photons. (From SimSET,
http://depts.washington.edu/simset/html/user_guide/randomsSimulation.html#overview.
With permission.)

Binning Module

The SimSET software automatically reports a variety of statistics on the
output photons to provide a general description of the simulation results.
Most users will want to further analyze the data in ways specific to their
research goals. The binning module can be used to create multidimensional
histograms of statistics. Binning can be performed using data from the PHG,
the collimator module, or the detector module, either on an event-by-event
basis while the other modules are running or afterwards from a photon history list. The user controls the binning using the binning parameters file.
Histograms can be formed over any combination of photons' detected spatial
position (i.e., line-of-response), number of scatters, and detected energy. For
PET, coincidences can also be binned by time-of-flight position and whether
the coincidence is a random event or not. The ranges and number of bins
for each dimension are set in the binning parameters file. Examples of sinograms produced by the binning module for unscattered and scattered photons from a SPECT simulation are shown in Figure 6.9. The resulting images,
reconstructed using filtered backprojection with a Hann filter, are shown
in Figure 6.10. The binning module's line-of-response binning options also
include binning by crystal number and binning appropriate for the 3DRP
[20] image reconstruction.

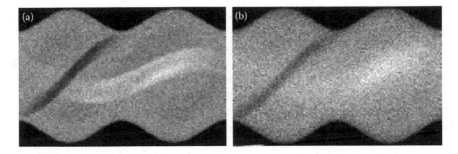

FIGURE 6.9
Examples of binned sinograms for unscattered (a) and scattered (b) photons in a simulation study. (From SimSET, http://depts.washington.edu/simset/html/user_guide/tutorials. html#analyzing_the_data. With permission.)

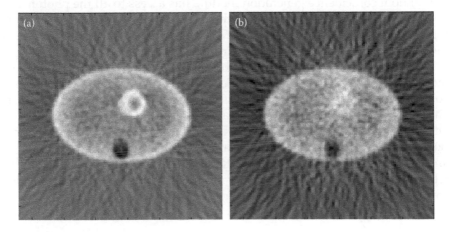

FIGURE 6.10
Reconstructed images from the binned sinograms in Figure 6.9. (From SimSET, http://depts. washington.edu/simset/html/user_guide/tutorials.html#analyzing_the_data. With permission.)

User Functions

SimSET provides several "user functions" where the user can intercept a photon or event during its processing and modify it, thus altering how it will be processed further downstream. In the distribution package these functions are empty (i.e., they do nothing), but they are called at critical points in decay/photon processing, providing a safe means of modifying or adding to SimSET. There are calls to user initialization and termination functions during SimSET's initialization and termination. These allow the user to set up read in instructions and input data, allocate/deallocate and initialize variables, report, and write out data, and incorporate other such functionality best performed before or after the decay-tracking loop. There are also calls to user functions just before each photon is processed in the collimator, detector, and binning modules, just before photons in each time window are processed and just before each event is written to the output history file in the random coincidence generation module, and just before the histogram is updated for each accepted event in the binning module. The user functions have access to all the parameters of the calling module. In addition, the user functions called before each photon is processed in the collimator, detector, and binning modules have access to all the tracking information for the current photon and decay; the user function called before the time-window processing in the random coincidence generation module has access to all the photon and decay information for all the events in the time window; the user function called before each event is written to the output history file in the random coincidence generation module has access to all the photon and decay information for the event; and the user function called before the histogram is updated for each accepted event in the binning module has access to all the photon, decay, and binning information for the event.

The user functions have many possible uses, from creating completely new modules to small changes in the processing of photons. For instance, one could create an electronics module including dead time and pulse pileup as part of the random coincidence generation module, or one could implement new types of collimators or detectors as collimator or detector user functions. Smaller changes might include tomograph-specific detected energy and position algorithms in the detector module user functions or Fourier-rebinning (FORE) [21] in the binning module user function, both of which could be implemented as binning user functions.

Efficiency

The main factors that determine the computational efficiency of the PHG are the geometry of the tomograph being simulated and the voxelization of the attenuation object. The portion of the object being imaged that falls within the tomograph FOV and the solid angle subtended by the tomograph limit the efficiency of conventional simulations. Events occurring or scattering outside the FOV are simulated as they might subsequently scatter into the FOV. Similarly, events within the FOV that start or scatter in directions that cannot be detected are tracked.

The computational cost of these obstacles can be reduced using IS (or variance reduction) techniques. The techniques we have adapted include (1) forced detection; (2) stratification; and (3) weight windows [10,22–24].

The details of our approach have been published [11,25], and the theory is discussed in Chapter 2.

We define any photon that reaches the target cylinder within the acceptance angle (Figure 6.2) to be productive. For an unscattered photon to be detected, the ray defined by its starting location and direction must intersect the target cylinder within the tomograph or collimator's acceptance angle (though this does not guarantee the photon will be detectable). Similar conditions for detectability can be set at a scattered photon's last interaction point. To prevent wasting compute time simulating events that will not be detected, the simulated decay locations and initial photon direction cosines are stratified by object slice and axial direction cosine, that is, more photons are simulated in slices and angles that are likely to produce detections. The productivity table stratifies the possible slices and axial angles and assigns "productivities" for each slice/angle bin. Different productivities are computed for scattered and unscattered events. Figure 6.11 illustrates the stratification of the emission angles. A short simulation with stratification disabled is used to compute the productivity table.

IS techniques require that each photon be assigned a weight that indicates how many photons it represents in the real world. The weight is adjusted at every IS step. For example, when stratifying photons' starting slice/angle, photons from oversampled slice/angles are given lower weights than photons from undersampled slice/angles. The PHG uses a "productivity table" to tell where the greatest number of "starts" should take place and to determine the photon weights.

The principle of forced detection is illustrated in Figure 6.12. As a photon is being tracked, a forced detection attempt is made with a copy of the photon after each interaction. The copy is forced to scatter somewhere along its path and then projected until it leaves the object. The scatter position and angle are chosen such that the photon hits the target cylinder within the acceptance angle, and thus is productive. The copy photon's weight is adjusted at each step by the probability of the forced action, that is, the probability that a scatter would occur somewhere "visible" from the target cylinder, the probability that the resulting scatter would be in a productive direction, and the probability that the photon would reach the target cylinder without another interaction. Once tracking of the photon copy is complete, tracking of the original photon continues until it leaves the object or its energy falls below user-specified minimum tracking energy, at which point the photon is discarded. The technique is implemented in the PHG with the use of precomputed probabilities that a photon with a specific direction and angle will scatter into a selectable range of outgoing angles. In particular, the forced-detection tables convert from the coordinates used by the Klein–Nishina formula (incoming energy and scatter angle) to the coordinates needed to implement forced detection (incoming energy, outgoing z-direction cosine, and sine of outgoing azimuthal angle).

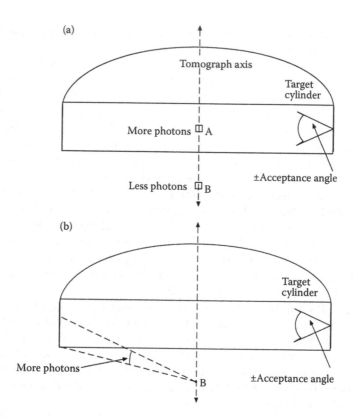

FIGURE 6.11

Stratification. (a) More decays are simulated at locations where they are likely to lead to detected photons than at locations where they are not. For instance, if voxels A and B have the same activity, more decays will be simulated in voxel A than B: decays in the field-of-view are more likely to be detected. (b) Once a decay location is selected, the photon direction is also stratified. Photon directions that will intersect the target cylinder within the acceptance angle are sampled more heavily than the ones that will not. (From SimSET, http://depts.washington.edu/simset/html/user_guide/phg.html#figure3. With permission.)

When creating sinograms or other binned output from importance-sampled events, instead of counting the events in each bin, one must sum the weights of the events. A list of N detected events with varying weights is worth less (i.e., statistics computed from them will have a greater variance) than a list of N uniformly weighted detected events by a multiplicative factor between 0 and 1 (see Chapter 2). This is called the quality factor (QF) and is used in the calculation of the computational figure-of-merit (CFOM) [11]:

$$\text{CFOM} = \frac{(N \times \text{QF})}{(\text{CPU seconds})} \tag{6.2}$$

For conventional simulations, CFOM gives the number of events produced per CPU second. For simulations using IS, it gives a similar figure adjusted

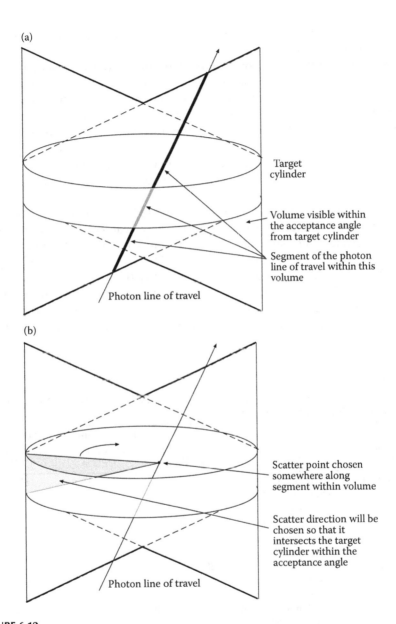

(a)

Target cylinder

Volume visible within the acceptance angle from target cylinder

Segment of the photon line of travel within this volume

Photon line of travel

(b)

Scatter point chosen somewhere along segment within volume

Scatter direction will be chosen so that it intersects the target cylinder within the acceptance angle

Photon line of travel

FIGURE 6.12
Forced detection. (a) When a photon passes through the volume visible from the target cylinder, a copy of the photon is made. (b) This copy is then forced to scatter somewhere in this volume. The photon direction after the scatter is chosen so that it will hit the target cylinder within the acceptance angle. Once the forced detection is complete, regular tracking of the original photon is resumed. (From SimSET, http://depts.washington.edu/simset/html/user_guide/phg.html#figure4. With permission.)

by the QF. In general, the CFOM improves significantly with the use of stratification and forced detection.

Increasing the number of voxels in the attenuation object usually increases the time spent tracking each photon: every time a photon crosses into a new voxel, the voxel attenuation value must be looked up, and the distance and free-path lengths to the next voxel calculated. An exception to this occurs when using stratification: increasing the number of slices may actually improve the efficiency of the stratification and thus of the simulation as a whole. However, once the voxelization has been chosen, the heterogeneity of activity and attenuation within the voxelized object has no effect on computation time.

Efficiency results for simulations of two 3D PET scanners are given below both with and without IS (see Table 6.1). Three water-filled cylinders with uniform activity were simulated: a 20 cm diameter, 15 cm long cylinder; a 20 cm diameter, 70 cm long cylinder; and a 30 cm diameter, 70 cm long cylinder. The 15 cm long cylinder was simulated with three different voxelizations: as a $1 \times 1 \times 1$ voxel object (SimSET truncates any attenuation/activity outside the object cylinder, so this becomes an exact representation of a cylinder); as a single axial slice with 128×128 voxels transaxially; and as

TABLE 6.1

Comparison of Computational Figure-of-Merit (CFOM) for Simulations with and without Importance Sampling (IS)

Detector Type	Object	Voxelization $(X \times Y \times Z)$	CFOM without IS (Events/ CPU s)	CFOM with IS (Events/ CPU s)
Cylindrical	20 cm diameter, 15 cm long water cylinder, uniform activity	$1 \times 1 \times 1$	1510	6790
Cylindrical	20 cm diameter, 15 cm long water cylinder, uniform activity	$\mathbf{128} \times \mathbf{128} \times 1$	507	2256
Cylindrical	20 cm diameter, 15 cm long water cylinder, uniform activity	$128 \times 128 \times \mathbf{15}$	451	2173
Cylindrical	20 cm diameter, **70 cm long** water cylinder, uniform activity	$128 \times 128 \times 70$	97	388
Cylindrical	**30 cm diameter**, 70 cm long water cylinder, uniform activity	$128 \times 128 \times 70$	58	198
Block	30 cm diameter, 70 cm long water cylinder, uniform activity	$128 \times 128 \times 70$	44	139

Note: Bold face indicates changes in the simulation from one line to the next.

15 1 cm thick axial slices, each 128×128 voxels transaxially. The 70 cm long phantoms were represented as 70 1 cm thick axial slices, each 128×128 cm transaxially.

Both of the simulated tomographs had a 70 cm diameter, 15 cm long patient FOV. The target cylinder had a radius of 35 cm, and extended from −10 to 10 cm axially; the acceptance angle was set to 90° (i.e., no photons were rejected due to the acceptance angle). The collimator simulation starts with a patient tunnel, a 0.5 cm thick, 20 cm long cylinder of polystyrene. Then, at either end of the FOV (starting at ±7.5 cm axially) there are two cylindrical lead endplates, 2.5 cm thick with inner radius 35.5 cm and outer radius 43.8 cm. The final radial layer in the collimator simulation is a cylindrical tunnel between the endplates and detectors, a 0.2 cm thick layer of polystyrene.

We used two different detector simulations. The first was a solid cylindrical detector, 3 cm thick BGO crystal, inner radius 44.14 cm, running from −7.5 to 7.5 cm axially. The second used four rings of block detectors. The structure of the blocks is shown in Figure 6.13; each ring consisted of 70 blocks, evenly spaced around the ring at 44 cm inner radius. Cylindrical detector simulations were run for all the phantoms; the block detector simulation was run only for the 30 cm diameter cylindrical phantom.

The IS results used stratification and forced detection. These results are from version 2.9.1 of the software. All simulations were run on one CPU of a Macintosh computer with dual 2.3 GHz PowerPC G5 CPUs.

The simulation results demonstrate some of the above-mentioned effects on sensitivity. IS leads to a 3–5-fold increase in efficiency, depending on the simulation. The first three lines show how the simulation efficiency decreases as the number of voxels in the object increases: approximately a factor of three as the voxelization changes from $1 \times 1 \times 1$ to $128 \times 128 \times 15$,

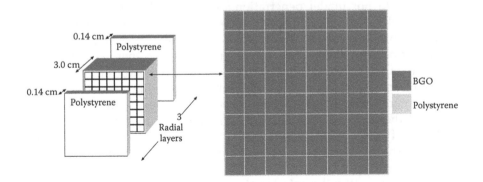

FIGURE 6.13
Block detector for efficiency tests. The block detector used in the efficiency tests consists of an 8×8 array of BGO crystals. The crystals are separated by 0.006 cm layers of polystyrene. On the axial and transaxial sides of the block, there is a 0.014 cm layer of polystyrene wrap. Radially, there are 0.14 cm layers of polystyrene before and after the crystal array. (From SimSET, http://depts. washington.edu/simset/html/user_guide/detector_modelling.html#figure7. With permission.)

with the importance-sampled simulation slowed by a slightly smaller factor. In particular, finer axial sampling had very little impact on the efficiency of the importance-sampled simulation.

The addition of activity outside the tomograph FOV greatly reduced simulation efficiency, as did the expansion of the water cylinder from 20 to 30 cm diameter. In both cases, the decrease was greater for the importance-sampled data: this was mainly due to a decrease in the QF.

Using block detectors further decreased the efficiency by 23% for the nonimportance-sampled simulation and 30% for the importance-sampled simulation. The bulk of the decrease was due to a sensitivity difference of 20%, not surprising given that the block detector tomograph has 12% less BGO than the cylindrical tomograph.

Future Development

We are constantly upgrading and extending the SimSET software, both to add functionality and to improve ease of use, efficiency, and flexibility of the current capabilities. We have recently added a photon-tracking simulation of block detectors such as those used in most PET tomographs, time-of-flight PET binning, and random coincidence simulation. We next plan to add photon-tracking and angular-response-function [26] versions of the SPECT collimator, improve simulation efficiency with improved IS, and increase the flexibility of the object description. We also plan to take more of an open-source approach to our software development.

SimSET currently offers only geometric collimation for SPECT. Geometric collimators do not model penetration, scatter, or the production of lead x-rays (k-shell fluorescence [27]) in the collimator. In some simulation studies, these omissions have little impact on the results. However, penetration and scatter become increasingly important for higher-energy photons; lead x-rays are important to account for in studies where dead time is an important factor and in simulations of thallium studies, which has a photopeak near the energy of lead x-rays. A photon-tracking simulation can accurately model all these effects, though such simulations are very inefficient: the great majority of photons tracked will not make it through the collimator to detectors. Another possibility is to use angular response functions: detailed photon-tracking simulations of the collimator are done for photons of different energies hitting the detector at different angles and the results summarized in empirical parametric form. Penetration, scatter, and lead x-rays are all included as possible outcomes, though their detected positions are approximated. Collimator simulations using angular response functions are considerably faster than photon-tracking simulations, but do not solve the problem that most photons tracked through the object are neither detected nor produce lead x-rays.

This is where improved IS can help. Our current IS algorithms were designed before we had collimator or detector simulations: the binning module processed all photons reaching the target cylinder within the acceptance angle. Now, for PET simulations, the acceptance angle must be set to 90° to avoid eliminating photons that could, conceivably, scatter in the collimator and be detected. As a result, many photons that the current IS regards as productive never make it to the detector. We plan to implement new algorithms that define productivity based on photons reaching the detector. We have previously prototyped this concept [28] and found that it could double SimSET's efficiency. We believe that combining this change with IS in the collimator and fictitious interaction tracking [29] will lead to even bigger efficiency gains.

We are also looking at ways to increase SimSET's flexibility without significantly impacting its ease of use or efficiency. One restriction that has made it difficult to use SimSET to study novel detector applications is the requirement that the object being imaged must lie entirely within a right circular cylinder (the object cylinder, described above) and the scanner entirely outside the same cylinder. For instance, in simulations of PEM, the object needs to be able to fill the rectangular box between the detectors. We plan to expand SimSET's object geometry to include both elliptical cylinders and rectangular boxes as acceptable object volumes. We will, however, maintain the requirement that the scanner be entirely outside the chosen object volume: this distinction is one main source of SimSET's efficiency advantage over high-energy physics packages.

Finally, we are planning to take a more open-source approach to SimSET development. SimSET has always been in the public domain, but very few of the extensions written by users have been incorporated into SimSET. Often, the authors are not even willing to share them: like much of the simulation software that was written before SimSET, the extensions are not carefully enough designed and written to be supported. As is often the case, the authors did not have the time to do so. To circumvent this problem, we are planning to offer more formal support to users writing extensions. We hope that providing our expertise will both allow such authors to work faster and to produce extensions that are well enough designed, written, and documented to be incorporated into the SimSET code base.

Acknowledgments

The work by the authors presented in this chapter was supported by PHS grants CA42593 and CA126593.

References

1. Nelson WR, Hirayama H, Rogers DWO. *EGS4 Code System: Stanford Linear Accelerator Center*, Menlo Park, CA (USA), 1985.
2. Forster RA, Cox LJ, Barrett RF et al. MCNP (Tm) Version 5. *Nuclear Instruments and Methods in Physics Research Section B: Beam Interactions with Materials and Atoms.* 2004;213:82–86.
3. Allison J, Amako K, Apostolakis J et al. Geant4 developments and applications. *IEEE Trans. Nucl. Sci.* 2006;53(1):270–278.
4. Anson C, Harrison R, Lewellen T. On using formal software engineering techniques in an academic environment. *Proceedings of the Annual International Conference of the IEEE Engineering in Medicine and Biology Society,* Seattle, WA, 1989.
5. Metz CE, Atkins F, Beck RN. The geometric transfer function component for scintillation camera collimators with straight parallel holes. *Phys. Med. Biol.* 1980;25:1059–1070.
6. Harrison RL, Laymon CM, Vannoy SD, Lewellen TK. Validation of the SPECT features of a simulation system for emission tomography. *Conference Records of the IEEE Nuclear Science and Medical Imaging Conference,* San Diego, CA, 2001.
7. Harrison R, Vannoy S, Kaplan M, Lewellen T. Slat collimation and cylindrical detectors for PET simulations using SIMSET. *Conference Records of the IEEE Nuclear Science and Medical Imaging Conference,* Lyon, France, 2000.
8. Harrison RL, Gillispie SB, Lewellen TK. Design and implementation of a block detector simulation in SimSET. *Conference Records of the IEEE Nuclear Science and Medical Imaging Conference,* Honolulu, HI, 2007.
9. Harrison RL, Gillispie SB, Alessio AM, Kinahan PE, Lewellen TK. The effects of object size, attenuation, scatter, and random coincidences on signal to noise ratio in simulations of time-of-flight positron emission tomography. *Conference Records of the IEEE Nuclear Science and Medical Imaging Conference,* San Diego, CA, 2006.
10. Kahn H. *Use of Different Monte Carlo Sampling Techniques*, Rand Corporation, Santa Monica, CA, 1955.
11. Haynor D, Harrison R, Lewellen T. The use of importance sampling techniques to improve the efficiency of photon tracking in emission tomography simulations. *Med. Phys.* 1991;18:990.
12. Zubal IG, Harrell CR, Smith EO, Rattner Z, Gindi G, Hoffer PB. Computerized three-dimensional segmented human anatomy. *Med. Phys.* 1994;21:299–299.
13. Segars WP, Tsui BMW. Study of the efficacy of respiratory gating in myocardial SPECT using the new 4-D NCAT phantom. *IEEE Trans. Nucl. Sci.* 2002;49(3):675–679.
14. Segars WP, Tsui BMW, Frey EC, Johnson GA, Berr SS. Development of a 4-D digital mouse phantom for molecular imaging research. *Mol. Imag. Biol.* 2004;6(3):149–159.
15. Palmer MR, Brownell G. Annihilation density distribution calculations for medically important positron emitters. *IEEE Trans. Med. Imag.* 1992;11(3):373–378.
16. Cullen DE, Hubbell JH, Kissel L. EPDL97: The evaluated photon data library,'97 Version. *Lawrence Livermore National Laboratory Report No. UCRL-50400.* 1997.

17. Tsui BMW, Gullberg G. The geometric transfer function for cone and fan beam collimators. *Phys. Med. Biol.* 1990;35:81–93.
18. Thompson C, Moreno-Cantu J, Picard Y. PETSIM: Monte Carlo simulation of all sensitivity and resolution parameters of cylindrical positron imaging systems. *Phys. Med. Biol.* 1992;37:731–749.
19. Badawi R, Kohlmyer S, Harrison R, Vannoy S, Lewellen T. The effect of camera geometry on singles flux, scatter fraction and trues and randoms sensitivity for cylindrical 3D PET—A simulation study. *IEEE Trans. Nucl. Sci.* 2000;47(3):1228–1232.
20. Kinahan P, Rogers J. Analytic 3D image reconstruction using all detected events. *IEEE Trans. Nucl. Sci.* 1989;36(1):964–968.
21. Defrise M, Kinahan P, Townsend D, Michel C, Sibomana M, Newport D. Exact and approximate rebinning algorithms for 3-D PET data. *IEEE Trans. Med. Imag.* 2002;16(2):145–158.
22. Booth TE, Hendricks JS. Importance estimation in forward Monte Carlo calculations. *Nucl. Technol./Fusion;(United States).* 1984;5(1):90–100.
23. Brown RS Hendricks J. Implementation of stratified sampling for Monte Carlo applications. *Trans. Am. Nucl. Soc.; (United States).* 1987;97:245–248.
24. Cramer S, Gonnord J, Hendricks J. Monte Carlo techniques for analyzing deep penetration problems. *Nucl. Sci. Eng.* 1986;92:280–288.
25. Haynor D, Harrison R, Lewellen T et al. Improving the efficiency of emission tomography simulations using variance reduction techniques. *IEEE Trans. Nucl. Sci.* 1990;37(2):749–753.
26. Song X, Segars WP, Du Y, Tsui BMW, Frey EC. Fast modelling of the collimator detector response in Monte Carlo simulation of SPECT imaging using the angular response function. *Phys. Med. Biol.* 2005;50:1791–1804.
27. McGuire E. K-shell Auger transition rates and fluorescence yields for elements Ar-Xe. *Phys. Rev. A.* 1970;2(2):273–278.
28. Harrison R, Dhavala S, Kumar P et al. Acceleration of SimSET photon history generation. *Conference Records of the IEEE Nuclear Science and Medical Imaging Conference*, Norfolk, VA, 2002.
29. Rehfeld N, Stute S, Apostolakis J, Soret M, Buvat I. Introducing improved voxel navigation and fictitious interaction tracking in GATE for enhanced efficiency. *Phys. Med. Biol.* 2009;54:2163.

7

The SIMIND Monte Carlo Program

Michael Ljungberg

CONTENTS

The Monte Carlo code SIMIND [1] simulates a clinical SPECT scintillation camera and can easily be modified for almost any type of calculation or measurement encountered in SPECT imaging. The SIMIND system consists of two main programs, CHANGE and SIMIND. The CHANGE program provides a menu-driven method of defining the system that will be simulated and to write these parameters to an external file. The actual Monte Carlo simulations are made by the program SIMIND that reads data from the input files, created by CHANGE, and after calculation writes results to the screen or to different data files. In this way, several input files can be prepared in a command file for submission to a batch queue, a convenient way when working with Monte Carlo simulations. Figure 7.1 shows a flowchart that describes a different file structure for the SIMIND and the CHANGE program.

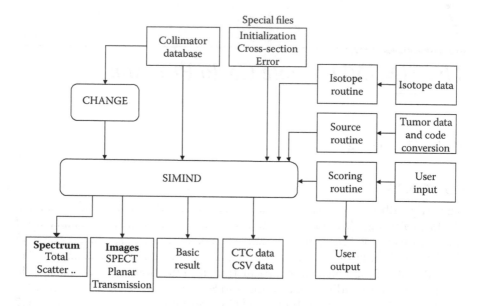

FIGURE 7.1

A schematic flowchart describing the input and output files used and created by the SIMIND code.

SIMIND includes a logical flag system that provides the user with different turn on/off features, such as SPECT simulation, simulation of interactions in the phantom, adding a Gaussian-based energy resolution, and so on without the need to redefine the input variables or the input file since the flags can be controlled from the command line.

A standard file format is sometimes needed to import simulated data to commercial imaging processing systems easily. SIMIND uses the Interfile V3.3 for an intermediate file format. This format is defined by a list of key-value pairs located in the header of an image data file [2] or as a separate file. Although DICOM has become a standard in image communication, many vendors still have support for the Interfile format.

The entire code has been written in Fortran-90 and includes versions that are fully operational on Linux 32 and 64 bit system versions and Windows 32 and 64 bits (Intel Visual Fortran Compiler and Lahey LF90 Compiler). Most of the main code structure is similar for all the operating systems, but, in cases where the operating system becomes unique, additional information on the code as it pertains to the specific system is provided. The code is strict Fortran-standard with only two system-dependent routines, namely a getarg function that reads the command line arguments from the operating system and a spawn command routine that calls the operating system and performs certain commands. These two routines are, however, often included by different compiler vendors in their system libraries. No imaging routines or reconstruction programs are provided.

The Fortran-90 language allows for a dynamic allocation of computer memory which is very convenient since arrays, such as image matrices and energy pulse-height spectrum, can be defined at execution and from definitions made by the CHANGE program. Dynamical allocated arrays thus significantly reduce the need for recompilation.

Basic Algorithm

The Monte Carlo program SIMIND is based on the use of uniformly distributed random numbers for modeling the process of radiation transport. In short, the code works as follows: photons are emitted from a simulated activity distribution in the phantom and are followed step by step toward the scintillation camera. Since the details of the photon history are not lost as is true in practical imaging, important parameters that are not accessible by measurements (for instance, the numbers of scatter interactions in the phantom for a particular photon history or scatter order, scatter angles, imparted energy, etc.) can always be deduced during the simulation. Several variance-reduction techniques, such as forced interaction and detection, important sampling, and limited direction emission, have been employed to significantly increase the efficiency of the simulation. The introduction of variance-reduction methods requires the use of a proper calculated photon history weight (PHW) in order to avoid bias in the result. See Chapter 2 for more details about variance-reduction methods.

Differential cross-section tables for the photoelectric interactions, coherent, incoherent scattering, and pair production have been generated in steps of 0.5 keV up to 3 MeV by the XCOM program [3]. Form factors and scattering functions used in the sampling of coherent scattering and in the correction for bounded electrons are taken from Hubbell et al. [4,5]. Included in the cross-section table are also the discontinuities close to the K-shell which is an effect that is of importance for material with relatively high atomic number.

If a photon is passing through the collimator, then it is followed explicitly in the scintillation crystal volume until it is absorbed or escapes the crystal. Characteristic x-ray photons, emitted from a photoelectric absorption, can be simulated. When the photon energy is above the energy threshold 2 mc^2, then pair production will be included in the possible interaction types. The imparted energy is scored at each interaction location and the x- and y-coordinate for the image is calculated from the centroid of these energy depositions.

A layer with a different material can be defined around the scintillator volume to simulate a protecting cover. Photon interactions are simulated in this volume but with no scoring of imparted energy. In most imaging situations of low-energy photons, the backscattering of photons from the light-guide and photomultipliers is not considered important. However, when simulating radionuclides emit multiple high-energy photons, such as ^{131}I, these backscatter photons can contribute significantly to the imparted energy recorded in an energy window in the lower energy region. To include this effect in the simulation, a volume can be defined behind the crystal with a thickness that is assumed to be equivalent to the scattering equipment behind the NaI(Tl) crystal. If a photon penetrates the crystal, it is followed in this volume until an escape. If the photon here is backscattered into the NaI(Tl) crystal, it is continued in this volume.

SPECT simulation can be made for both circular and noncircular orbits. Projections are stored as floating-point matrices, since variance-reduction techniques are used. Simulations of 360° and 180° rotation modes can be selected with an arbitrary start angle. A noncircular camera orbit is in its simplest form a trajectory defined as an ellipsoid for which the ratio between the major and minor axis is specified in CHANGE. When using voxel-based phantoms, the orbit can also be automatically calculated to mimic auto-contouring. Here, a center-to-camera distance (CTC) is calculated for each projection angle by a ray-tracing method using the density distribution. A constant value can be added to the calculated CTC distances mimicking a safety distance. The CTC data can be saved in a data file for further use in, for example, a reconstruction program. As a third alternative, discrete CTC values for each projection angle can be read from a data file.

Statistical variations in the energy signal are simulated by convolving the imparted energy in the crystal from every photon history "on-line" with an energy-dependent Gaussian function with an FWHM that depends as function on $1/\sqrt{E}$. Here, it is required that the user provides a reference value of the FWHM at 140 keV. The convolution procedure also affects the spatial location of each interaction site. If the simulated energy signal is within a predefined energy window, the apparent position of the event is calculated from the centroid for the imparted energy in the scintillation crystal and is stored in the projection matrix.

"Delta Scattering" Method

This sampling method for a photon path length is a statistical method involving fictitious interaction that produces unbiased estimates and is efficient when simulating photon transport in a heterogeneous volume of interest. The basic idea is that a photon path length, d_i, is sampled from

$$d_i = \frac{\ln(R_i)}{\mu(Z, h\nu)_{max}} \tag{7.1}$$

where R is a uniform distributed random number and μ_{max} is the largest possible value of the attenuation coefficient within the volume of interest. The photon is then moved a distance d_i to a new point. The ratio between the attenuation coefficient at this location and the μ_{max} is calculated. This ratio will always be less than or equal to unity. Then, if a sampled random number R_{i+1} is less than this ratio, the current position of the photon will be defined as the endpoint. Otherwise, the photon is continued a distance d_{i+1} sampled from Equation 7.1 and the test procedure is repeated with a new random number R_{i+2}. For large matrices, this method is more efficient than explicit ray-tracing methods based on calculating the exact path length within each voxel.

Interactions in the Phantom

A photon history in the phantom starts by sampling a decay position within the source volume. The photon history is then split into one primary photon history and a number of scattered photon histories dependent on the selected number of scatter orders. For the primary case, the photon is "forced" to penetrate the phantom without any subsequent interactions to continue toward the camera in a direction within an acceptance angle determined (normally defined from the collimator-hole properties). The PHW (the probability for the photon to travel the simulated path) is multiplied by the probability for escape without any interactions which are calculated by accumulating the attenuation coefficients in discrete step along the photon path to the border of the phantom. The photon is then followed until termination in the crystal or other compartments. The control is then returned to the position of the decay in the phantom in order to continue simulating the "split-scattered" histories.

When simulating the "scattered part" of the photon, an isotropic direction is initially sampled at the decay point. A photon path is then calculated using the delta-scattering method, described above. At the end of the sampled path (if inside the phantom), the scattered photon is split into two parts. The first part is directed toward the camera by applying a Compton scattering in a forced direction. This scattered photon is then forced to escape the phantom without interaction and is then followed until termination. The control is returned to the interaction point in the phantom. The type of interaction for the second part of the scattered photon is properly sampled to be either Compton scattering or coherent scattering and this photon is followed in the phantom by the delta-scattering sample, as described above. This simulation procedure is repeated until the total number of selected scattering orders has been simulated. For each photon history, this splitting method thus always results in one primary photon history and a number of scattered photons dependent on selected maximum scattering order. This method improves the overall efficiency of the simulation because fewer histories are needed to be simulated.

The calculation of the path length from an interaction point to the boundary is thus made in two different ways, depending on the type of phantom being used. For simple intrinsic phantoms, the path length is analytically calculated from equations describing ellipsoids, cylinders, and rectangular phantoms. For voxel-based phantoms, a combination of the step-by-step method and the delta-scattering method is used.

Voxel-based computer phantoms are internally defined by integer matrices where a matrix cell describes the density (or a value proportional to the activity concentration) in a particular location in the object. The reason SIMIND is working with density distribution and not with attenuation distribution is that the attenuation is a function of photon energy. If the maps were scaled to attenuation values, then one would, in principle, have needed a set of 3D maps for all possible photon energies. Instead, the attenuation coefficient is calculated from the product of the density taken from the voxel maps, and tabulated mass-attenuation coefficients, $\mu(h\nu,Z)/\rho$, stored as differential cross-section coefficients (in steps of 0.5 keV) as a function of material and energy. Two different phantom materials can be selected with the main purpose of differentiation between bone and muscle tissue. The selection of which material will be used depends on a predefined threshold value of the density.

Collimator Simulation

Included in the CHANGE program is a large database that defines the parameters for most of the commercial collimators available for both SPECT and planar imaging. These are easily defined in CHANGE or at the command line level by a unique collimator code. Four different collimators are presently available in the program.

Collimator Based on a Geometrical Transfer Function

This collimator routine has been developed in collaboration with Dr. Eric Frey, Johns Hopkins Medical Center, Baltimore, USA. The collimator model is based on an analytic formulation of the geometric response for parallel- [6] and converging-beam [7] collimators. The weight of the incoming photon is adjusted according to its direction and to a related geometrical response function for different types of collimators. None of the photons are rejected, so this collimator routine is computationally very efficient. However, no effect from septum penetration and collimator scatter is included.

Collimation Including Penetration and Scatter

The simulation and validation of this collimator has been described in the work by Ljungberg et al. [8]. The computation is based on the delta-scattering method and the collimator algorithm includes both simulation of penetration effects and scattering in the collimator septa. A drawback with this colli-mator algorithm is, however, that the path length for low-energy photons becomes very small because the μ_{max} is calculated for lead. Therefore, this collimator should be used mainly in applications where penetration effects really are important.

Parallel-Hole Collimator by Forced Convolution

This is a recent development that uses a forced-convolution method to simu-late a collimator. Technically, this is a scoring routine (more information on this type of routine is given below). In this routine, the ordinary collima-tor is turned off and the photons are forced toward the crystal in the direc-tion parallel to the normal of the crystal surface. If the absorbed energy in the crystal is within the energy window, the (x,y) position of each history is stored in a set of image matrices, where each matrix represents the events coming from photons with the last interaction points defined as a plane par-allel to the crystal at a certain distance from the lower surface of the crystal. Thus, for each photon, the distance between the surface and the last point of interaction (or point of emission) needs to be calculated. A typical thick-ness for such a plane is 5 mm. After all photon histories have been simulated for a particular projection angle, then each of these images is convolved by a Gaussian function with a distance-dependent width to simulate the col-limator resolution at that distance. The sum of all convolved images is then calculated to obtain the final projection image.

Pinhole Collimation

The pinhole collimator is also based on the delta-scattering sample. The user needs to define four parameters in addition to the collimator thickness d_{coll}. The first is the wedge angle α. The second and third parameters, denoted d_A and d_B in Figure 7.2, relate to the position where the front wedge and the back wedge are defined. The fourth parameter is the collimator-to-detector distance, d_{det}. Thus, if $d_A + d_B = d_{coll}$ then the collimator will be of a wedge type. If not, then it will be of a channel type.

Comparison of the Four Collimators

Figure 7.3 shows a ^{99}Tc^m planar cold-spot simulation of a source defined as a thin circular-shaped sheet parallel to the surface of the camera and located at a distance of 11 cm using each of the four collimators described above. The same number of histories (1,634,900) has been used. For the first three collimators, the source was the same and the circular source plane was 11 cm in radius. For the pinhole collimator simulation, due to the magnification effect, the radius of the circular source was reduced from 11 to 2.6 cm and the source-to-collimator distance was reduced from 10 to 5 cm. The pinhole

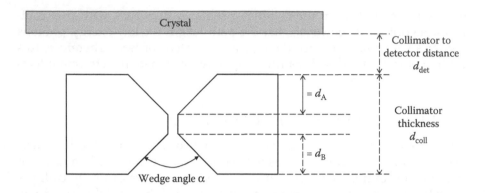

FIGURE 7.2
A schematic image showing the geometric setup of the pinhole collimator. If d_A equals d_B, then the pinhole is of a wedge type. Otherwise, it will be of a channel type.

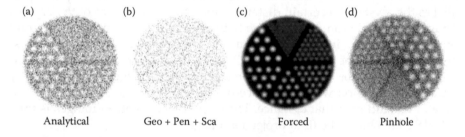

FIGURE 7.3
Four simulations using the same number of histories and (a) the standard collimator based on analytical weighting, (b) the collimator that allows for geometrical collimation, septal penetration, and Compton scatter in the septa, (c) the "forced convolution" collimator, and (d) the pinhole collimator. The source represents a thin layer of activity with a circular shape.

collimator was wedge-shaped with a 1 mm diameter circular hole and the pinhole-to-detector distance was set to 30 cm.

Transmission Imaging

Included in SIMIND is a possibility to simulate transmission studies (TCT) with SPECT with either a parallel-hole collimator or a fan-beam collimator [9]. The difference between SPECT and TCT is mainly in that the source in this type of acquisition mode is located outside the patient in a position opposite to the camera head relative to the patient.

Figure 7.4 shows schematically the geometry for the transmission simulation. In SIMIND, the phantom needs to be defined as voxel matrices and is

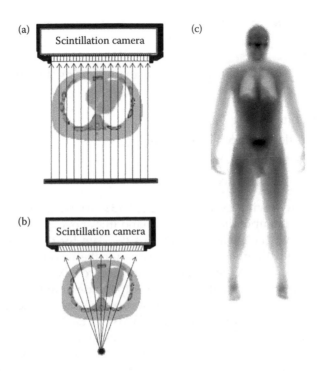

FIGURE 7.4
The left column shows two schematic images of (a) the transmission geometry based on a flood source (a) and a fan-beam geometry (b). The right image (c) shows a result obtained from a transmission simulation of the XCAT phantom using the flood source geometry.

located between the transmission source and the camera. An emitted transmission photon is traced into the phantom matrices and is checked whether or not it will strike the phantom. If true, the entrance matrix cell is calculated and the simulation is continued as has been described in the previous chapter. If not true, the photon is regarded as being a primary photon and is continued toward the camera without interaction in the phantom. The maximum number of scatter orders is here also necessary to define. The transmission source can be a line or a flood source at different distances from the phantom. The emission angle for the transmission photon can be selected to be forced into a cone or a fan (which matches a transmission measurement where the transmission source is collimated with a slit in the axial direction). The output of the transmission simulation is transmission projection and/or blank projection (without the object in site). The latter can also be correlated to the transmission projection. The advantage with this option is that artifacts in the reconstructed images due to large deviation in the logarithm calculation can be avoided.

User Interface

When executing a SIMIND command at the operating system prompt, the syntax consists of the following parts:

- The program name
- An input file as created by CHANGE
- An output file base name
- Optional control switches

Consider the following example: `c:\simind> simind point point1/ 01:140/fa:12`. The input file in the command line above, `point.smc`, defines the file created by CHANGE that contains the data for the particular simulation. If not explicitly given, the base name for the results, in this example `point1`, will be equal to the input file name. Since SIMIND creates several result files, all files have usually the same output base name but are separated by different file extensions. The user can at run time change the input parameters, originally created by the program CHANGE. Each value, defined by an index number in CHANGE, can be overridden by giving a corresponding control switch at run time. The advantage with this option is that only one data file (e.g., `point.smc`) is necessary for multiple simulations and it is very easy to set up multiple simulations in a script file. In the above example, the value for index 01 in CHANGE that defines the photon energy will be changed to 140 keV and flag 12, that determine if a Gaussian energy resolution should be added, is set to false. Below is shown an example of a Windows script file for a simulation where `point.smc` is the input file and a point source is positioned at various depth by data given by switch/18: for two photon energies (140 and 364 keV). The switch/SC: determines the maximum number of scatter orders in the phantom. Switch 53 sets the type of collimator routine (0 = analytical collimator, 1 = collimator allowing for scatter and penetration).

```
C:\user> simind point/01:140/sc:6/53:0/18:10.00 point1a
C:\user> simind point/01:364/sc:6/53:1/18:10.00 point1b
C:\user> simind point/01:140/sc:6/53:0/18:0.000 point2a
C:\user> simind point/01:364/sc:6/53:1/18:0.000 point2b
C:\user> simind point/01:140/sc:6/53:0/18:-10.0 point3a
C:\user> simind point/01:364/sc:6/53:1/18:-10.0 point3b
```

Programming SIMIND

The native SIMIND program can simulate different types of detector parameters and create images, planar images or SPECT projections, and energy pulse-height distributions. To take advantage of the Monte Carlo code fully, however, some programming is necessary. Generally, this can be a major task if the user is not familiar with the particular source code. Furthermore, problems will occur in keeping track of different versions if the author of the main code makes modifications. An option to link a user-written subroutine has therefore been included. This has the feature for separating the code, written by a user, from further distributions of SIMIND. By communicating with the main code using subroutines and global common blocks rather than adding source codes to the main program, the user can access all important variables without requiring an extensive knowledge of the actual SIMIND code. The advantage is thus the ability to write special routines for different simulations and simply link the compiled versions of these to the main program when needed. The user can write three basic types of subroutines: the score routine, the source routine, and the isotope routine.

Score Routine

A call is made to the scoring routine at several distinctive stages during the simulation of a photon history which enables the user to decode an integer flag and thus determines where the photon is "located" at the time when the call was made. An integer variable is assigned a value depending on the stage of the photon history. The call to the subroutine from SIMIND is of the form call score(iopt,lun), where lun is a logical unit number used for output printing to the screen or the result file and iopt is an integer number that provides information to scoring routines in which stage in the photon history simulation the call to the scoring routine was made. In a scoring routine, further calculation can be made based on the value of the iopt variable. The values of iopt and the corresponding stages for the call in the code are listed in Table 7.1.

TABLE 7.1

The Different Stages in the Photon History Where the Scoring Routine in Called

IOPT	Stage in the Photon History
0	After reading data from the *.smc file created by the CHANGE program
1	Before starting the main loop of the program
2	After generation of the decay point X_0, Y_0, Z_0 and the direction. If phantom is selected then after escape from the phantom. If SPECT is selected then after the phantom/source has been rotated
3	After "n" photon interactions in the crystal
4	After the photon history has been terminated
5	After a complete simulation of an SPECT projection
6	After simulating the total number of photon histories
7	During the printout, so the scoring routine can provide results to the main result file
8	After an interaction in the phantom. The call is made after calculation of new cross-section data and direction cosines but before calculating the coordinates for the new interaction point

Consider the following routine where the imparted energy at each interaction point, stored in the variable BIDRAG, is accumulated and the average energy for each photon history is calculated using the variable PHOTON, which is the number of histories per projection.

```
Subroutine score(iopt,lun)

Include 'simind.fcm'          ! Connect to simind through
                               ! common block
Real sum,average

If(iopt.eq.0) then            ! Just after reading smc values
        Sum = 0               ! Initialize the counter
Elseif(iopt.eq.3) then        ! After each interaction
        Sum = sum + bidrag    ! Integrate energy
Elseif(iopt.eq.6) then        ! All histories terminated
        Average = sum / photon ! Calculate average
Elseif(iopt.eq.7) then        ! Print result in *.res file
        Write(lun,10),average ! using the lun variable
10      Format(' Average was found to be :',f12.3)
   End if
End
```

The following scoring routines are included in the main program:

Scattwin: This routine is used for simulations of multiple-energy window acquisitions. Energy windows are defined from a file and space is allocated at run time. For each energy window, projection images will be stored for the "air" case where no interactions occur in the phantom, "tot" including all events and "sca" including only those events coming from photon scattered in the phantom. The routine also outputs a detailed energy spectrum separated into components of primary events and scatter events of different scatter orders. The user can easily add energy windows by simply adding new lines to the text file defining the upper and lower energy window threshold and an optional flag that can allow for summation of counts obtained from one of the other energy windows.

Penetrate: This routine has been written for investigation of the different components in an image. The final image here is separated into 18 components where each image shows the event coming from histories that have experienced different types of interaction. For example, those photons that scattered in the phantom and penetrated a septa but that were not scattered behind the crystal will form one image component. The usefulness of this routine is in optimization studies where the relative contributions need to be quantified.

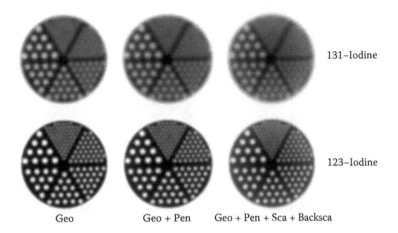

131–Iodine

123–Iodine

Geo   Geo + Pen   Geo + Pen + Sca + Backsca

FIGURE 7.5
Examples of the ability to separate the events in an image into components are displayed for two radionuclides. The left column shows images of events from only geometrical collimated photons. The middle column displays images of events where septum penetration in the collimator is also included. The right column displays images of events where, in addition to geometrical and penetration events, scatter in the collimator and backscattering of photons in the compartment behind the crystal compartment are allowed. It is clear that these unwanted additional events degrade the image quality and therefore may need to be compensated for.

An example of such a simulation is shown in Figure 7.5 where there is a simulation of a plane consisting of a uniform source of ^{131}I with an HE collimator (upper row) and ^{123}I (lower row) with an LEGP collimator. Both cases include cold-spots of different diameters located as six pie shapes. Note that this is not a reconstructed image. The left column shows events coming from geometrically collimated photons. The middle row shows geometrical collimation plus penetration and collimator-scatter events. The right row (C) shows in addition to (B) also events that come from photons backscattered in the compartment behind the crystal. Figure 7.6 shows corresponding energy pulse-height distributions.

Forced Collimator: The forced collimator routine is described in detail in the collimator section. The routine is aimed to produce a fast simulation for those cases where interactions in the collimator and the crystal are not necessary. The explicit simulation of each photon passing the collimator is here replaced by a distance-dependent convolution using a Gaussian function.

List-mode: A scoring routine written to allow for a list-mode storage which means that important information about each history is stored in a binary file for further processing. An advantage of this feature is that no new simulation is needed when, for example, the energy window setting needs to be redefined. By this feature, it is

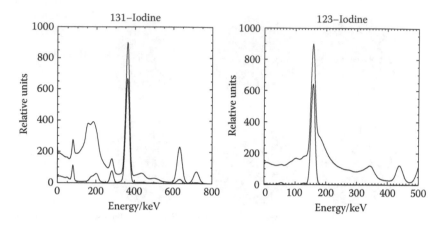

FIGURE 7.6

Energy pulse-height distribution for iodine-131 (left) and iodine-123 (right) showing events from geometrical collimated photons only (lower distribution) and all events, including events coming from septum penetration, scattering in the collimator, and backscattered photons (upper distribution).

also possible to link SIMIND calculation to own-written software. One such application could be more explicit simulation of light propagation in the crystal and in light-guides.

Source Routine

The aim of this routine is simply to give SIMIND a Cartesian coordinate x_0, y_0, z_0 that describes where the decay occurred. The user can then write source routines that are more complex than the geometries included in the SIMIND (i.e., cylinders and spheres). Provided with the SIMIND code are source routines simulating simple source shapes, such as spheres and cylinders. For simulation of more complex source distributions, a routine is used to generate decays according to the voxel values in a set of image matrices where the location of an individual matrix cell is scaled to a (x,y,z) location.

Also included is a multiple-source routine used to set up a simulation consisting of several sources of different shapes. Each source is defined in a text file by a line that defines (a) the radius in (x,y,z) coordinates, (b) the location of the center point, and (c) a relative activity concentration (~MBq/cc). It is also possible to define a background activity concentration by a switch/BG and thereby be able to simulate a cold-spot case. The shapes of the sources can be elliptical, cylindrical, hexagonal, or conic.

More complex sources can be simulated by using a source map. This map is a 2-byte integer matrix where each pixel value shows the number of decays to be simulated and the relative pixel location indicates the location for those decays. A user can then by an external program define any type of source geometry. An example of such a simulation is shown in Figure 7.3 where a map, created by a CT study of the elliptical Deluxe rod insertion (Data Spectrum, Inc.), was made and the image used to sample decays.

Included in SIMIND is also access to an anthropomorphic voxel-based computer phantom, for example, the phantoms developed by Zubal et al. [10] and Segars et al. [11]. These source routines use digital images to simulate a source distribution. The images can be simple user-created integer matrices or images of known code-based formats and characteristics such as the different Zubal phantom or images obtained from the MCAT, NCAT, or XCAT software. To make it easy for the user, a conversion table can be specified that together with a code-based phantom such as the Zubal phantom is used to create a 3D density array and a 3D source distribution. Code-based phantoms are created in such a way that each voxel value represents a unique code used to identify a particular organ or structure to which the voxel belongs. An example of such a conversion table is shown as a part in Table 7.2.

The first row is the tissue description, the second column is the code, and the two remaining columns are the density in units of g/L and a relative value of the activity in the voxel. The advantage of this method is that it is very easy to change the simulated distribution for a particular coded phantom file.

It is also possible to add tumors to an activity distribution that reflect a normal distribution. The syntax for the tumor definitions are the same as for the "multiple source" routine, described above. However, here the initially defined voxel values are replaced by the relative activity concentration defined in this tumor file.

TABLE 7.2

Some Examples of Structures in the XCAT Phantom with Related Codes, Densities and Relative Activity Concentrations

Organ/Structure	Code	Density g/dm³	Relative Activity Conc.
hrt_myocard LV	1	1060	75
hrt_myocard RV	2	1060	75
hrt_myocard LA	3	1060	5
hrt_myocard RA	4	1060	5
hrt_blodpool LV	5	1060	1
hrt_blodpool RV	6	1060	1
hrt_blodpool LA	7	1060	1
hrt_blodpool RA	8	1060	1
Body	9	1000	1
Liver	10	1060	50
Gall bladder	11	1026	20
Lung	12	260	1
Stomach wall	13	1030	1
Stomach content	14	1060	1
Kidney	15	1050	10
Spleen	16	1060	15

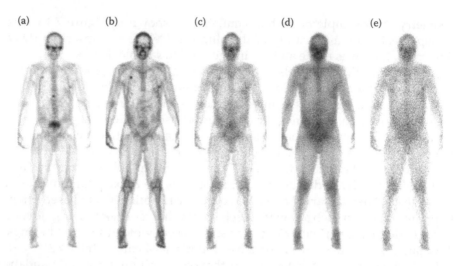

FIGURE 7.7
Examples of simulations using the scattwin routine. The photon energy is 140 keV. The energy window for image (a)–(c) is 126–154 keV. Image (a) represents a case with no photon interactions in the phantom. This image is not achievable in a real measurement. Image (b) shows a case where attenuation and scatter have been allowed in the phantom. This case represents a real measurement. Image (c) shows the distribution of scatter events that is included in image (b). This scatter distribution represents what is compensated for in various scatter correction methods. Image (d) shows events obtained using a wide scatter window below the photopeak window (92–126 keV) and image (e) shows events obtained in a narrow energy window (120–126 keV). (d) and (e) represent images used in energy-window-based scatter compensation methods.

Figure 7.7 shows some examples of simulations using the XCAT phantom in combination with the scattwin routine.

Isotope Routine

This routine is created to provide SIMIND with a value of the photon energy. Such a routine can include decay schemes of various complexities. A logical flag in SIMIND controls when to stop the calls to the isotope routine. The difference between this technique and the alternative of running several independent simulations of different photon energies and summing the results is the addition of summation pulse-pile up events due to simultaneous multiple photon emission.

Future Developments

Currently, the code is being modified so that it is able to simulate pixelated solid-state detectors. In these camera systems, each pixel acts like a single detector. The generation of an event is made by collecting charges created in the detector. This collection of charges depends on the voltage settings and the energy spectrum therefore looks quite different as compared to scintillation detectors, such as NaI(Tl), and has a pronounced low-energy tail close to the photo-peak. The ability to simulate multiple pinhole configurations will also be added.

Documentation

A online manual and also some tutorials can be found on Lund University's Department of Medical Radiation physics webpage: http://www2.msf.lu.se/simind

References

1. Ljungberg M, Strand SE. A Monte Carlo program simulating scintillation camera imaging. *Comp. Meth. Progr. Biomed.* 1989;29:257–272.
2. Todd-Pokropek A, Cradduck TD, Deconinck F. A file format for the exchange of nuclear medicine image data: A specification of Interfile version 3.3. *Nucl. Med. Commun.* 1992;13:673–699.
3. Berger MJ, Hubbell JR. *XCOM: Photon Cross-Sections on a Personal Computer.* Washington DC: National Bureau of Standards, 1987.
4. Hubbell JH, Veigle JW, Briggs EA, Brown RT, Cramer DT, Howerton RJ. Atomic form factors, incoherent scattering functions and photon scattering cross sections. *J. Phys. Chem. Ref. Data.* 1975;4:471–616.
5. Hubbell JH, Øverbø I. Relativistic atomic form factors and photon coherent scattering cross section. *J. Phys. Chem. Ref. Data.* 1979;8:69–105.
6. Metz CE, Atkins HL, Beck RN. The geometric transfer function component for scintillation camera collimators with straight parallel holes. *Phys. Med. Biol.* 1980;25:1059–1070.
7. Tsui BMW, Gullberg GT. The geometric transfer function for cone and fan beam collimators. *Phys. Med. Biol.* 1990;35(1):81–93.
8. Ljungberg M, Larsson A, Johansson L. A new collimator simulation in SIMIND based on the delta-scattering technique. *IEEE Trans. Nucl. Sci.* 2005;52(5):1370–1375.
9. Ljungberg M, Strand SE, Rajeevan N, King MA. Monte Carlo simulation of transmission studies using a planar sources with a parallel collimator and a line source with a fan-beam collimators. *IEEE Trans. Nucl. Sci.* 1994;41(4):1577–1584.
10. Zubal IG, Harrell CR. Computerized three-dimensional segmented human anatomy. *Med. Phys.* 1994;21:299–302.
11. Segars WP, Sturgeon G, Mendonca S, Grimes J, Tsui BMW. 4D XCAT phantom for multimodality imaging research. *Med. Phys.* 2010;37(9):4902–4915.

8

GATE: A Toolkit for Monte Carlo Simulations in Tomography and Radiation Therapy

Irène Buvat, Didier Benoit, and Sébastien Jan

CONTENTS

GATE, which stands for Geant4 Application for Emission Tomography, is an opensource and freely distributed Monte Carlo simulation tool. The origin of GATE can be traced back to a workshop organized in July 2001 by Irène Buvat in Paris, France, and attended by several research groups sharing a common interest in Monte Carlo simulations. The focus was to brainstorm about the future of Monte Carlo simulations in nuclear medicine. Indeed, the dedicated Monte Carlo codes developed for positron emission tomography (PET) and single photon emission computed tomography (SPECT) suffered from a variety of drawbacks and limitations in terms of validation, accuracy, and support [1] and did not meet users' requirements. Accurate and versatile general-purpose simulation codes such as Geant3 [2], EGS4 [3], MCNP [4], and Geant4 [5] were available. These codes were quite appealing as they included well-validated physics models, geometry modeling tools, and efficient visualization utilities. However, modeling PET or SPECT acquisitions using these codes required a large effort. After discussions, the need to develop a simulation toolkit that would combine the best of both worlds became clear. The aim was to have a dedicated Monte Carlo platform for emission tomography including specific features relevant in that context (modeling of decay kinetics, of detector dead time, of detector and patient movements), and to benefit from the versatility and support of general-purpose simulation tools. At that time, object-oriented technology appeared to be the best choice to ensure high modularity of the code. Therefore, the consensus was to select the simulation toolkit developed in C++ by the Geant4 Collaboration and to warrant long-term support by sharing the developments among many research groups.

Soon after, the specifications of the code were defined and four groups (Laboratory of High Energy Physics in Lausanne, Geant4 Low Energy Electromagnetic Physics working group, Laboratory of Corpuscular Physics in Clermont Ferrand, and Electronics and Information System lab at the University of Ghent) started coding. In mid-2002, these four groups decided to create the OpenGATE collaboration (http://www.opengatecollaboration. org). The objective of this collaboration was to develop, improve, test, and maintain GATE. The collaboration grew very rapidly until it reached 20 members from Europe, North America, and Asia. Based on the work of this collaboration, the GATE simulation tool was first publicly released in May 2004 [6]. Since then, 18 new versions have been released, and a second collaboration paper describing the extended features of GATE has been published [7].

Initially, GATE was dedicated to the modeling of planar scintigraphy, SPECT, and PET acquisitions. Over the years, the functionalities of the platform have been greatly enhanced. In 2011, GATE version 6 makes it possible to simulate planar scintigraphy, SPECT, PET, and computed tomography (CT) acquisitions, but also radiation therapy experiments, including conventional radiotherapy with photon or electron beams, as well as proton and ion therapy treatments. These latest developments have been stimulated by the increasing integration of radiotherapy and imaging, either to customize treatment at best by accounting for metabolic information [8] or for patient monitoring (e.g., [9–11]).

In the section "Overview of GATE," an overview of the GATE code and main concepts is presented. The core physics models in GATE derived from those of Geant4 are described in the section "Physics of GATE." The design of a GATE simulation is explained in the section "Designing a GATE Simulation." The actual running of a simulation, depending on the available platform, is described in the section "Running a GATE Simulation." The section "Current Applications of GATE" reviews the different systems that have already been modeled in GATE, while the section "Conclusion" summarizes the main applications of GATE in imaging and radiation therapy research.

Overview of GATE

Basic Architecture of the Code

GATE consists of several hundred C++ classes. These classes are organized in a layer architecture (shown in Figure 8.1), consisting of a Geant4 kernel surrounded by three layers: the "core layer," the "application layer," and the "user layer." The core layer includes classes defining the basic mechanisms available in GATE (e.g., basic geometry definition, radioactive source management, physical processes, time management, visualization management). Some of these classes are used in all Geant4-based simulations, while others are specific to GATE, such as the GATE virtual clock to manage time-dependent phenomena (e.g., radioactive decay, or biokinetics of a tracer).

The "application layer" contains upper-level classes based on the core layer classes and specific to the imaging or radiation therapy applications (e.g., design of specific volume shapes, design of specific volume motions). The range of functionalities available in GATE can be increased by adding new classes in this layer, without changing the overall structure of the code.

The "user layer" manages the mechanisms through which simulations can be run using macrofiles, without any C++ coding. Each class of GATE

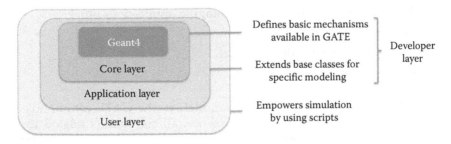

FIGURE 8.1
Structure of the GATE code in four layers.

provides a dedicated extension to a command interpreter class, so that the end-user of GATE does not have to perform any C++ programming. A simulation can be entirely described by a list of instructions (or command lines) that can be gathered in a macrofile.

As GATE is opensource, interested developers can customize the code and introduce their own functionalities by adding classes in the core or application layers.

Mechanism of Macros in GATE

In GATE, a simulation is designed as a list of command lines. Each command performs a particular function using one or more parameters, and all functions in GATE can be accessed through a command line. The command lines can be either entered interactively at the prompt appearing after starting GATE, or can be gathered into so-called macro files (files with a .mac extension). This principle of macros is based on the Geant4 messenger, which makes it possible to run a Geant4 application without having to edit and recompile the C++ code.

Macros are thus ascii files in which each line contains either a command or a comment (lines starting by #) (Figure 8.2). Macros can be executed from the command line in GATE (right after the prompt) or by passing the macrofilename as a parameter when starting GATE, or when being called by another GATE macro.

A macro or a set of macros (as one macro can call another macro) should include all commands describing the simulation set-up in the correct order. A simulation set-up consists of the following steps:

1. Defining the visualization parameters
2. Defining a scanner (imaging applications) or a beam (radiotherapy applications) geometry
3. Defining a phantom geometry
4. Selecting the physics processes
5. Initializing the simulation
6. Selecting the detector model (in imaging applications only)
7. Defining the source(s)
8. Specifying the data output format
9. Starting the acquisition

To facilitate the design of a GATE simulation, the GATE commands are organized as a tree structure. For instance, all commands related to the source definition (step 7 above) start by "/gate/source." As all commands can also be entered interactively, the building of the simulation set-up

```
#/vis/disable
/control/execute visu.mac

/gate/geometry/setMaterialDatabase ../../GateMaterials .db

#        WORLD
/gate/world/geometry/setXLength 150. cm
/gate/world/geometry/setYLength 150. cm
/gate/world/geometry/setZLength 150. cm

/control/execute camera.mac
/control/execute phantom.mac
/control/execute physics.mac

#        INITIALIZE
/gate/run/initialize

/control/execute digitizer.mac

#        SOURCE
/control/execute sources.mac

#        OUTPUT
/gate/output/root/enable
/gate/output/root/setFileName benchmarkPET
/gate/output/root/setRootCoincidencesFlag 1
/gate/output/root/setRootdelayFlag 1

#        RANDOM
# JamesRandom Ranlux64 MersenneTwister
/gate/random/setEngineName MersenneTwister
/gate/random/setEngineSeed auto

#        START
/gate/application/setTimeSlice  120. s
/gate/application/setTimeStart    0. s
/gate/application/setTimeStop   240. s
/gate/application/startDAQ
```

FIGURE 8.2
Example of macro in GATE. Note that this macro calls five other macros: camera.mac, phantom.mac, physics.mac, digitizer.mac, and sources.mac.

(especially of the detector and of the phantom) can be controlled on line, through a sophisticated visual rendering. Tracks of simulated particles can also be visualized on line for debugging or educational purpose as is shown in Figure 8.3.

To understand how a GATE simulation is designed, a number of concepts specific to Geant4 or GATE have to be well understood. The most important of these concepts are defined below.

Important Concepts in GATE

Systems

A *system* is a key concept in GATE. It provides a template defining the hierarchical structure of scanner geometry and is compatible with one or several output formats. This notion of *system* is relevant because many scanners share the same components. For instance, in PET, most scanners consist of

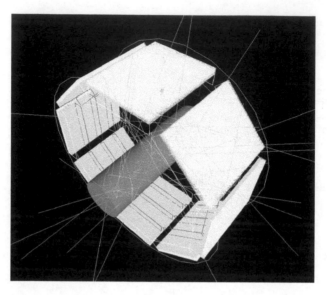

FIGURE 8.3
Example of visualization of a PET system in GATE with tracks of particles. The scanner consists in eight detector heads, each composed of 100 × 100 LSO crystals (light grey) and 100 × 100 BGO crystals (white). A cylindrical phantom is located in the field of view of the scanner. The grey lines correspond to photon tracks.

several rings, each ring being composed of several scintillator blocks, each block consisting of several small crystals. Similarly, in SPECT, gamma cameras are usually composed of a continuous or pixelated crystal and a collimator. Therefore, the same *system* can be used to simulate different scanners sharing the same geometrical features. Nine *systems* are available in GATE, called "scanner," "SPECTHead," "cylindricalPET," "PETscanner," "ecat," "ecatAccel," "CPET," "OPET," and "CTscanner." These *systems* can be used to simulate most of the existing tomographic imaging devices. The "SPECTHead" *system* is appropriate to model most gamma cameras. The "CTscanner" *system* is appropriate to model a simple CT scanner. The "cylindricalPET," "ecat," "ecatAccel," "CPET," "PETscanner," and "OPET" *systems* are appropriate to model PET imaging devices, each with its own specificities. Lastly, the "scanner" *system* is the most flexible and can be considered for any unconventional imaging device not compatible with the other available *systems*. The choice of an appropriate *system* mostly depends on the geometric shapes of the different components of the imaging device to be simulated (gantry, scintillator) and on the output format in which the user is willing to store the simulation results. For instance, only the "OPET" and the "cylindricalPET" *systems* support the list mode format as an output format.

Sensitive Detectors and Actors

A *sensitive detector* is a virtual entity that has to be defined to store information about interactions of a particle with matter. *Sensitive detectors* must be attached to geometrical volumes in which the user wants to track and record interactions. GATE stores information related to the particle interactions only in volumes that are attached to a *sensitive detector*. Two types of *sensitive detectors* should be distinguished: the *crystal sensitive detector* (*crystalSD*) and the *phantom sensitive detector* (*phantomSD*). *CrystalSD* is used to record information regarding interactions taking place inside some volumes of the scanner, such as energy deposit, position of interaction, type of interaction, and so on. *PhantomSD* is used to detect and tally Compton and Rayleigh interactions taking place in the field of view of the scanner.

An *actor* is similar to a *sensitive detector* but has extended capabilities. In addition to storing information about particle interaction, it can act on particles. In other words, *actors* can actually modify the simulation. Many *actors* are available in GATE. For instance, a *DoseActor* measures the energy deposited in a given volume while the *KillActor* stops the tracking of particles when they reach a given volume. The so-called *filters* can be combined with *actors* to modify the behavior of *actors*. *Filters* perform tests on the particle properties (e.g., energy, type, or position) and trigger the actor only if the test output is true. For instance, a *filter* can be used to calculate a dose distribution only for electrons with energy greater than a certain threshold. Developers can add new types of actors and filters using an easy mechanism.

Digitizer

The purpose of what is called the *digitizer* in GATE is to simulate a scanner electronic readout scheme. GATE uses Geant4 to generate particles and transport them through different materials, modeling the physical interactions between particles and matter. The information generated during this transportation process is used by GATE to produce the data observed by the scanner. The *digitizer* models the series of operations turning an interaction in a detector element (called a *hit* and characterized by a set of observables such as energy, position, momentum, and time of interaction) into a signal as delivered by the scanner. This series of operations includes modeling of the detector response and of the processing performed by the front end electronic (FEE). The *digitizer* is actually made of several modules, among which (nonexhaustive list, a comprehensive list can be found online in the GATE user documentation):

- The *hit adder*: This module sums the energy deposited by multiple *hits* into a *sensitive detector*, to produce a signal called a *pulse*. For instance, a gamma ray interacting within a scintillation crystal can yield *hits* corresponding to several Compton scattering followed by a photoelectric absorption. The *hit adder* sums the energy deposited by each *hit* within a *CrystalSD* to yield a *pulse*. The *pulse* position is the energy-weighted centroid of the *hit* positions, and the time associated with the *pulse* is that of the first *hit* within the *sensitive detector*. If a particle interacts in several *sensitive detectors*, the *hit adder* generates a list of *pulses*, one for each *sensitive detector*.

- The *pulse reader:* This module adds the *pulses* together within a user-defined group of *sensitive detectors*. This yields a new *pulse* of energy equal to the total energy deposited in the group of *sensitive detectors*. The position of this *pulse* is set to that of the *pulse* from the *hit adder* that has the greatest energy.

- The *energy blurring* module: Different energy blurring models are available, including simple Gaussian blurring of the energy value associated with a *pulse*, Gaussian energy blurring with a different energy resolution for each crystal in the detector block or Gaussian energy blurring with different energy resolutions for different volumes (for instance for detectors including several layers of different scintillation crystals). Modules are also available for spatial blurring and temporal blurring to model the limited time resolution of the scanners.

- The *dead time* modeling module: Two models of dead time are available: paralyzable and nonparalyzable dead times. These models can be implemented event by event during a simulation. The detailed method underlying these models can be found in [12].

- The *energy window* setting: For both PET and SPECT applications, the *digitizer* is used to set the acceptance energy window, by defining its lower and upper limit.

Many more *digitizer* modules are available and a complete list can be found in the online documentation of GATE.

In PET simulations, the *digitizer* is also used to sort coincidences among the *pulses* that are detected. The *pulses* are actually called *singles* before entering the so-called *coincidence sorter* module. The *coincidence sorter* module identifies pairs of coincident *singles* in the *singles* list. Whenever two or more *singles* are detected within a coincidence window, these *singles* are grouped to create a coincidence. Two methods can be used to identify coincidences within GATE. In the first method, when a *single* is detected, it opens a coincidence window, and a second single occurring during the duration of the window yields a coincidence. As long as the window opened by the first *single* is not closed, no other *single* can open another coincidence window. In the second method, all *singles* open their own coincidence window, and a logical OR is made between all the individual *singles* to find coincidences. The duration of the coincidence window can be set by the user, as well as the number of blocks that should separate the two crystals in which *singles* have been detected for the coincidence to be validated.

Each *single* is stored with an event ID number, which uniquely identifies the decay from which the *single* comes from. If two event IDs are not identical in a coincidence event, the event is defined as a random coincidence. Similarly, a flag identifies Compton-scattered events. The Compton scatter flag can be used to differentiate true from scattered coincidences.

GATE also offers the possibility to specify an offset value for the coincidence sorter so that prompts and/or delayed coincidence lines can be simulated.

When more than two *singles* are found in a coincidence window, GATE can process them using nine different rules. These rules go from ignoring all coincidences (killAll policy) to storing all pairs of *singles* separated by at least the number of blocks specified by the user (takeAllGoods policy).

Once the coincidences are identified, further processing can be applied to mimic count losses that may occur due to dead time, limited bandwidth of wires, or buffer capacities of the I/O interface. For PET scanners using a delayed coincidence line to detect randoms, data corresponding to prompts and delayed coincidences can be processed by a unique *coincidence sorter*. For instance, a prompt coincidence can produce dead time that hides the detection of a delayed coincidence. The prompt coincidence events can also saturate the bandwidth, so that some random events are lost.

Overall, appropriate setting and parameterization of the *digitizer* is a key step to properly reproduce the behavior of a certain imaging system over a range of count rates.

Data Output

GATE supports eight types of output format for imaging applications: ASCII, binary, ROOT (http://root.cern.ch/drupal/), Interfile, sinogram format, ECAT7 format (native file format from CPS Innovations, Knoxville, USA for their ECAT scanner series), List Mode Format (LMF) developed by the Crystal Clear Collaboration, and CT image format. Some of these formats are only supported by some *systems* (see the section "Systems") so the user should carefully choose the *system* as a function of the output format he is interested in.

In GATE, it is also possible to separate particle tracking into several steps. For instance, one can run the simulation until the particles exit the phantom and then store the phantom tracking particle histories in a ROOT file. This ROOT file can then be subsequently used as an input for a second simulation modeling only tracking in the detector. This is the so-called *phase space* approach, in which intermediate results are stored in *phase space* files. This approach is especially useful to test several detector settings, with always the same object in the field of view. In that case, it is useless to repeat the propagation of events in the phantom. This step of the simulation can be run only once and the resulting file can then be input as many times as needed in different models of detector. *Phase space* data can be stored either in a ROOT format or in the IAEA format [13]. In the *phase space* approach, the name of the particle, its kinetic energy, its position along the three axes, its direction along the three axes, and the weight of each particle are stored. Unlike the IAEA format, the ROOT format also stores the name of the volume in which the particle was produced and the name of the process that produced the particle. To save space, it is possible to select only the pieces of information to be stored in the *phase space* files.

In dosimetry applications, a *dose output* module should be used so as to output a dose map in the phantom. The output file is then a binary file containing dose expressed in Gy. In addition, a binary file containing the uncertainty on the dose values is also stored.

Time Management

GATE can pretty easily manage time-dependent phenomena, such as patient respiratory and cardiac motions, detector motion, collimator motion in radiotherapy applications, or tracer kinetics. Introducing time-dependent phenomena in GATE requires (1) defining the movements of the phantom and/or detector components; (2) describing the radioactive source motions if any; and (3) setting the starting and stopping times of the acquisition. The Geant4 tool requires the geometry to be static during a simulation. However, GATE takes advantage of the fact that the typical duration of a single event is very short when compared to the motions of interest in imaging and radiotherapy applications. Motions are thus synchronized with the evolution of the source activities by subdividing the acquisition time frames (typically of the order of minutes or hours) into small *time slices*. At the beginning of each *time slice*, the geometry is updated according to the predefined movements. During each *time slice*, the geometry is static and the simulation of the particle transport proceeds. Within the *time slices*, the source can only decay so that the number of events decreases exponentially from one *time slice* to the next, and within the *time slices* themselves. Simulations can be split into *time slices* with durations varying from one *time slice* to another.

Physics of GATE

Radioactive Sources

A source in GATE is defined by a particle type (e.g., radionuclide, gamma, and positrons), the particle position, direction, energy, and activity. The decay time of a radioactive source is usually obtained from the Geant4 database, but can also be set by the user. Multiple sources can be defined with independent properties.

GATE includes two modules dedicated to PET. One uses the von Neumann algorithm [14] to randomly generate the positron energy according to the measured β+ spectra. This method is faster than using the decay scheme of radionuclides used by Geant4. The β+ spectra of commonly used radionuclides (11C, 15O, and 18F) have been parametrized in GATE based on the Landolt–Börnstein tables [15]. The second module models the acollinearity of the two annihilation photons, which is not accounted for in Geant4. In GATE, acollinearity is modeled using a 0.58° full width at half maximum (FWHM) Gaussian blur. This width corresponds to experimental values measured in water [16].

In GATE, the source distribution can be described either using an analytical description or using a voxelized description, that is using maps of activity distribution. In addition to the radioactive source distributions, the attenuation properties should be defined using the same approach (analytical or voxelized). Patient images can even be used to define the source distribution and attenuation properties to reproduce highly realistic acquisitions [17]. In that case, emission images are converted into activity levels, and an analogous translator converts the voxel value of the voxelized attenuation map into a specific material.

Functionalities are also available in GATE to define and manage dynamic voxelized phantoms, that is, phantoms in which the activity distribution or the attenuation map or both vary with time [18].

Principle of Particle Propagation in GATE

In GATE, the typical history of an event is as follows:

1. A particle is generated with its parameters, such as initial type, time, momentum, and energy.

2. An elementary *step* is calculated. A *step* corresponds to the trajectory of a particle between discrete interactions (i.e., photoelectric, Compton, pair production, etc.). During a *step*, the changes to particle's energy and momentum are calculated. The length of a *step* depends on the type of particle, the propagation materials, and the nature of interaction. The calculation of step length is complex and is fully described in the Geant4 documentation.

3. If a *step* occurs within a volume corresponding to a *sensitive detector*, the interaction information between the particle and the material is stored. Relevant information may include the deposited energy, the momentum before and after the interaction, the name of the volume where the interaction occurred, and so on. This set of information is referred to as a *hit* (see the section "Digitizer").

Stages 2 and 3 are repeated until the energy of the particle becomes lower than a predefined value, or the particle position goes outside a predefined region. The entire series of *steps* is a simulated trajectory of the particle and is called a *track* in Geant4.

Physics Models

In Geant4, each physics process is described by a model and a corresponding cross-section table. Several models are sometimes available for a given physics process. All Geant4 physics models and cross sections below 10 GeV are available in GATE, including models describing the transport of optical photons and hadronic interactions. The set of processes and associated options selected by the user is named the *physics list*. GATE does not provide more processes or datasets than the ones available in Geant4, but integrates them for imaging and radiation therapy-related applications. Some examples of *physics lists* are provided with the GATE documentation. The Geant4 "option 3" *physics list* (Geant4 Physics Reference Manual), which has been designed for high-precision simulations such as those needed for medical applications, is available in GATE. A material database file contains all material parameters required by Geant4 to calculate the interaction cross sections. This material database can be edited by the user.

Electromagnetic Processes

Electromagnetic processes are used to simulate the electromagnetic interactions of particles with matter. The electromagnetic physics package manages electrons, positrons, γ-rays, x-rays, optical photons, muons, hadrons, and nuclei.

In GATE V6, three packages of models and associated cross sections are available to simulate electromagnetic processes: the *standard package*, the *low-energy package,* and the *PENELOPE package*. Models and cross sections are based on theoretical calculations and on experimental data. The *standard package* relies on the parameterization of experimental data (see Geant4 Physics Reference Manual) and is supposed to be appropriate for processes between 1 keV and 100 TeV. Using this package, photoelectric effect and Compton scatter can be simulated at energies above 10 keV. Under 100 keV however, relative errors on the cross sections are higher than 10% and can rise up to 50%. The *low-energy package* uses directly experimental data [19] and is supposed to properly describe electromagnetic processes occurring between 250 eV and 100 GeV. It models photon and electron interactions and includes Rayleigh scattering. The *PENELOPE package* is based on the PENELOPE code (PENetration and Energy LOss of Positrons and Electrons), version 2001 [20] and models electromagnetic processes between 250 eV and 1 GeV. The electron multiple-scattering algorithms of PENELOPE have not been implemented yet in GATE.

The energy loss of photons is due to discrete interactions only. The energy loss of electrons, however, is separated into continuous energy loss (soft collisions) and secondary electron production (hard collisions or knock-ons). The production threshold (often referred to as "cut") is defined as the minimum energy E_{cut} above which secondary particles are produced and tracked. When the electron energy E is lower than E_{cut}, E decreases only by continuous energy loss. When E is greater than E_{cut}, secondary electrons are produced. This threshold E_{cut} is actually set as a range cut and is internally converted into energy, to avoid the dependence on material.

Hadronic Processes

Hadronic processes describe the interactions between energetic hadrons/ions and target nuclei. At low energies, incomplete fusion may occur in central collisions and elastic or inelastic scattering in peripheral collisions. At higher energies, central collisions produce fragmentation into several lighter fragments or nucleons, while in peripheral collisions, one can distinguish the so-called participant and spectator regions. In Geant4, fusion, inelastic scattering and fragmentation processes are included in the inelastic process type. Inelastic hadronic interactions allow simulation of the cascade, pre-equilibrium, and de-excitation stages of ions interacting with matter. These processes produce prompt-gamma radiations and β+ emitters among other particles. Each process is described through models and sets of interaction data. The different Geant4 models available (Bertini, binary cascade, quantum molecular dynamics model, low-energy parameterized model) offer different levels of complexity for the simulations.

Models can be selected as a function of particle type, energy, and material. Several interaction datasets may be available for each hadronic process. Full validation of the hadronic physics models available in Geant4 for dosimetry and imaging applications is still needed, although some results are already available [21,22].

Optical and Scintillation Processes

Optical photons are at the origin of the detected signal in scanners involving scintillators, such as used in SPECT and PET. Although simulating all optical photons created using a SPECT or PET acquisitions would be unpractical due to the too large number of optical photons to be simulated, it can still be useful to accurately model the physics of optical photon creation and propagation when interested in the optimization of a single detector. GATE can thus simulate the scintillation process, the scintillation photon propagation through the crystal to the light detector and the conversion of the scintillation photons into electronic signals, based on the models available in Geant4 [23]. These models account for the absorption and scattering of optical photons within dielectric materials, for the reflection, transmission, and absorption at various types of material interfaces (dielectric–dielectric, dielectric–metal, etc.) as a function of the surface treatment (optically polished, ground, etched). When using these models, a large number of parameters need to be set: light yield, energy resolution, emission spectrum, and decay characteristics of the scintillator; refractive index, absorption length, scattering length and surface roughness of the crystal; and reflection and absorption properties of the light sensor and any other materials surrounding the crystal. These parameters either can be obtained from published data or have to be determined experimentally as described, for instance, in [24]. Examples of simulations of the transportation of optical photons with GATE have been published [25,26].

Designing a GATE Simulation

Building a Macro

The easiest way to design a GATE simulation, be it for imaging or radiation therapy application, is to start from the examples or the benchmarks provided with the release of GATE. These examples indeed give the general template of what a simulation macro should look like, and these templates can then be modified and customized to a specific experimental set-up. GATE is currently provided with three benchmarks (one in PET, one in SPECT, and one for radiation therapy applications) and a large number of examples. The role of the benchmarks is to be run every time a new release is produced so as to check the integrity of the GATE installation and the consistency of the results with the results obtained when testing the software on a broad variety of hardware. Unlike the benchmarks, the examples are not provided with "benchmark results" and their only purpose is to help the user define a macro corresponding to his own application.

The PET benchmark simulates a whole-body scanner that does not correspond to any existing system. It consists of eight detector heads arranged in an 88 cm diameter and 40 cm axial length octagonal cylinder. Each head is made of 400 detector blocks and each block is a 5×5 array of dual-layer LSO–BGO crystals. The heads are equipped with partial septa and can rotate in a step-and-shoot mode. A cylindrical phantom including F18 and O15 activity is set inside the PET scanner. The duration of the simulated acquisition is 4 min. The acquisition is divided into two 2 min frames so that after the first frame, the gantry rotates by 22.5°. Only coincidences are recorded with a coincidence window set to 120 ns. The standard electromagnetic package of Geant4 is used. To speed up the simulation, x-rays and secondary electrons are not tracked. Approximately 3.7×10^7 decays occur during the simulated acquisition. About 7×10^5 coincidences are recorded and written in a ROOT file. Based on the ROOT output, figures and plots are calculated to check the correct execution of the simulation. The benchmark results include (1) the total number of generated events and detected coincidences, (2) their spatial and time distributions, (3) the fraction of random and scattered coincidences, and (4) the average acolinearity between the two annihilation gammas. The average values and standard deviations of these figures of merit obtained in 12 different configurations (varying by the operating system and its version) are provided so that the user can check the validity of his results.

Similarly, a SPECT benchmark models a fictitious gamma camera with four heads, equipped with a parallel hole collimator and scanning a phantom filled with Tc99m undergoing a translation motion on a table. The simulated acquisition consists of 64 projections (16 projections per head), acquired along a circular orbit. The physics processes are modeled using the low-energy electromagnetic processes package. Rayleigh, photoelectric, and Compton interactions are turned on while the gamma conversion interactions are turned off. To speed up the simulation, x-rays are tracked until their energy fall under 20 keV while secondary electrons are not tracked. Compton events occurring in the phantom, collimator, backcompartment, shielding, and table are recorded. A Gaussian energy blur is applied to all events detected in the crystal, using an energy resolution of 10% at 140 keV. The limited spatial resolution of the photomultipliers and associated electronic is modeled using a Gaussian blur with a standard deviation of 2 mm. Only photons detected with an energy between 20 and 190 keV are stored. The benchmark average results are also provided for the user to check the integrity of the software installation.

Although these two benchmarks do not correspond to any real imaging systems, they are comprehensive enough to be used as templates to design specific PET and SPECT simulations. The radiotherapy benchmark consists of a proton beam and a photon beam interacting in a $40 \times 40 \times 40$ cm^3 water phantom. The proton beam energy is 150 MeV while the photon beam energy is 18 MV. Three hundred and fifty thousand protons and two million photons are generated. The simulated energy deposition profile is automatically produced by the analysis code provided with the benchmark together with a file including the number of events, tracks, and steps so that the user benchmark results can be compared with the OpenGATE collaboration results.

In addition to these benchmarks, examples are provided to serve as additional templates corresponding to different experimental set-up. These examples include macros for performing an optical simulation, for modeling a CT system, and for modeling the kinetic of tracers in a phantom.

In radiotherapy applications, ten examples are also provided, among which models for various photon, proton, and carbon ion therapy applications. These examples illustrate the many features useful in radiation therapy applications, such as how to output a 3D dose distribution map, how to use phase spaces, or how to use a patient CT image in the simulation.

As in imaging applications, these examples can be used as templates to help in the design of specific radiotherapy set-ups.

Acceleration Options

The high accuracy of GATE simulations is at the cost of a moderate computational efficiency. Fortunately, a number of options can be used to speed up GATE simulations. These include

- The use of the regular navigator and fictitious interaction approaches appropriate for SPECT and PET [27]
- The tabulated modeling of the detector response in SPECT which circumvents the tracking of particles within the collimator/detector pair [28]
- Compressed voxel strategies to optimize memory and CPU use [29]
- The conventional splitting and Russian roulette variance-reduction techniques (Chapter 2)
- A cloning principle to accelerate CT simulations [7]

In addition, the so-called cuts can be used to tune the compromise between simulation accuracy and efficiency. *Cuts* are criteria that specify when the tracking of a particle should be stopped or when secondary particles should not be produced. The user can define four types of *cuts* to limit the tracking of particles and speed up the simulation: the maximum total track length, the maximum total time of flight, the minimum kinetic energy of the particle, and the minimum remaining range. When a particle reaches any of these *cuts*, its remaining energy is deposited locally. Eliminating the secondary particles with an initial energy below the production threshold can also greatly increase the computational efficiency.

For the reason that low-energy processes generate more secondary particles than standard energy processes, *cuts* affect simulation speed more strongly when applied with the *low-energy package*. Turning off the production of electrons, x-rays and delta-rays by setting high thresholds can substantially reduce the simulation time for a typical simulation of a PET scanner, without affecting the accuracy of the simulation in terms of single or coincidence photon countings. The influence of different *cut* values has been described for macro-dosimetry using carbon ions [30] and protons [31], but also for micro-dosimetry [32].

Other options (stepping function, maximum step size, stepping algorithm) [5,19] and (Geant4 Physics Reference Manual) are also available for controlling the compromise between accuracy and CPU time.

Running a GATE Simulation

GATE simulations can be run on a single PC, or on a cluster of computers, or on a grid. When using a single PC, GATE can be run interactively using macrofiles or in a batch mode.

To reduce the overall computing time of GATE simulations and when a cluster of computers is available, GATE can be run in cluster mode [33,34]. This mode consists of splitting the job using a job splitter, running the actual simulations on several CPU, and merging the resulting files through a file merger. The inputs to the job splitter are the GATE scripts, parameters, and command-line options. A generator of pseudo-random numbers provides statistically independent seeds for the set of macrofiles. The job splitter produces a split file that contains all information about the partitioned simulation to enable the merging of the output files. The output merger uses as input the ROOT output files resulting from the parallelized simulations and merges them based on the split file content. The cluster implementation is fully automatic and does not require interaction from the user.

GATE is also compatible with the deployment on a grid infrastructure. GATE has already been deployed on the EGEE grid [33], and functionalities to run GATE simulations on distributed resources have been developed.

Finally, GATE is available as a virtual machine, under the vGATE name. This virtual machine can be run on any host machine, with any operating system (Linux, Windows, MacOS) provided that the Virtual Box program (http://www.virtualbox.org) is installed. This version of GATE makes it possible for the user to run his first simulation very quickly, without having to install GATE, Geant4, and all associated libraries step by step.

Systems Already Modeled and Validated with GATE

A large number of imaging systems have already been modeled using GATE. Table 8.1 summarizes the imaging systems modeled by members of the OpenGATE collaboration as well as associated references. Regarding the simulation of radiation therapy systems, only the model of the Elekta Precise Linac with 6 MV photon beam has been described so far given that such modeling has been made possible only since the GATE V6 version released in 2010. Figure 8.4 illustrates a number of systems already simulated with GATE.

Current Applications of GATE

Given its broad scope, GATE has been and is still used for a large variety of applications. Only a few examples will be listed here to give some insight into the type of studies that can benefit from using GATE. GATE is widely relied on for studying prototypes of imaging devices and optimizing some of their components (for instance, [33,37–44]). It can also be taken advantage

TABLE 8.1

Imaging Systems Already Modeled by Members of the OpenGATE
Collaboration

Imaging Modality	Imaging System	Reference
PET	ECAT EXACT HR+ Siemens	[65]
	ECAT HRRT Siemens	[66]
	HiRez Siemens	[67,68]
	Allegro Philips	[69]
	GE Advance Discovery LS GEMS	[70]
Small-animal PET	MicroPET P4, Concorde	[71]
	MicroPET Focus 220 Siemens	[71,72]
	Mosaic Philips	[73]
SPECT	IRIX Philips	[74]
	AXIS Philips	[75]
	DST Xli GEMS	[49,76]
	Millenium VG Hawkeye GEMS	[49]
	Symbia T Siemens	[28]
Prototype	Solstice Philips	[75]
	LSO/LuYAP phoswich PET	[77]
	Atlas	[38,40]
	CsI(Tl) SPECT	[37]
	OPET	[39]
	ClearPET/XPAD small-animal PET/CT	[78]

PET EXACT ECAT HR+ (Siemens)   SPECT DST XLI (GE)

PET Allegro (Philips)   Micro-TEP Focus 220 (Siemens)

FIGURE 8.4
Examples of systems already simulated with GATE (see Table 8.1 for references).

of to better understand the response of the imaging system in CT [45], in PET [46–48], and in SPECT [49].

In tomographic reconstruction, the system matrix to be inverted is most often calculated analytically, or using simplified simulations. GATE can actually also be used to calculate this system matrix so that physical effects for which there is no easily tractable analytical model can be accounted for. This has been already performed both in SPECT [50,51] and in PET [42,52].

GATE simulations are also often performed to assess reconstruction algorithms (e.g., [53,54]) and quantification methods. A large number of papers fall into this kind of applications (e.g., [55–57]) and it is expected that more and more evaluation studies will rely on highly realistic Monte Carlo simulated data, which have the advantage of closely mimicking real data while offering a full control of all phantom and scanner parameters.

GATE is also used for dose calculation in internal dosimetry [58–62]. Several codes dedicated to dosimetry applications (e.g., MCNP [4], EGSnrc [63] have already been well validated and are widely used. The room for GATE is thus rather in those applications involving both dosimetry and imaging, as GATE is currently the only integrated Monte Carlo simulation platform supporting radiation therapy simulations and imaging simulations.

Conclusion

GATE is a simulation tool resulting from a collaboration effort. In addition to the leading role of the OpenGATE collaboration in the development of GATE, its evolution is also largely driven by user demands and contributions. To that end, a number of tools are available to get the GATE users largely involved in the development, validation, and documentation of the code. The source code is available to any user; a GATE user mailing list is open to post question and share expertise on GATE; and the documentation is a wiki to which any user can contribute. GATE user meetings are also organized to strengthen links and collaborations within the GATE user community.

GATE future developments include a possible extension of the code so that it could also model optical imaging experiments (such as fluorescence and bioluminescence) given the growing interest in coupling optical imaging with micro-PET for instance [64]. New approaches to speed up GATE simulations are also under developments, taking advantages of GPU architectures or hybrid methods coupling the Monte Carlo methods with analytical calculations.

References

1. Buvat I, Castiglioni I. Monte Carlo simulations in SPET and PET. *Q. J. Nucl. Med.* 2002;46(1):48–61.
2. Brun R, Bruyant F, Maire M, McPherson AC, Zanarini P. *GEANT3 Technical Report* CERN;1987.
3. Bielajew AF, Hirayama H, Nelson WR, Rogers DWO. *History, Overview and Recent Improvements of EGS4.* Otawa: National Research Council;1994.
4. Briesmeister JF. *MCNP—A General Monte Carlo N-particle Transport Code LANL.* Los Alamos: Los Alamos National Laboratory;1993.
5. Agostinelli S, Allison J, Amako K et al. GEANT4—A simulation toolkit. *Nucl. Instrum. Meth. Phys. Res. A.* 2003;506:250–303.
6. Jan S, Santin G, Strul D et al. GATE: A simulation toolkit for PET and SPECT. *Phys. Med. Biol.* 2004;49(19):4543–4561.
7. Jan S, Benoit D, Becheva E et al. GATE V6: A major enhancement of the GATE simulation platform enabling modelling of CT and radiotherapy. *Phys. Med. Biol.* 2011;56(4):881–901.
8. Ford EC, Herman J, Yorke E, Wahl RL. 18F-FDG PET/CT for image-guided and intensity-modulated radiotherapy. *J. Nucl. Med.* 2009;50(10):1655–1665.
9. Crespo P, Shakirin G, Fiedler F, Enghardt W, Wagner A. Direct time-of-flight for quantitative, real-time in-beam PET: A concept and feasibility study. *Phys. Med. Biol.* 2007;52(23):6795–6811.
10. Knopf A, Parodi K, Paganetti H, Cascio E, Bonab A, Bortfeld T. Quantitative assessment of the physical potential of proton beam range verification with PET/CT. *Phys. Med. Biol.* 2008;53(15):4137–4151.
11. Nishio T, Ogino T, Nomura K, Uchida H. Dose-volume delivery guided proton therapy using beam on-line PET system. *Med. Phys.* 2006;33(11):4190–4197.
12. Knoll GF. *Radiation Detection and Measurement.* New York: John Wiley & Sons; 1979.
13. Capote R, Jeraj R, Ma CP et al. *Phase-Space Database for External Beam Radiotherapy:* IAEA Technical Report INDC(NDS)-0484, Vienna, 2006.
14. Von Neumann J. 12. Various techniques in connection with random digits-Monte Carlo methods. *Nat. Bureau Standards AMS.* 1951;12:36–38.
15. Behrens H, Janecke J. *Numerical Tables for Beta-Decay and Electron Capture.* Vol 4. Berlin: Springer;1969.
16. Iwata K, Greaves RG, Surko CM. γ-Ray spectra from positron annihilation on atoms and molecules. *Phys. Rev. A.* 1997;5:3586–3604.
17. Stute S, Carlier T, Cristina K, Noblet C, Martineau A, Hutton B, Barnden L, Buvat I. Monte Carlo simulations of clinical PET and SPECT scans: Impact of the input data on the simulated images. *Phys. Med. Biol.* 2011;56:6441–6457.
18. Descourt P, Segars WP, Lamare F et al. Real Time NCAT: Implementing real time physiological movement of voxellised phantoms in GATE. *IEEE Medical Imaging Conference Record* 2005;5:3163–3165.
19. Apostolakis J, Giani S, Maire M, Nieminen P, Pia MG, Urban L. *Geant4 Low Energy Electromagnetic Models for Electrons and Photons.* CERN-OPEN-99-034, INFN/AE-99/18; 1999.

20. Salvat F, Fernández-Varea JM, Acosta E, Sempau J. PENELOPE: A code system for Monte Carlo simulation of electron and photon transport. Paper presented at: AEN-NEA; 5–7 November 2001; Issy-les-Moulineaux, 2001.
21. Chen Y, Ahmad S. Evaluation of inelastic hadronic processes for 250 MeV proton interactions in tissue and iron using Geant4. *Radiat. Prot. Dosimetry.* 2009;136: 11–16.
22. Bohlen TT, Cerutti F, Dosanjh M et al. Benchmarking nuclear models of FLUKA and GEANT4 for carbon ion therapy. *Phys. Med. Biol.* 2010;55(19):5833–5847.
23. Levin A, Moisan C. A more physical approach to model the surface treatment of scintillation counters and its implementation into DETECT. *IEEE Medical Imaging Conference Record* 1996;2:702–706.
24. van der Laan DJ, Schaart DR, Maas MC, Beekman FJ, Bruyndonckx P, van Eijk CW. Optical simulation of monolithic scintillator detectors using GATE/GEANT4. *Phys. Med. Biol.* 2010;55(6):1659–1675.
25. van Der Laan DJ, Maas MC, Schaart DR, Bruyndonckx P, Léonard S, van Eijk CWE. Using Cramer–Rao theory combined with Monte Carlo simulations for the optimization of monolithic scintillator PET detectors. *IEEE Trans. Nucl. Sci.* 2006;53:1063–1070.
26. Janecek M, Moses WW. Simulating scintillator light collection using measured optical reflectance. *IEEE Trans. Nucl. Sci.* 2010;57:964–970.
27. Rehfeld NS, Stute S, Apostolakis J, Soret M, Buvat I. Introducing improved voxel navigation and fictitious interaction tracking in GATE for enhanced efficiency. *Phys. Med. Biol.* 2009;54(7):2163–2178.
28. Descourt P, Carlier T, Du Y et al. Implementation of angular response function modeling in SPECT simulations with GATE. *Phys. Med. Biol.* 2010;55(9): N253–N266.
29. Taschereau R, Chatziioannou AF. Compressed voxels for high-resolution phantom simulations in GATE. *Mol. Imag. Biol.* 2008;10(1):40–47.
30. Zahra N, Frisson T, Grevillot L, Lautesse P, Sarrut D. Influence of Geant4 parameters on dose distribution and computation time for carbon ion therapy simulation. *Phys. Med. Biol.* 2010;26(4):202–208.
31. Grevillot L, Frisson T, Zahra N et al. Optimization of GEANT4 settings for proton pencil beam scanning simulations using GATE. *Nucl. Instrum. Meth. Phys. Res. B.* 2010;268:3295–3305.
32. Frisson T, Zahra N, Lautesse P, Sarrut D. Monte-Carlo based prediction of radiochromic film response for hadrontherapy dosimetry. *Nucl. Instrum. Meth. Phys. Res. A.* 2009;606:749–754.
33. Staelens S, De Beenhouwer J, Kruecker D et al. GATE: Improving the computational efficiency. *Nucl. Instrum. Meth. Phys. Res. A.* 2006;569:341–345.
34. De Beenhouwer J, Staelens S, Kruecker D et al. Cluster computing software for GATE simulations. *Med. Phys.* 2007;34(6):1926–1933.
35. Staelens S, Vunckx K, De Beenhouwer J, Beekman F, D'Asseler Y, Nuyts J, Lemahieu I. GATE simulations for optimization of pinhole imaging. *Nucl. Instrum. Meth. A.* 2006;569:359–363.
36. Boisson F, Bekaert V, El Bitar Z, Wurtz J, Steibel J, Brasse D. Characterization of a rotating slat collimator system dedicated to small animal imaging. *Phys. Med. Biol.* 2011;56:1471–1485.

37. Lazaro D, Buvat I, Loudos G et al. Validation of the GATE Monte Carlo simulation platform for modelling a CsI(Tl) scintillation camera dedicated to small-animal imaging. *Phys. Med. Biol.* 2004;49(2):271–285.
38. Chung YH, Choi Y, Cho G, Choe YS, Lee KH, Kim BT. Characterization of dual layer phoswich detector performance for small animal PET using Monte Carlo simulation. *Phys. Med. Biol.* 2004;49(13):2881–2890.
39. Rannou F, Kohli V, Prout D, Chatziioannou A. Investigation of OPET peformance using GATE, a Geant4-based simulation software. *IEEE Trans. Nucl. Sci.* 2004;51:2713–2717.
40. Chung YH, Choi Y, Cho GS, Choe YS, Lee KH, Kim BT. Optimization of dual layer phoswich detector consisting of LSO and LuYAP for small animal PET. *IEEE Trans. Nucl. Sci.* 2005;52:217–221.
41. van der Laan DJ, Maas MC, de Jong HWAM et al. Simulated performance of a small-animal PET scanner based on monolithic scintillation detectors. *Nucl. Instrum. Meth. Phys. Res. A.* 2007;571:227–230.
42. Vandenberghe S, Staelens S, Byrne C, Soares E, Lemahieu I, Glick S. Reconstruction of 2D PET data with Monte Carlo generated system matrix for generalized natural pixels. *Phys. Med. Biol.* 2006;51:3105–3125.
43. Visvikis D, Lefevre T, Lamare F, Kontaxakis G, Santos A, Darambaravan D. Monte Carlo based performance assessment of different animal PET architectures using pixellated CZT detectors. *Nucl. Instrum. Meth. Phys. Res. A.* 2006;569:225–229.
44. Boussion N, Hatt M, Lamare F et al. A multiresolution image based approach for correction of partial volume effects in emission tomography. *Phys. Med. Biol.* 2006;51(7):1857–1876.
45. Chen Y, Liu B, O'Connor JM, Didier CS, Glick SJ. Characterization of scatter in cone-beam CT breast imaging: Comparison of experimental measurements and Monte Carlo simulation. *Med. Phys.* 2009;36(3):857–869.
46. Stute S, Benoit D, Martineau A, Rehfeld NS, Buvat I. A method for accurate modelling of the crystal response function at a crystal sub-level applied to PET reconstruction. *Phys. Med. Biol.* 2011;56:793–809.
47. Zeraatkar N, Ay MR, Ghafarian P, Sarkar S, Geramifar P, Rahmim A. Monte Carlo-based evaluation of inter-crystal scatter and penetration in the PET subsystem of three GE Discovery PET/CT scanners. *Nucl. Instrum. Meth. Phys. Res. A.* 2011;659(1):508–514.
48. Torres-Espallardo I, Rafecas M, Spanoudaki V, McElroy DP, Ziegler SI. Effect of inter-crystal scatter on estimation methods for random coincidences and subsequent correction. *Phys. Med. Biol.* 2008;53(9):2391–2411.
49. Autret D, Bitar A, Ferrer L, Lisbona A, Bardies M. Monte Carlo modeling of gamma cameras for I-131 imaging in targeted radiotherapy. *Cancer Biother. Radiopharm.* 2005;20(1):77–84.
50. Lazaro D, El Bitar Z, Breton V, Hill D, Buvat I. Fully 3D Monte Carlo reconstruction in SPECT: A feasibility study. *Phys. Med. Biol.* 2005;50(16):3739–3754.
51. El Bitar Z, Lazaro D, Breton V, Hill D, Buvat I. Fully 3D Monte Carlo image reconstruction in SPECT using functional regions. *Nucl. Instr. Meth. Phys. Res. A.* 2006;569:399–403.
52. Aguiar P, Rafecas M, Ortuno JE et al. Geometrical and Monte Carlo projectors in 3D PET reconstruction. *Med. Phys.* 2010;37(11):5691–5702.

53. Vandenberghe S, Daube-Witherspoon ME, Lewitt RM, Karp JS. Fast reconstruction of 3D time-of-flight PET data by axial rebinning and transverse mashing. *Phys. Med. Biol.* 2006;51(6):1603–1621.
54. Martineau A, Rocchisani JM, Moretti JL. Coded aperture optimization using Monte Carlo simulations. *Nucl. Instrum. Meth. Phys. Res. A.* 2010;616:75–80.
55. Tylski P, Stute S, Grotus N et al. Comparative assessment of methods for estimating tumor volume and standardized uptake value in (18)F-FDG PET. *J. Nucl. Med.* 2010;51(2):268–276.
56. Assié K, Dieudonne A, Gardin I, Véra P, Buvat I. A preliminary study of quantitative protocols in Indium 111 SPECT using computational simulations and phantoms. *IEEE Trans. Nucl. Sci.* 2010;57:1096–1104.
57. Dewalle-Vignion AS, Betrouni N, Lopes R, Huglo D, Stute S, Vermandel M. A new method for volume segmentation of PET images, based on possibility theory. *IEEE Trans. Med. Imaging.* 2010;30(2):409–423.
58. Visvikis D, Bardies M, Chiavassa S et al. Use of the GATE Monte Carlo package for dosimetry applications. *Nucl. Instrum. Meth. Phys. Res. A.* 2006;569:335–340.
59. Taschereau R, Chow PL, Chatziioannou AF. Monte carlo simulations of dose from microCT imaging procedures in a realistic mouse phantom. *Med. Phys.* 2006;33(1):216–224.
60. Taschereau R, Chatziioannou AF. Monte Carlo simulations of absorbed dose in a mouse phantom from 18-fluorine compounds. *Med. Phys.* 2007;34(3):1026–1036.
61. Thiam CO, Breton V, Donnarieix D, Habib B, Maigne L. Validation of a dose deposited by low-energy photons using GATE/GEANT4. *Phys. Med. Biol.* 2008; 53(11):3039–3055.
62. Maigne L, Perrot Y, Schaart DR, Donnarieix D, Breton V. Comparison of GATE/ GEANT4 with EGSnrc and MCNP for electron dose calculations at energies between 15 keV and 20 MeV. *Phys. Med. Biol.* 2011;56(3):811–827.
63. Kawrakow I, Rogers DWO. *NRCC Report PIRS-701*, National Research Council of Canada, Ottawa; 2001
64. Li C, Yang Y, Mitchell GS, Cherry SR. Simultaneous PET and multispectral 3-dimensional fluorescence optical tomography imaging system. *J. Nucl. Med.* 2011;52(8):1268–1275.
65. Jan S, Comtat C, Strul D, Santin G, Trébossen R. Monte Carlo simulation for the ECAT EXACT HR+ system using GATE. *IEEE Trans. Nucl. Sci.* 2005;52:627–633.
66. Bataille F, Comtat C, Jan S, Trébossen R. Monte Carlo simulation for the ECAT HRRT using GATE. *IEEE Medical Imaging Conference Record* 2004;4:2570–2574.
67. Michel C, Eriksson L, Rothfuss H, Bendriem B, Lazaro D, Buvat I. Influence of crystal material on the performance of the HiRez 3D PET scanner: A Monte-Carlo study. *IEEE Medical Imaging Conference Record* 2006;4:1528–1531.
68. Rehfeld NS, Vauclin S, Stute S, Buvat I. Multidimensional B-spline parameterization of the detection probability of PET systems to improve the efficiency of Monte Carlo simulations. *Phys. Med. Biol.* 2010;55(12):3339–3361.
69. Lamare F, Turzo A, Bizais Y, Le Rest CC, Visvikis D. Validation of a Monte Carlo simulation of the Philips Allegro/GEMINI PET systems using GATE. *Phys. Med. Biol.* 2006;51(4):943–962.
70. Schmidtlein CR, Kirov AS, Nehmeh SA, et al. Validation of GATE Monte Carlo simulations of the GE Advance/Discovery LS PET scanners. *Med. Phys.* 2006; 33(1):198–208.

71. Vandervoort E, Camborde ML, Jan S, Sossi V. Monte Carlo modelling of singles-mode transmission data for small animal PET scanners. *Phys. Med. Biol.* 2007;52(11):3169–3184.
72. Branco S, Jan S, Almeida P. Monte Carlo simulations in small animal PET imaging. *Nucl. Instr. Meth. Phys. Res. A.* 2007;580:1127–1130.
73. Merheb C, Petegnief Y, Talbot JN. Full modelling of the MOSAIC animal PET system based on the GATE Monte Carlo simulation code. *Phys. Med. Biol.* 2007; 52(3):563–576.
74. Staelens S, Vandenberghe S, De Beenhouwer J et al. A simulation study comparing the imaging performance of a solid state detector with a rotating slat collimator versus a traditional scintillation camera. *Proc. SPIE Med. Imag. Conf.* 2004;5372:301–310.
75. Staelens S, Koole M, Vandenberghe S, D'Asseler Y, Lemahieu I, Van de Walle R. The geometric transfer function for a slat collimator mounted on a strip detector. *IEEE Trans. Nucl. Sci.* 2005;52:708–713.
76. Assie K, Gardin I, Vera P, Buvat I. Validation of the Monte Carlo simulator GATE for indium-111 imaging. *Phys. Med. Biol.* 2005;50(13):3113–3125.
77. Rey M, Vieira JM, Mosset JB et al. Measured and simulated specifications of the Lausanne ClearPET scanner demonstrator. *IEEE Medical Imaging Conference Record* 2005;4:2070–2073.
78. Nicol S, Karkar S, Hemmer C et al. Design and construction of the ClearPET/ XPAD small animal PET/CT scanner. *IEEE Medical Imaging Conference Record* 2009:3311–3314.

9

The MCNP Monte Carlo Program

Per Munck af Rosenschöld and Erik Larsson

CONTENTS

Monte Carlo N-Particle (MCNP) is a Monte Carlo code package allowing coupled neutron, photon, and electron transport calculations. Also, the possibility of performing heavy charged particle transport calculations was recently introduced with the twin MCNPX code package. An arbitrary three-dimensional problem can be formulated through the use of surfaces defining building blocks ("cells") that are assigned density, material, and relevant cross-section tables. The source can be specified as point, surface, or volumes using generic or as a phase/space file.

Neutron calculations can be assigned cross-section using free gas or $S(\alpha,\beta)$ models. Neutron criticality calculations can be performed. Photon interactions include both coherent and incoherent scattering, photoelectric absorption with fluorescent emission, as well as pair production. Electron transport is Class I, and includes production of secondary particles including knock-on electrons, Auger electrons, Bremsstrahlung photons, K x-rays, and annihilation photons.

The MCNP code is used in multiple areas of research due to the flexible geometry/source specification, and a range of detector ("tallies") and variance-reduction techniques. The code has a pulse height tally for simulating detector response, for example, for simulating photon spectra measurement using a high-purity germanium detector. Multiple tallies can be used in a single problem.

The code packages can be run under Windows, Unix, and Linux operating systems. The system is supported and updated by the Los Alamos National Laboratory in California. The code requires no programming skill per se; not even a compiler is needed as executables are delivered with the code. However, the problem at hand needs to be specified in accordance with how the code was written with respect to geometry, physics, and radiation source, which requires some experience to do correctly and in particular how to interpret.

The expert staff at the Los Alamos Laboratory provide regular training courses at multiple sites worldwide at the beginners and at the advanced level in use of the MCNP code.

Brief History and Code Availability

One of the first applications of the computerized Monte Carlo technique was neutronic calculations for the development of atomic bombs. Scientists at the Los Alamos Laboratory during the World War II were actively pursuing the potential use of computers in making multiple statistical calculations feasible. The initial formulation of the Monte Carlo technique was performed already in the late 1940s but the technique evolved along the improvement of computers and programming languages. The first version of the MCNP code was written in Fortran in the 1960s at Los Alamos Laboratory, at that point the code could be used to perform three-dimensional neutron transport calculations. The possibility of performing gamma transport calculations was added in the 1970s, at which point the code was called "MCNG" (Monte Carlo Neutron Gamma). This was shortly followed by the generalization to perform photon transport calculations as well as criticality calculations. The code changed name to "MCNP," at first this stood for Monte Carlo Neutron Photon and later Monte Carlo N-Particle.

During the 1980s, the MCNP code got updated again and tally plotting and the generalization of source specification. In the 1990s, the version 4 of the MCNP code got released which increased the flexibility and use through the integration of the Integrated TIGER Series (ITS) Monte Carlo system for electron transport calculations. With that update, the MCNP code arguably got much of the functionality and form of the present version.

The version 5 of the MCNP code includes various small enhancements, for instance improved functionality of the geometry specification and the plotter. In addition, parallelization of problems was introduced, which allows for easier reduction in statistical uncertainties within a given time-frame if multiple processors are available for use.

A parallel work was initiated in the mid-1990s on the so-called MCNPX code package. The code built on MCNP version 4B, with its neutron–photon–electron transport capabilities, and the LAHET code built for high-energy particle transport calculations. The MCNPX code includes the standard geometry and physics modules—albeit not the latest version—but with an extended use allowing for transport calculations of several other particles. For instance, the MCNPX code allows for proton, antiproton, pion, helium, and neutrino calculations.

The MCNP and MCNPX codes are available from *Radiation Safety Information Computational Center* (RSICC), Oak Ridge, Tennessee (http://rsicc.ornl.gov). The MCNP and MCNPX codes were previously available from OECD Nuclear Energy Agency (NEA), Issy-les-Moulineaux, France (http://www.nea.fr), free of charge. However, this distribution channel of the code has presently been indefinitely suspended. Given that the codes include the possibility of neutron criticality calculations, the distribution is restricted and supervised by the RSICC, and the distribution is presently, for example, limited to certain countries and users. In this chapter, we will limit the discussion to the standard version of MCNP; however, all of the things below apply also to MCNPX but the text below does not deal with some of the features available only in MCNPX.

Code Structure

In this chapter, a brief overview of how the code is set up and run is presented. The intent is to give an idea of the basic concepts of the code—more details are found in the manual.

Running MCNP

The code is invoked through using a DOS prompt in Windows and a c-shell in Linux/Unix. There are basically two ways the code can be run, either by starting a new calculation or by restarting an old calculation ("continue run"), the latter is a practical way to improve the statistics of a calculation.

In the command line, the user calls for specific functions in the code as well as the input file. The input file is basically a standard text file. The file needs to adhere to a specific format, or else the code will stop immediately without performing any calculations. Each line in the input file must start with a "card," each card has a special meaning to the code. A card simply refers to a single line in the input file; a card can be up to 80 characters long. There are three types of cards: "cell," "surface," and "data" cards.

Typically, each card has a set of options ("flags") that can be set in one of more ways in order to run the code in the desired fashion. For instance, the "mode" card has the flags "n," "p," and "e," specifying whether the code be run simulating neutrons, photon, and/or electrons.

The program makes use of three files in each calculation; the input file, the runtpe file, and the output file. The input file needs to have precisely the following format:

```
Message line (optional)
Blank line
One line problem title card
Cell cards
. . .
Blank line
Surface cards
. . .
Blank line
Data cards
Blank line (optional)
```

The output file contains information on the problem run. The amount of information provided can be steered by the "print" data card, where the output information specified. Users should be aware, however, that by requesting a lot of information on the print card tends to slow down the code. The information may include, for instance, the number of particle tracks entering and leaving certain parts of the geometry and number of reactions of a certain kind.

In case the problem was incorrectly set up, the code may have detected that in one of the many integrity tests performed at the start, and then the output file contains information on what has gone wrong. The output file also tends to contain a lot of warning messages of various kinds. The user is advised to carefully review the output file to find out how the problem has run and if the results can be trusted. It is typically very useful to run the code with a low number of histories first and then examine the result and then consider if corrections to the variance reduction, geometry, and so on can be advantageous. An important aid in the judgment of to what extent the results can be trusted is provided in the output file, where 10 statistical tests are performed on each tally. These tests are key to understand how the problem is performed and in terms of trustworthiness of results. Specifically, the tests are shown as a function of source particle histories, which tends to indicate if the problem is converging properly or not. Poor use of variance-reduction methods causing the calculations to fluctuate is a common reason for erratic behavior of the tally as indicated by the statistical tests. The tally fluctuation is used by the code to derive the statistical uncertainty of the tally results, for example, one standard deviation of the photon fluence across a surface.

Geometry Specification

The geometry of a problem is made up of surface cards and cell cards. The surface cards enable specifying planes, cones, spheres, cylinders, ellipsoids, hyperboloid, paraboloid, and elliptical or circular torus defined by coordinates and parameters allowing for arbitrary positioning and size in space. There are also so-called macrobodies that are really just a set of surfaces combined into a volume, for instance, a box or cylinder.

The cells (or "geometry blocks") are specified with a material number (see below), density in grams per cubic centimeter and a series of surface cards that are used to define and limit the cell in space. The cell is specified in terms of its relation or direction toward a certain surface or set of surfaces. The set of surfaces are then combined using Boolean operators. In reality, users tend to make mistakes in the geometry specification of the problem. Therefore, it is highly recommendable to plot the geometry before running the problem as this may indicate any errors. Another useful way of testing the correctness of the geometry specification is to run the particle transport with all cell densities set to zero—if particles enter a region with an ambiguous geometry specification they will get lost and the run will discontinue. The output file provided by the code will provide detailed information concerning lost particles direction and coordinates.

The geometry also offers the possibility of specifying repeated structures ("lattices"), which can be useful in for instance simulations of detector systems. The geometry specification capabilities in MCNP are fairly flexible and very complex geometries can be described using a combination of simple surfaces. In our experience, it typically requires some trial and error to set up a problem except very trivial ones even for an experienced user; the Visual Editor can help in this respect (described below).

Data Cards

The rest of the cards that are needed to specify a given problem in MCNP are lumped together into the "data card" section. Several of the data cards refer back to the cell or surface cards and set properties or parameter relative to the way these should be used in the problem. There are a considerable number of optional data cards that offer a great deal of possibilities with respect to specifying a problem. The following sections will give some account of the data cards (e.g., tallies and variance reduction), but the reader is advised to read the manual in order to see the full potential of the program.

The mandatory data cards involve the mode card, importance card (or weight windows, see below), source card, tally cards, material cards, and problem cut offs (such NPS—number of particle histories). The mode card specifies which particles are simulated in the problem; in the MCNP5 code this is neutrons, photons, and electrons in any combination. If a particle type is not simulated as specified on the mode card, then the particles are considered locally absorbed if created. The importance card, allows the user to instruct the code to spit particles upon entering certain parts of the geometry. This is useful for increasing the number of particles in the interesting/ important part of a problem, typically near a detector, and reducing them in the less important part of the geometry.

The source card is needed to specify a generic point, surface, or volumetric radiation source. The source can emit one type of particles, the directionality, the kinetic energy, and position can be specified as well as biased, if needed. Results from the simulation are normalized per source particle; therefore one can easily combine the results of runs of the same geometry with multiple types of source particles.

In addition to the generic source, the user can also write particles entering a surface or cell to a file, this file can be called for in the simulation and used as a source file. Finally, the code can be run using a criticality source with neutron emitters. Each problem needs to have at least one tally specified; typically giving the fluence of particles in a cell or across a surface (tallies are covered in some detail below).

The material cards are often requiring some extra care in its specification. The material cards are specified as one or a number of natural elements with their relative atomic or weight fraction as well as the cross-section table to use. In particular, the selection of neutron cross sections, the user is advised to select a cross section with the highest accuracy/detail but also that includes generation of secondary gammas in case this is of interest in the calculation. It can be worthwhile to perform multiple calculations varying the selection of cross section in order to substantiate the results.

Finally, the problem needs a definite end at which point the calculation will terminate as based on the number of particle histories followed or the time spent in the calculation. The user can also interrupt the program in order to stop the calculation.

Physics, Tallies, and Detectors

As mentioned before, the presuccessor of MCNP was originally developed to model the physical behavior of neutrons for criticality in materials. Since then, MCNP has been under constant development, and the MCNP5 [1,2] can track neutrons, photons, electrons, and positrons. The integration of LAHET (Los Alamos High-Energy Transport) and MCNP4B led to the MCNPX code. The MCNPX 2.6 [3] package can track almost all particles at almost all energies. The code utilized nuclear cross sections and physical modules for particles or energy where data are not tabulated.

The MCNP code package includes an extensive set of neutron interaction cross sections. The data set available contains interaction cross-section data evaluations provided by multiple researchers and institutions worldwide. Typically, the data set has been compiled based on both theoretical physics models and fitted to experimental data. The cross-section tables include up to and above 100,000 data points, which indicate to some extent how fine-resolved the resonance peaks are. The cross sections provided by the U.S. initiative called the ENDF—a joint group of U.S. industry and research institutions—provided a standard for the format of presentation used by the other groups. Data sets, provided for each isotope, differ, for example, in the number of data points, tolerance for which it was generated, and with respect to tabulation of secondary gammas. Note also that simplifications with respect to the level of details in the gamma production is evident; gamma lines may have been lumped together into average bins, and so on.

Cross sections are provided in discreet and continuous format. Presently, computer speed is not that limiting as a factor in MCNP calculations; therefore, it is generally not a good idea to use the discreet cross sections. Other than that, it can be difficult to provide advice on the best cross-section data to use. However, using the most recent data, and in coupled a neutron/photon calculation, cross-section data including gamma production, appears to be reasonable to favor.

Photons

The latest photon cross-section tables were introduced in MCNP in 2002. These are processed from the ENDF/B-VI.8 library and include incoherent, coherent, photoelectric effect, and pair production cross sections for energies from 1 keV to 100 GeV and for atomic numbers Z between 1 and 100. Data of photoatomic and atomic relaxation are extracted from the EPDL97 library [4]. Also included are coherent form factors, incoherent scattering functions, fluorescence data [5], and the bound-electron momentum [6]. These bonded electrons can add or subtract energy to an incoherent scattered photon resulting in a so-called Doppler energy broadening of the photon. All photon physics is controlled with the phys:p card.

MCNP allows for two photon transport interaction models, a simple and a detailed mode. The simple model is intended for high-energy physics and is the default mode for energies above 100 MeV. This model ignores Thomson scattering and fluorescent photons from photoelectric absorption. Also for coherent scattering, the electrons are assumed to be free and no form factors are used. The detailed photon interaction models include Thomson scattering and florescent photons. It assumes that electrons are bonded and uses form factors for incoherent and coherent scattering.

There is the option to run photon transport only (Mode P), where electron transport is turned off or coupled-photon electron transport (Mode P E). The photon physics models in these two modes are the same; it is the generation of Bremsstrahlung and annihilation photons from secondary electrons and positrons that differs. For the photon only mode, two options are available. The default option includes a thick target Bremsstrahlung (TTB) model, in which electrons and positrons are produced but immediately slowed to rest so that Bremsstrahlung and annihilation photons are generated directly at the interaction point of the photon. The second option omits Bremsstrahlung photons completely but includes the annihilation photons. In the coupled photon electron mode (Mode P E), secondary electrons and positrons are banked for later transport and can then produce Bremsstrahlung photons along its trajectory or annihilation photons at the end of the positron track. The big gain in running photon mode is to avoid the time-consuming electron transport and thereby greatly reduce the simulation time.

Introduced in MCNP5, there is now the option to simulate photonuclear events, although the default mode is off. A photonuclear event means that the photon can be absorbed by the nucleus by the giant dipole resonance absorption mechanism or the quasi-deuteron absorption mechanism. Once a photon has been absorbed, one or several secondary particles, such as gamma-rays, neutrons, protons, deuterons, tritons, helium-3 particles, alphas, and fission fragments, can be emitted, assuming the photon energy is high enough. Photonuclear events have a resonance peak of about 5–20 MeV.

Electrons

Simulation of electrons is fundamentally different from noncharged particles such as neutrons and photons in that sense that noncharged particles undergo few interactions with a free path between whereas electrons are affected by the Coulomb force resulting in many but small interactions with matter. To make for realistic long simulations, MCNP uses a method where many small interactions are reduced into larger steps, a method called condensed histories [7]. MCNP is built around the Integrated Tiger Series (ITS) 3.0 code [8], which is based on the ETRAN code [9]. It is a class-I electron transport code and utilizes precalculated step-lengths. The physics is controlled with the phys:e card where it is possible to control if electrons will produce photons or the opposite, that photons will produce electrons. It is also the card where you control the production of knock-on electrons, Bremsstrahlung, and electron-induced x-rays. If an alternative energy loss-straggling method is wanted, this should be specified in the 18th entry in the debug information card dbcn.

MCNP uses a class-I algorithm for collisional energy loss and a class-II algorithm for radiative losses. The Goudsmit–Saunderson [10] theory is used to calculate angular deflections and the energy loss fluctuations are calculated with the Landau theory [11] with the Blunck and Leisegang [12] enhancements for binding effects. These models require that the energy loss is small compared to the electron energy. Therefore, major steps are chosen so that the average energy loss is 8.3%. To better calculate the electron trajectory, the major energy steps are divided into a number of substeps, m. For geometries where a major step is large compared to the distance between the cell surfaces, the default value of m does not typically need to be increased. However, for smaller geometries, the number of substeps could influence the accuracy of the electron trajectory. It is recommended that an electron undergoes at least 10 substeps in a cell before leaving. Raising the number of substeps of course also increases the simulation time.

At the MCNP initialization, parameters for energy loss are precalculated in discrete energies on an energy grid $E_1, E_2, \ldots, E_{n-1}, E_n, E_{n+1}$, where each energy bin represents a new major energy step. At the start of a new energy step, the energy of the electron E lies between two energies E_n and E_{n+1} and which electron cross-section data to assign the electron are chosen by an energy-indexing logic. The default algorithm is the MCNP style, also called bin-centered, and which assigns cross section to the electron with energy E in a group n, so that $E_n > E \geq E_{n+1}$, saying that the cross sections calculated for energy E_n are used throughout the whole energy step. Another option is the so-called ITS style $(E_{n-1} + E_n)/2 > E \geq (E_n + E_{n+1})/2$ (chosen by dbcn 17j 1) and is in style with what is used in the ITS 3.0 code. This method assigns the cross sections of the upper boundary that are closest to the electron energy (nearest boundary). This difference causes the MCNP-style energy-indexing algorithm to assign electron cross-section data, which are on average half a group higher than in the ITS-style algorithm. The ITS method is the most successful of the two methods but for historical reasons, the MCNP style remains as default [13]. However, the ITS method does not work perfectly. When an electron crosses a surface, the energy step is interrupted and if there are many surfaces for the electron to pass, the energy loss and angular deflection can be affected. Introduced in MCNP 5 1.40 is a new logic for electron transport (chosen by dbcn 17j 2), that does not rely on the precalculated energy grid but calculates transport parameters at every substep. This is a much more robust method and allows for a more precise simulation in small cells. The drawback is that this new algorithm is much more time consuming than the two previous ones.

Tallies

Tally cards are used to gather the information that the user wants to extract from the simulation. There are a seven standard tallies denoted F1–F8 that collects data of surface flux, cell flux, point flux, surface current, pulse height, energy deposition, heating, fission energy deposition, and so on. These tally options are summarized in Table 9.1. Several tallies can be used in the same problems and up to 999 tally cards of the same tally type can be used. The tallies can be further subdivided into bins of, for example, energy, cosine, or time. At the end of a simulation, the tallies are normalized to be per source particle, except if changed with a tally modifier or for criticality simulations (per weight). The tally results undergo 10 statistical test of tally convergence, including the variance of the variance (VOV). If a statistical test is not fulfilled, the user will be notified of this. All tally results, together with the statistical errors and the results of the statistical tests, will be printed in the output file. The user can choose to get the tally in a special file format called mctal for easier data extraction.

There are several ways that a tally could be modified. By using an asterisk in front of the tally card, the units are multiplied with the energy. A tally

TABLE 9.1

Summary of Tallies Available in MCNP5/MCNPX

Mnemonic	Tally Description	Fn Units	*Fn Units
F1: <pl>	Current integrated over surface particles	MeV	MeV
F2: <pl>	Flux averaged over a surface	Particles/cm^2	MeV/cm^2
F4: <pl>	Flux averaged over a cell	Particles/cm^2	MeV/cm^2
F5a:N or F5a:P	Flux at a point or ring detector	Particles/cm^2	MeV/cm^2
F6: <pl>	Energy deposition averaged over a cell	MeV/g	jerks/g
+F6	Collision heating	MeV/g	N/A
F7:N	Fission energy deposition averaged over a cell	MeV/g	jerks/g
F8: <pl>	Energy distribution of pulses created in a detector	Pulses	MeV
+F8: <pl>	Deposition charge	N/A	
FMESHn: <pl>	MESH tally (MCNP5), where n is the tally number	Particles/cm^2	MeV/cm^2
	Only tally type 4 allowed		
(R/C/S) MESHn: <pl>	MESH tally (MCNPX). Rectangular, cylindrical, or spherical grid. Mesh tally types n 1–4		
	Type 1: Average flux, fluence, or current		
	Type 2: Scores source origin point of particle <pl>		
	Type 3: Energy deposition		
	Type 4. Scores track contribution to point detectors or DEXTRAN spheres (*)		
FIP/IP	Pinhole image radiography tally, MCNP5/MCNPX		
FIR/TIR	Transmission radiography tally, rectangular. MCNP5/MCNPX		
FIC/TIC	Transmission radiography tally, cylindrical. MCNP5/MCNPX		

could be multiplied with energy or time response functions, where the ith bin is multiplied with a multiplier m_i. Particles may be flagged when passing a specific surface or cell in the geometry, using the tally cell and surface flagging cards, CF and SF, resulting in a double tally printout. Tallies could be changed with the special treatment card FT that could for instance add a Gaussian energy distribution to the energy bins to simulate a detector energy response or separate the tally result depending on particle type.

The pulse height tally F8 is different to the other tallies in that sense that it stores the total energy or charge deposition in a cell per source particle, including from secondary particles. This implies that it is important to model the whole event correctly as particles with lower energy may affect the result. For the tallies above, it is only important to model the events so that expectance values of the macroscopic variables are correct.

A special kind of tally group is the radiography tallies. These are tallies which allow the user to visualize the geometry problem like it was an x-ray transmission image or a pin-hole image of a geometry containing the source.

A third type of tally is the MESH-tallies, which are tallies scored over a 3D mesh that is independent of the problem geometry. The mesh can be in rectangular, cylindrical, or spherical form. These are track-length tallies and the mesh itself does not interfere with the transport of particles. This type of tally could also be visualized in 2D plots where the results are superimposed on top of the problem geometry. Plots could be produced during simulations using the MCPLOT function or when simulation is interrupted. It can also be displayed later by the MCNP plot procedure or in the MCNP Visual Editor (VISED). Below follows a short description of the 10 statistical tests the tallies undergo at the end of the simulation.

Tally Mean, x:
1. The mean must exhibit, for the last half of the problem, a nonmonotomic behavior (no up or down trends) as the number of histories N increases.

Relative Error, R:
2. R must be <0.1 and <0.05 for point detectors.
3. R must decrease monotonically with N for the last half of the problem.
4. R must decrease as $1/pN$ for the last half of the problem.

Variance of the Variance, VOV:
5. The magnitude of the VOV must be <0.1 for all types of tallies.
6. VOV must decrease monotonically for the last half of the problem.
7. VOV must decrease as $1/N$ for the last half of the problem.

Figure of Merit, FOM:
8. FOM $\equiv 1/(R^2T)$ must remain statistically constant for the last half of the problem.

9. FOM must exhibit a nonmonotonic trend in the last half of the problem.

Tally PDF, f(x):

10. The SLOPE n in x^{-n} of the 25–201 largest history scores x are determined and must be greater than 3.

Although not a tally card, the PTRAC (particle track) card creates an output file that contains a list of the particle tracks and includes data of particle type, position of interaction, interaction type, energy, time, and weight. The user can use some options to filter what information to be written to the file, otherwise this file easily becomes very large.

Variance Reduction

Probably one of the strong points of the MCNP code is the large number and creative methods for variance reduction available for use. The basic premises of all variance-reduction techniques is that the "particle weight" is adjusted in proportion to the bias introduced, for example, if a particle is selected for a game of Russian roulette and survives, the particle weight is adjusted taking into account the risk of the particle getting killed. The particle weight is used in the calculation of a contribution to a tally.

The complete list of variance-reduction techniques in MCNP are Cell Importance, Energy Splitting and Roulette, Time Splitting and Roulette, Photon Weight, Exponential Transform, Vector Input, Forced Collision, Weight Windows (Energies, Times or Cell-Based), Weight Window Generation, Superimposed Importance Mesh for Mesh-Based Weight Window Generator, Detector Contribution, DXTRAN Contribution, and Bremsstrahlung Biasing. Either the Cell Importance or the Cell-Based Weight Windows are required in the problem. Many of the variance-reduction techniques are optional in use and can be used in combination with other techniques. The use of multiple techniques, and the advanced ones, is probably for the advanced user only, considering that interaction may lead to unforeseeable or even erroneous results. In general, the use of variance reduction is cautioned. When an incorrect use of variance reduction occurs, it can sometimes be noticed by inspecting the output file and by looking at the Tally output, which may behave erratically in the statistical tests performed in MCNP. For example, the user might specify the variance-reduction technique such that events that are important for the problem are not sampled adequately, which results in infrequent appearance of particles that impact the tally results greatly (i.e., "high-weight particles").

What follows is a very brief description of the variance-reduction techniques available in MCNP; full details are beyond the scope of this chapter.

Cell Importance

Each cell is assigned a value that tells the code if the particles should be split or killed using Russian roulette, upon leaving and entering it, depending on the respective values. A relatively smooth transition between low and high importance is preferable, where division of the problem into more cells might be warranted in order to accommodate that.

Energy Splitting

The card offers the opportunity to perform splitting/Russian roulette based on particle energies.

Time Splitting

Using the card, the user can split or perform Russian roulette based on time intervals from the starting point of the emission of source particles.

Photon Weight

The card is used to control the weight of photons generated in a given cell in order to make sure the code is not spending time on particles of inconsequential weight.

Exponential Transform

The card makes use of a deterministic transport of particles from one part to another in the geometry. Typically, that process is simulated with the Monte Carlo (statistical) approach, but the card offers the possibility of combining the two. The exponential transform method is only available for photons and electrons

Vector Input

The card is used to define the direction of which the exponential transform (above) is working.

Forced Collision

The forced collisions card makes photons and neutrons interact in a given cell with absolute certainty. The technique is useful in the situation with a small chance of interactions occurring. Forced collision can be used with neutrons and photons.

Weight Windows (Energy, Time, or Cell-Based)

The card is an elaborate way of controlling the particle weight across the geometry of a problem. Divisions can be made toward energy, time, and type of particle, with multiple and separate cards for each. Basically, the weight window method divides particle weight into three ranges (i.e., "windows") where particles in the lowest range are played Russian roulette upon and the survivors are upregulated in weight into the mid-range. Particles in the mid-range are ignored as far as the weight window card is concerned. Particles in the high range are split one or more times so that their weight is maintained in the mid-range (window). A smooth transition between cells is warranted in order to avoid excessive use of particle splitting and Russian roulette, which tends to slow down the speed of the problem converging.

The weight window card is in fact very a practical and efficient method for controlling the weight of particles simulate, and its use is recommended. If the problem is very simple, the cell importance and weight windows might work equally well.

Weight Window Generation

The parameters of the weight window card are not always intuitive to set manually. Therefore, the generator offers a practical solution to the problem. When the generator is turned on, the weight windows are adjusted based on the relative contribution of the cells of the problem. Particles tend to have lower weight close to the important region (equals more particle tracks). The use implies specifying one tally to be the primary in the problem for which the weight window setting will be optimized. The use also causes calculations to go slower, and the user is recommended to run with the generator turned on for a while, and afterwards review and manually adjust the settings if needed before running the final calculation.

Superimposed Importance Mesh for Mesh-Based Weight Windows

The technique offers the opportunity to disconnect the weight window card from the geometry, that is, the cells. The mesh is defined in parallel to the geometry with the sole purpose of controlling particle weight in various parts of the problem.

Detector Contribution

The card can be used to increase the speed of running a given problem by reducing, or even completely eradicating, the contribution of particles to the total of a tally. Use of the technique can be warranted in case the probability of contribution to a tally is insignificant from a certain part of a problem.

DXTRAN Contribution

The variance-reduction technique is particularly useful in case a tally is located such that scattering toward it constitutes a low probability event, such as a problem in which the user is only interested in particles scattered in a small angular interval [14]. The use of DXTRAN concept is nontrivial and too complex to be covered in any detail here. In short, DXTRAN is somewhat akin to forced collisions and exponential transform; at each time a collision occurs, a special "DXTRAN particle" is created that is forced to scatter in the direction and travel to the surface of a predefined sphere (the "DXTRAN sphere"). Preferably, the sphere is covering the tally of interest. The user is able to define several such DXTRAN spheres, though this is typically inadvisable.

Bremsstrahlung Biasing

Bremsstrahlung tends to be of low energy but in a given calculation, the contribution from the high-energy photons may contribute proportionally higher to the tally result. The card therefore offers the possibility of biasing the Bremsstrahlung spectra in a way that is presumed to be more efficient.

Problem Visualization

There are a couple of programs in the MCNP package that are of great value when it comes to building and checking the geometry. Up until MCNP5, there were the geometry plotter and the MCPLOT tally and cross-section plotter, and now there is also the MCNP Visual Editor (VISED) [15]. The geometry plotter and the tally and cross-section plotter are programs based on X-window graphics, which require an X client when working in Windows environment.

The MCNP geometry plotter is a visual aid to the creation of the MCNP input files, and is used to produce plots of the geometry in any 2D plane (although it is not recommended to produce a plot plane that is identical to a surface). The geometry plotter is very useful to produce plots to verify that the surfaces and the combination into cells are correct and as intended. If surfaces appear red and dashed and cells are uncolored, there is a big chance that there is an error in the geometry. By default, different materials have different colors. The plotter can also label for instance surface number, cell number, particle importance in cells, lattice coordinates, densities, mass, and materials of the cells. The geometry plotter is started by typing the command mcnp5 IP i = *filename* in the c-shell/dos prompt, where I stands for initiate and P stands for plot.

From the geometry plotter or while running an MCNP file, it is possible to call for the MCPLOT tally and cross-section plotter. This plotter can be used for plotting cross sections (not while running MCNP) as a function of energy for the materials specified in the input file and to plot MCNP tallies. Tally plotting can be 2D-plots like the tally toward bins, for example, like a detector response to energy or 2D-plots with the tally result on top of the geometry with a color or as a contour. Such examples of tally plotting are given in Figures 9.1 and 9.2. MCPLOT is invoked by the MPLOT card in the input file, in the program interrupt mode or the command mcnp5 Z i = *filename* in the dos prompt/c-shell.

Attached with the MCNP5 package followed the MCNP Visual Editor (VISED). This is an interactive program where it is, like the geometry plotter, possible to produce 2D plots of the geometry and label the surface numbers, cell numbers, densities, importance, and so on. It also contains the features of the MCPLOT. It is also possible to make 3D views of the geometry, where it is possible to rotate or zoom in the geometry. By a mouse click on a surface, the surfaces can be made semitransparent or invisible. 3D ray tracing images, looking like an x-ray image of the geometry, can be generated. It is also possible to track particles (source, collisions, etc.) in the geometry as well as tally plotting (as in MCPLOT).

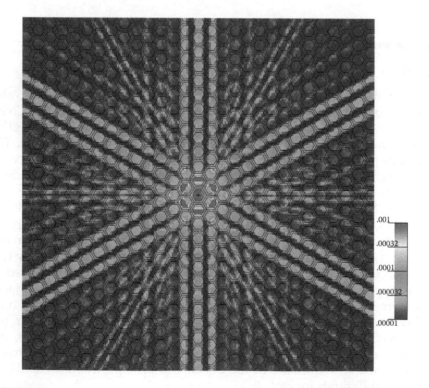

FIGURE 9.1
An image from the MCPLOT tally plotter showing the photon energy flux at the back of an MEGP collimator from a ^{131}I point source 10 cm in front of the collimator. The mesh tally *fmesh:4 and a rectangular mesh with a cubic mesh size of 1 mm^3 were used.

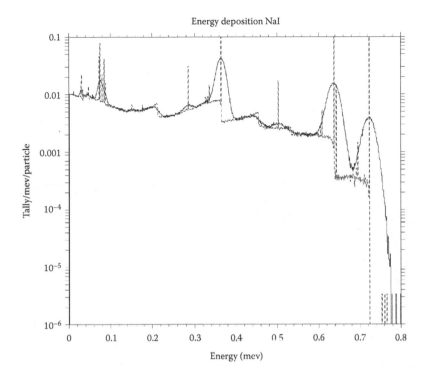

FIGURE 9.2

The energy spectra in a ½ in. thick NaI detector behind the collimator. The dashed line represents the true energy deposition in the crystal and the solid line represents a modeled energy resolution. The tallies used were the F8 tally with energy bins and the Gaussian energy broadening (GEB) option in the special treatment card FT. The geometry was developed with the hexagonal lattice (lat = 1) option, and each lattice element was filled with a universe describing the collimator hole and septa.

There are, beyond the plotting capabilities, a few extra features that could be very useful. Input files can be created in a graphical editor by choosing among tabs labeled surface, cell, source, and data. This can be a big help in the sometimes tedious process of creating the geometry in a text editor. In the surface tab, you can choose among the standard surfaces in MCNP and either draw the surfaces with the mouse or enter the parameters (the cell flags) in text boxes. To combine the surfaces into cells, each surface that constitutes the cell is highlighted with the mouse and then with a mouse click the senses are automatically determined and the cells are created. It is also possible to change or enter surfaces in a text editor. The source definition is made in the source tab and materials, mode, and importance are specified in the data tab. Another geometry feature in VISED is to convert 2D or 3D CAD files into MCNP format. Supported file formats are 2D.dxf and 3D.sat files. VISED will transform the surfaces recognized by the MCNP geometry package and create an input file of the surfaces and the description of

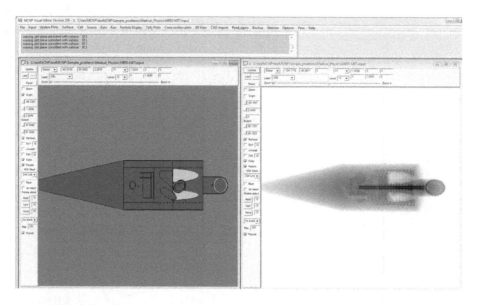

FIGURE 9.3

A screen dump of the Visual Editor, VISED, with plots of the MIRD-MIT phantom that is available and included in the MCNP5/MCNPX package. The left window shows a standard 2D plot much like in the geometry plotter. The right window shows a radiograph (ray length * cross section) of the geometry from a 50 keV source 100 cm above the body.

cells. The user then only has to add material and density to get the geometry. Figure 9.3 shows a screen-shot from the VISED program.

References

1. X-5 Monte Carlo Team. MCNP—A General Monte Carlo N-Particle Transport Code, Version 5, Volume I: Overview and Theory. Los Alamos National Laboratory 1987.
2. X-5 Monte Carlo Team. MCNP—A General Monte Carlo N-Particle Transport Code, Version 5, Volume II: User's Guide. Los Alamos National Laboratory 2003.
3. Pelowitz DB. MCNPXTM User's Manual. Los Alamos National Laboratory 2008.
4. Cullen DE, Hubbell JH, Kissel LD. EPDL97: The Evaluated Photon Data Library, '97 Version. Lawrence Livermore National Laboratory 1997.
5. Everett CJ, Cashwell ED. MCP Code Fluorescence-Routine Revision. Los Alamos National Laboratory 1973.
6. Biggs F, Mendelsohn LB, Mann JB. Hartree-Fock Compton profiles for the elements. *At. Data Nucl. Data Tables*. 1975;16(3):201–309.

7. Berger MJ. Monte Carlo calculation of the penetration and diffusion of fast charged particles. In: Alder B, Fernbach S, Rotenberg M, eds. *Methods in Computational Physics*, Vol. 1. New York: Academic Press; 1963: p. 135.

8. Halbleib J. Structure and operation of the ITS code system. In: Jenkins TM, Nelson WR, Rindi A, eds. *Monte Carlo Transport of Electrons and Photons*. New York: Plenum Press; 1988.

9. Seltzer SM, Berger MJ. An overview of ETRAN Monte Carlo methods. In: Jenkins TM, Nelson WR, Rindi A, eds. *Monte Carlo Transport of Electrons and Photons*. New York: Plenum Press; 1988: p. 153.

10. Goudsmit S, Saunderson JL. Multiple scattering of electrons. *Phys. Rev.* 1940;57:24–29.

11. Landau L. On the energy loss of fast particles by ionization. *J. Phys. (USSR)*. 1944;8:201.

12. Blunk O, Leisegang S. Zum Energieverlust schneller Elektronen in dünnen Schichten.*Z. Physik*. 1950;128:500.

13. Schaart RD, Jansen JT, Zoetelief J, de Leege PF. A comparison of MCNP4C electron transport with ITS 3.0 and experiment at incident energies between 100 keV and 20 MeV: Influence of voxel size, substeps and energy indexing algorithm. *Phys. Med. Biol.* 2002;47(9):1459.

14. Munck af Rosenschöld PM, Verbakel WF, Ceberg CP, Stecher-Rasmussen F, Persson BR. Toward clinical application of prompt gamma spectroscopy for *in vivo* monitoring of boron uptake in boron neutron capture therapy. *Med. Phys.* 2001 May;28(5):787–795.

15. Schwarz AL, Schwarz RA, Carter LL. MCNP/MCNPX Visual Editor Computer Code Manual, For Vised Version 22S. Los Alamos National Laboratory 2008.

10

The EGS Family of Code Systems

Scott J. Wilderman and Yoshihito Namito

CONTENTS

The electron gamma shower (EGS) Monte Carlo particle transport simulation code system was originally developed by researchers at the Stanford Linear Accelerator Center as a platform for modeling high-energy physics experiments, and through the release of version 3 of the code (EGS3) in 1978 [1], developmental work on EGS focused on modeling physics processes at very high energies. This changed with the release of EGS4 in 1985 [2], whose authors cited the "many requests to extend EGS3 down to lower energies" as a major reason for the creation of EGS4 [2]. Much of the impetus for this extension of the dynamic range of EGS was provided by medical physics investigators, who also contributed a great deal of the effort required to modify the code system for low-energy simulations.

After the release of EGS4, interest in the use of the EGS package as a general-purpose Monte Carlo platform exploded. A large number of researchers contributed improvements to the EGS4 models of both photon and electron transport, especially at low energies, and official updates to EGS4 were released periodically through version 4.3. After EGS4.3, development work on EGS split into two distinct paths led by different groups. The first, centered at the Ionizing Radiation Standards group of the National Research Council of Canada, released an updated version of EGS4 called EGSnrc in 2000 [3,4]. The second group, comprising EGS4 author Ralph Nelson from SLAC plus investigators at the Radiation Science Center, Advanced Research Laboratory at the High Energy Accelerator Research Organization (KEK) in Japan and at the University of Michigan, released EGS5 in 2005 [5]. It should be noted that even though the release dates of the two codes are given as 2000 and 2005, these dates are somewhat misleading. Both EGSnrc and EGS5 are distributed via the World Wide Web, which allows both the codes and the manuals for both systems to be updated almost continuously. Indeed, both NRCC-PIRS-701 and SLAC-R-730 were updated in 2010, and even though it was initially described as "the new EGS4 version" [3], the current version of EGSnrc retains only a small fraction of the initial EGS4 source code and its underlying electron transport algorithm bears scant resemblance to that in EGS4.

The remainder of this chapter will first briefly describe the low-energy models in EGS4 (including the extensions to the package following its initial release), which are relevant to both EGSnrc and EGS5. The report will then go on to discuss the models of EGSnrc and EGS5, focusing on their differences at energies relevant to nuclear medicine, especially their quite distinct approaches to electron transport. General information about the structure and use of the two programs will then be provided, followed by the description of some relevant applications of EGS to nuclear medicine problems.

Low-Energy Electron Transport in EGS4

The major problem in simulating particle transport at low energies with EGS3 was that default electron transport step sizes, though appropriate at higher energies (i.e., above 10–20 MeV), were so long that many of the approximations used in the basic transport modeling algorithm no longer held. Because it is necessary for understanding the deficiencies of the EGS4 and the motivation for the different approaches taken by EGSnrc and EGS5 in surmounting them, we present here a brief summary of the electron transport algorithm of EGS4.

Because electron scattering cross sections (both elastic and inelastic) are so very large, it is almost never practical to model each collision in an individual or analog fashion. Instead, the cumulative effect of all collisions occurring over a finite distance is treated as having occurred during a single step. This method is known as the "condensed history" approach [6], and is illustrated in Figure 10.1.

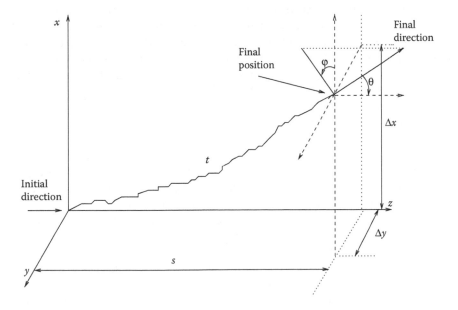

FIGURE 10.1
Schematic of a condensed history electron transport step.

Suppose an electron starting at the origin traveling along the z axis is scattered a large number of times as it traverses a pathlength t. When it reaches the end of that step, it will have lost energy ΔE, its position will be given by Δx, Δy, and s, and its direction by the angles θ and φ.

In a condensed history Monte Carlo simulation, only t (or ΔE, from which t is derived) is known (or assumed) initially, and s, Δx, Δy, θ, and ΔE (or t) are estimated. Mechanically, then, a condensed history step involves translating the particle's coordinates by Δx, Δy, and s, decrementing its energy by ΔE and updating its direction vector from computed values of θ and φ (which is random). Obviously, the accuracy of the simulation will be determined by the fidelity with which the calculated/estimated components of the simulated transport reflect the true transport. Typically, ΔE is determined from t (or t from ΔE) using the continuous slowing-down approximation. The angle θ is usually sampled from an analytical "multiple scattering" distribution function which models the aggregate scattering of an electron of a given energy traveling through a pathlength t. Values of Δx, Δy, and s are usually estimated from t, θ, and ΔE. The models and methods used by Monte Carlo pfogram in determining these values and in transporting partides through given t has come be known as the algorithm's "transport mechanics," a term coined by Bielajew and coworkers [7,8].

The EGS3 transport model used very long pathlengths t, employed an approximate high-energy theory in determining s, and ignored the lateral deflections Δx and Δy. All of these approximations are for the most part valid at very high energies where the average multiple scattering angle is small, even over large paths. At lower energies, where the elastic scattering cross section is not only orders of magnitude larger, but is also less preferentially forward, ignoring lateral deflection over long pathlengths leads to glaring inaccuracies. In addition, the multiple scattering distribution of EGS was derived in a small angle approximation, which is not valid at low energies for the long steps which were the EGS3 defaults.

These issues were first addressed in EGS3 by Rogers [9] in his paper "Low Energy Electron Transport With EGS." In that work, Rogers introduced a method for permitting user control of electron step size by the specification of both a maximum allowable fractional energy loss over a step due to continuous loss processes and also a maximum user-defined geometric step. These and other changes geared toward improving the accuracy of EGS at lower energies were incorporated into the initial release of EGS4.

By taking shorter steps at lower energies, the deleterious effects of ignoring the lateral deflection and employing an inappropriate model for computing the relationship between s and t were somewhat mitigated in the initial version of EGS4. But use of a constant fractional energy loss for determining t frequently led to simulations taking extremely small steps at the lowest energies of the problem, resulting in very long run times. Perhaps the most important modification to EGS4 after its initial release was the introduction of the "PRESTA" transport mechanics by Bielajew and Rogers [10], which permitted accurate simulation over much larger step sizes by including a treatment of the lateral deflections Δx and Δy (after Berger [6]) and by more accurately modeling s, among other things.

But even PRESTA had limitations. A perfect transport mechanics model will produce identical results regardless of the initial choice of t. Simulations using imperfect models yield different results for different choices of t, and converge to the result from a single-scattering model as t decreases (Larsen convergence [11]). As results from simulations using the PRESTA algorithm sometimes exhibited a fairly strong dependence on step size, improving upon the PRESTA transport mechanics was a focus of both EGSnrc and EGS5.

Models of Basic Physics Processes in EGSnrc and EGS5

The physics models of EGSnrc and EGS5, many of which were derived from or are related to those in EGS4 and so can sometimes be quite similar, are summarized here.

Electron Transport Models

As noted earlier, the basic electron transport approach in EGS is typically referred to as the condensed history model. More specifically, within the condensed history framework, the EGS codes treat collisions which result in the creation of energetic secondary particles as discrete interactions, an approach typically referred to as the "Class II condensed history model" [6]. Thus the primary components of the EGS system electron transport models are the cross sections used in modeling the discrete collisions, the stopping powers (which model the collective effect of the small energy loss collisions), the multiple scattering distributions (for the collective effect of the many elastic scattering collisions), and the transport mechanics (for piecing the other components together). The most substantive differences between EGSnrc and EGS5 lie in the multiple scattering distributions they use and in their respective treatments of transport mechanics, which will be covered separately in the next section.

Discrete Collisions with Atomic Electrons

Electron–electron collisions which result in the ejection of an energetic target electron (assumed to be initially at rest) are modeled in both EGSnrc and EGS5 by using the Møller cross section, as implemented by Bielajew and Rogers [12]. The analogous positron–electron interaction (Bhabha scattering) is likewise treated similarly in both codes, again following Bielajew and Rogers.

Electron Impact Ionization

Both EGS5 and EGSnrc provide options for explicit treatment of the effect of atomic binding on electron–electron collisions, which is important when it is desirable to model atomic relaxation subsequent to the creation of inner shell vacancies.

EGS5 follows the EGS4 modifications of Namito and Hirayama [13,14] and provides options for using any of the following cross sections for treating K-shell electron-impact ionization (EII):

1. Casnati [15,16]
2. Kolbenstvedt (original) [17]
3. Kolbenstvedt (revised) [18]
4. Jakoby [19]
5. Gryziński [20–22]
6. Gryziński (relativistic) [20–22]

When the option to treat EII explicitly is invoked in EGSnrc, by default the cross section used is the semiempirical formulation of Kawrakow [23], which permits treatment of any K-shell or L-shell vacancy above 1 keV and provides better agreement with the experimental data used by Casnati. The EGSnrc implementation explicitly corrects the differential cross section so as to preserve stopping powers. Options are also provided for using the cross sections of Casnati, Gryziński, or Kolbenstvedt, taken from the respective original sources.

Discrete Bremsstrahlung Interactions

Following the review article of Koch and Motz [24], below 50 MeV, both EGS5 and EGSnrc use Bethe-Heitler (first Born approximation) cross sections with empirical correction factors. EGS5 retains the EGS4 sampling scheme, while EGSnrc uses a new implementation which corrects some of the errors in the EGS4 method. In both codes, total cross sections are normalized so that their energy integrals (which are the radiative stopping powers) are consistent with those recommend in ICRU Report 37 [25]. This normalization, first implemented in EGS4 by Rogers et al. [26], can be quite significant at energies below a few MeV, where bremsstrahlung production is very small [27]. EGSnrc employs updated version of the normalization, using more densely spaced tables for improved interpolation.

EGSnrc supports two additional options for bremsstrahlung cross sections. The first is based on the NIST bremsstrahlung cross section database, which uses partial wave cross sections of Pratt et al. [28–30] below 2 MeV. The second option applies to electron–electron bremsstrahlung and is taken from an exact first Born approximation calculation by Tessier and Kawrakow [31].

In both EGS5 and EGSnrc, the angular distribution of ejected photons is taken from the Schiff formula of Koch and Motz [24]. EGS5 follows the EGS4 implementation of Bielajew et al. [32], while EGSnrc employs its own, new, faster method.

Two Photon Positron Annihilation

Both EGS5 and EGSnrc use the EGS4 prescription for treating positron annihilation, which is taken from pp. 268–270 of Heitler [33].

Stopping Powers

Stopping powers in both EGSnrc and EGS5 are derived from the EGS4 model which was based on the Bethe theory [34] as formulated by Berger and Seltzer [35]. Duane et al. [36] updated the EGS4 system so that its electron–electron and positron–election stopping powers would be identical to those in ICRU Report 37 [25], and both EGSnrc and EGS5 incorporate that work. Similarly, Rogers et al. [26] modified the EGS4 package so as to make its radiative stopping powers (for treating the aggregate effect of low-energy bremsstrahlung collisions) consistent with those in ICRU 37.

Because the collisional stopping power formalism relies on semiempirical parameters (the mean ionization energy, I, and density effect, δ) which may not be adequate for detailed simulations, especially at low energies and for high Z, both EGSnrc and EGS5 retain options from EGS4 for user input of either the parameters themselves or constants used in determining the parameters.

Multiple Elastic Scattering

EGS4 employed Moliére's [37] small-angle multiple scattering distribution as formulated by Bethe [38] in determining the polar scattering angle of an electron having undergone a larger number of elastic scattering collisions while traveling a given transport distance. The Moliére theory has several shortcomings, however. First, the underlying single scattering cross section on which it is based (the screened Rutherford cross section) is inadequate for modeling large-angle scattering at low energies in high-Z materials, as well as in the MeV range, when spin and relativistic effects are important. Second, some of the approximations in its derivation are not valid for pathlengths of <20 elastic mean free paths. This restriction is not always met in Monte Carlo simulations, and can lead to significant errors.

EGSnrc circumvents the limitations of the Moliére theory by employing a model due to Kawrakow [3], who added an exact treatment of energy loss and spin and relativistic elastic scattering cross sections to a framework originally devised by Kawrakow and Bielajew [39]. The underlying algorithm is based on modeling the multiple scattering distribution as the combination of the differential single scattering cross section and the multiple scattering distribution resulting from two or more elastic scatters. By default, EGSnrc uses in this model the exact partial wave cross section of Mott [40] as developed by Riley [41], using Hartee–Fock electron densities based on Desclaux [42]. Users may elect to use instead a screened Rutherford cross section.

EGS5 uses the EGS4 implementation of Moliére as its default multiple scattering distribution function and relies on its transport mechanics model, described in the next section, to get around the problems with Moliére at small pathlengths. For simulations in which the screened Rutherford cross section is inadequate, EGS5 provides the exact multiple scattering distribution of Goudsmit and Saunderson [43], using an exact partial wave cross section which includes virtual orbits at subrelativistic energies, spin and Pauli effects in the near-relativistic range, and nuclear size effects at higher energies taken from an unpublished work of Salvat (F. Salvat, private communication, 2000). In this approach, multiple scattering distributions are precomputed for a number of different step sizes at a number of energies for each material in the simulation, and the cumulative distribution function is fitted to a readily sampled form [44] over a fine grid of angles.

Photon Transport Models

Both EGSnrc and EGS5 model photoelectric absorption, Compton scattering, pair production, and Rayleigh scattering, and both also provide the ability to model the emission of secondary particles arising from atomic relaxation subsequent to inner shell ionization. The models used by each code in simulating all of these processes are described here.

Pair Production

For modeling the production of electron–positron pairs by scattering of energetic photons off atomic electrons, EGS5 uses the EGS4 formulas taken from the review article of Motz, Olsen, and Koch [45], with an empirical correction (taken from Storm and Israel [46] below 50 MeV). EGS5 also includes (as an option) Bielajew's modification of EGS4 [47] for determining the angular distribution of the electron and positron pair using either the leading order term in the Sauter–Gluckstern–Hull distribution (Equation 3D-2000 of Motz et al. [45]) or the Schiff angular distribution (Motz et al., Equation 3D-2003 [45]). Note that because of the use of the extreme relativistic approximation, the EGS4 implementation of the differential cross section can become quite inaccurate at photon energies near the pair production threshold (twice the electron rest mass).

EGSnrc employs the EGS4 methodology for determining the pair production cross section, but uses an updated sampling scheme and also contains an option for a user-provided correction at low energies. In addition, an option is available for using instead a library of differential cross sections based on an exact partial-wave analysis of Øverbø, Mork, and Olsen [48]. This tabulation explicitly treats asymmetry in the energy distribution of the electron–positron pairs, which can be quite significant for energies below 10 MeV.

EGSnrc also permits the explicit treatment of triplet production, which occurs when the atomic electron is ejected during a pair production interaction, using an implementation based on the derivation of Mork [49].

Compton Scattering

Although they take different approaches, both EGSnrc and EGS5 have extended the EGS4 treatment of Compton scattering by including options to model electron-binding effects, polarization, and Doppler broadening of the ejected electron energy spectra, which can have a pronounced impact at lower energies [50].

Electron-binding effects are treated by multiplying the Klein–Nishina cross section by the incoherent scattering function for either an atom as a whole or for each atomic subshell. EGS5 incorporates the work of Namito and Hirayama [51], which uses the Waller–Hartree theory [52] to compute tables of incoherent scattering functions. The EGS5 treatment of Doppler broadening of Compton electron energy spectra, which is significant at diagnostic nuclear medicine energies, is based on the impulse approximation model described in Ribberfors and Berggren [53] and originally implemented in EGS4 by Namito et al. [54], again using tabulated data. Explicit modeling of these effects (on a shell-by-shell basis) is available (as an option) in EGS5. Namito et al. [55] also introduced a method for modeling Compton (and Rayleigh as well) scattering of polarized photons into EGS4, and EGS5 has retained that methodology.

Even though both EGS5 and EGSnrc use methods based on the same impulse approximation theory, the implementation of binding effects and Doppler broadening in EGSnrc is quite different. EGS5 first samples the scattering angle from the singly differential cross section and then samples the shell number before determining the photon energy from the Compton profile. In contrast, EGSnrc first samples the shell number and then samples the scattering angle and momentum (from which the energy is computed) from the doubly differential cross section for the given shell, a method based on an analytical approximation for the Compton profiles suggested by Brusa et al. [56]. This detailed treatment is the default in EGSnrc, as is the generation of any fluorescent photons and Auger electrons which might result from the relaxation of a vacancy produced in an inner shell Compton collision. EGSnrc also provides an option for including radiative corrections.

Rayleigh Scattering

Both EGS5 and EGSnrc model Rayleigh scattering, taking the total cross sections from Storm and Israel [46], with atomic form factors derived from

Hubbell and Øverbø [57]. EGS5 follows the EGS4 methodology, while EGSnrc uses its own implementation. Both EGS5 and EGSnrc provide an option for user input of form factors, and EGS5 permits modeling of the effects of the incident photon polarization on coherent scattering, after Namito et al. [55].

Photoelectric Effect

EGS5 uses the PHOTX [58] library of total photoelectric cross sections, which was originally implemented in EGS4 by Sakamoto [59]. To ensure accuracy in the total cross section near subshell edge energies, EGS5 employs the LEM method of Namito and Hirayama [60], first developed for EGS4. Additionally, a method devised by Hirayama and Namito [61,62] for EGS4, based on the PHOTX data, for determining the subshell of interaction in a photoelectric absorption, has been incorporated into EGS5. EGSnrc uses the total photo-absorption cross section library of Storm and Israel [46] (as did EGS4), though provision is made to allow users to use any data they wish. Cross sections from the Evaluated Photon Data Library (EPDL) [63] and XCOM [64] are provided with the package. To resolve interactions into subshells, EGSnrc first uses cross section values fit to data taken from XCOM to determine the element involved, and then uses precomputed approximate interaction probability data to select the subshell within the given atom. Both EGSnrc and EGS5 employ the EGS4 method of Bielajew and Rogers [65], based on the theory of Sauter [66], for determining the angle of emitted photoelectrons.

Atomic Relaxation

Standard EGS4 did not create or transport fluorescent photons or Auger electrons emitted in atomic relaxation processes, which can be important in certain nuclear medicine applications, such as detector modeling. Both EGS5 and EGSnrc have remedied this deficiency. The work of Hirayama and Namito [62], used in EGS5 to resolve photoelectric absorptions into subshells, also provides data and methodology for modeling generalized atomic relaxation of K- and L-shell vacancies, and has been incorporated into EGS5. EGS5 models 20 possible transitions (those with intensity of at least 1% of the L_{α_1} transition for Fermium ($Z = 100$)) during the relaxation of L-shell vacancies. EGSnrc explicitly models the relaxation of all K- and L-shell vacancies with binding greater than 1 keV, and treats M- and N-shell vacancies in an average way.

Electron Transport Mechanics of EGSnrc and EGS5

The schematic representing the condensed history electron transport algorithm shown earlier (Figure 10.1) actually presents an incomplete picture of

Class II condensed history transport mechanics. In practice, the step could have been interrupted by a large energy loss collision at any point, and the distance traveled t could range from zero to the selected step size. When such "catastrophic" collisions occurred in EGS4, the step size was simply reduced to the distance traveled prior to the collision and the transport mechanics were then applied. Note that this could sometimes lead to very small steps, for which the Moliére distribution may not be valid. Another potential problem in the implementation of condensed history schemes arises in that since electrons continuously lose energy through low transfer collisions during a transport step, the cross section for catastrophic collisions changes between the initial step and collision, thus introducing an inaccuracy in the collision distance.

Transport near to and along the boundary between two materials presents an even more difficult problem. Because the actual path is meandering, there exists the possibility that the particle will travel in media on both sides of the boundary for a portion of its path during a single step. To get around this problem, Bielajew and Rogers introduced a boundary crossing algorithm as part of the PRESTA [10] update to EGS4. In that work, the total path t was limited to the lesser of the maximum step for which the Moliére theory (which was derived using a small angle approximation) is valid and the perpendicular distance to the nearest boundary (this ensures that an electron will not cross a region boundary during a step). Since a particle would never reach a boundary under an exact implementation of such a scheme, a restriction was placed on the degree of truncation of the step due to the perpendicular distance limit, based on the minimum distance for which Moliére is valid. When this restriction was invoked, the particle was transported exactly to the boundary, where multiple scattering was applied but the lateral displacements Δx and Δy were ignored.

While the PRESTA boundary crossing algorithm represented a decided improvement, it was not able to fully overcome all of the limitations inherent in the fundamental algorithmic structure of the EGS4. Foote and Smyth demonstrated that applying the multiple scattering deflection exactly at a region's bounding surface can lead to singularities in simulated electron fluences [67]. In addition, it was found that once energy loss was incorporated, the algorithms in PRESTA slightly underestimated both lateral and longitudinal displacement [68]. Because of the deficiencies of EGS4/PRESTA models of boundary crossing and lateral and longitudinal displacement, both EGS5 and EGSnrc have completely replaced the EGS4 transport mechanics scheme.

EGSnrc Transport Mechanics

Because the multiple scattering distribution of EGSnrc is accurate at any pathlength, there is no minimum limit for steps taken near boundaries. It was found, however, that once an electron is within three mean free paths of a boundary, it is more efficient to treat each elastic scattering event individually than to sample the multiple scattering distribution, and so EGSnrc switches to single scattering mode near boundaries. (An option is also provided to run entire simulations in single scattering mode.)

To treat lateral and longitudinal displacements, EGSnrc uses a method based on randomly splitting each condensed history step into two substeps, determining and then combining the multiple scattering angles over each substep, and then computing the displacements Δx, Δy, and s by an accurate model of Kawrakow [3], based on a method from Kawrakow and Bielajew [69]. Kawrakow's model preserves both first- and second-order spatial moments (e.g., $\langle z \rangle$, $\langle z^2 \rangle$, $\langle r^2 \rangle$) to better than 0.1% if the fractional electron energy loss is <20–25%. To handle the issue of cross sections for discrete interactions which vary over transport steps, EGSnrc employs the well-known method of "fictitious cross sections," in which the cross section over a step is taken to be a constant which is given by the sum of the actual cross section and a cross section for a fictitious event, in which no scattering occurs. To improve the efficiency of this technique (which is exact) at low energies (when the discrete interaction cross section can vary by orders of magnitude over a step, yielding a preponderance of fictitious events), EGSnrc uses the cross section per unit energy loss (the cross section divided by the stopping power).

EGS5 Transport Mechanics

EGS5 takes a completely different approach to boundary crossing and transport mechanics, generalizing the random hinge methodology of PENELOPE [70] into a "dual random hinge" method, in which energy loss and multiple elastic scattering are fully decoupled [71]. The multiple scattering random hinge is illustrated in Figure 10.2. Instead of being transported by Δx, Δy, and s in a single step, an electron is transported along its direction vector in two steps. After being transported through the first step (the hinge point), it is deflected, with θ determined from the multiple scattering distribution for electrons traveling the full path t. It is then transported linearly through the remaining distance. It can be shown that if the hinge points are taken at distances which correspond to randomly distributed fractions of K1, the "scattering strengths," (defined as the integral over t of the first transport mean free path, which is commonly referred to as the "scattering power"), near second-order spatial moments of the transport equation are preserved over long steps [69]. No explicit lateral displacement treatment is required.

Random energy loss hinges are similarly applied in EGS5, superimposed on the multiple scattering hinges. Electrons traveling through an energy loss step ΔE are transported until they travel a distance after which they have lost a randomly chosen fraction of ΔE. This distance defines the energy hinge point, at which the electron energy is decremented by the full ΔE. The particle is then transported the distance required to lose the remaining fraction

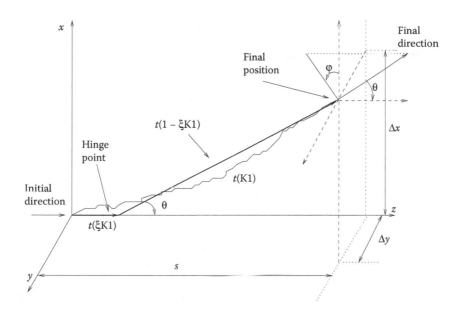

FIGURE 10.2
Schematic of a random hinge electron transport step.

of ΔE. Note that the electron is considered to travel with constant energy during each part of the hinge: its initial energy prior to the hinge, and its final energy through the residual portion of the hinge. Note also that as the hinges are decoupled, energy hinges occur sometimes before and sometimes after multiple scattering hinges. Since the most recently updated energy is used when computing the variables required to calculate the deflection, there is an implicit averaging of these variables. This implicit energy averaging also applies to the variation in the discrete cross sections over the hinge steps, though the averaging in this case is not exact.

In addition to permitting long transport steps, this method can be formulated so as to permit transport across boundaries between regions of differing media. Because the electrons are traveling linearly and at constant energy between hinge points, when a boundary is encountered, the particle can simply be momentarily stopped, and the distances remaining to the next hinges updated from the remaining scattering strength and energy loss, using the stopping and scattering powers of the new medium. This precludes the need to compute a perpendicular distance to the nearest boundary to limit multiple scattering steps. Such computations can be extremely expensive for boundaries defined by higher-order surfaces.

The random hinge method involves transporting all electrons in their initial direction before any deflection is imposed. Thus, problems which have a strong spatial or directional sensitivity to events occurring prior to the first multiple scattering hinge (such as deep penetration shower simulations) may exhibit a hinge artifact not present in EGS4, since EGS4 always imposed multiple scattering prior to discrete events and boundary crossings. To overcome this problem, a mechanism has been introduced into EGS5 to force small multiple hinges (typically the smallest value of K1 for which Moliére is valid) at the beginning of primary electrons histories. Subsequent steps are increased in size until the default hinge step is reached.

EGSnrc and EGS5 Code Packages

One of the most significant features of the EGS system is that it completely decouples the particle transport algorithm (germane to every possible application) from the source, geometry, and tallying processes, which can be unique to each specific application. The main advantages of this approach are that the program is very efficient, as it is not encumbered by having to support a multitude of options in an attempt to treat all possible applications, and that it is completely flexible with respect to the types of problems to which it can be applied. The main disadvantage with this approach is that each problem requires its own set of source code for modeling primary source particle generation, computing geometric distances to boundary, and tallying and scoring. These problem-specific source codes, which can be non-trivial for some applications, must be installed by the user (hence they are commonly referred to as "user codes").

To guide users through the steps necessary to construct and run applications which exploit the various features of the package, EGS releases have always included a set of tutorial programs. In addition, starting with EGS4, system releases have also included examples of user codes required to attack more sophisticated problems, as well as generalized user codes which support specific geometries (Cartesian grids and cylindrical, for example) and so can be readily adapted by users in lieu of writing their own source code for problems with conforming geometries. Additionally, both EGSnrc and EGS5 provide ancillary programs and packages with visual interfaces which permit users to quite readily construct complex geometries. Further, the more recent releases of EGSnrc provide two additional packages which greatly assist in the development of user codes. In 2006, a multiplatform package, EGSnrcMP, was released, which includes a graphical interface which permits user to access, compile, and run EGSnrc user codes interactively. And as of 2009, releases of EGSnrc have included a set of C++ class libraries from which user codes for modeling complex problems can readily be constructed.

Though it is not the intent of this chapter to reproduce the respective EGS5 and EGSnrc user manuals, in this section, we will summarize the procedures for setting up and running the programs and review some of the requirements, options, and features of both systems.

Installation and Structure of Packages

EGS was developed for command-line operating systems, especially Unix/ Linux, with the compilation of source code and the running executable performed using a series of scripts. While EGS5 retains this basic structure as its default configuration, the current version of EGSnrc (beginning in 2006 with the release of EGSnrcMP [72]) is structured quite differently. While the command line-based structure still exists in EGSnrc (though the scripts used in EGS4 for compiling and running user code have been replaced by the Unix/ Linux make utility), it can be completely hidden from users who elect to run a generic graphical interface, egs_gui, provided with the package. This interface permits users to completely control execution, including specification of materials, compilation of user code, generating of input data files, and so on.

Both EGSnrc and EGS5 can also be run under Windows. EGSnrcMP exploits the functionality of "make" in creating a facility for installing and running EGSnrc on Windows platforms while retaining all the features of the Unix/Linux EGSnrc environment. Though not part of the main EGS5 distribution, KEK supports a publicly available method for running EGS5 on Windows which retains the basic EGS5 command-line structure, employing the Windows program "junction.exe" to mimic Linux/Unix symbolic links. Both systems are installed by downloading archives from the respective websites:

http://irs.inms.nrc.ca/software/egsnrc/download.html

and

http://rcwww.kek.jp/research/egs/egs5.html

The EGSnrc package can be installed either by executing a self-extracting setup and installation wizard available from the EGSnrc website, or by downloading and executing setup scripts which control extraction, installation, and compilation (if necessary) of the files in the archive. EGSnrc can be set up for single users or to allow multiple users to access the same code base. As mentioned above, users can elect to use the EGSnrcMP GUI to select, set up, compile, and run EGSnrc jobs without ever-running scripts from a command line. Alternatively, compilation of EGSnrc source codes can still be performed from the command line using the make utility, with simulations run by simply invoking the name of the executable.

EGS5 is tailored primarily for single users, and so "installation" is a simple matter of extracting the EGS5 files from the archive. A single script, "egs5run," controls the compilation and execution of EGS5, and the user is required to modify this script to specify the directory containing the EGS5 source, compiler, and compilation options. At this point, it must be noted that both EGSnrc and EGS5 are proprietary, copyrighted systems, though free licenses are granted for use for educational and noncommercial research purposes.

Preparing Material Data Files with PEGS

As alluded to earlier, to preclude extensive computations at run time, much of the data required by EGS5 and EGSnrc is computed prior to simulations being run (though it should be noted that in its current incarnation, EGSnrc computes most electron physics data on the fly). In EGS, these preparatory computations are performed by PEGS. Historically, PEGS was provided as a self-contained program so that data files generated for a given material could be examined by users interested in investigating the physics data associated with the material (such as stopping powers, cross sections, etc.) and also reused in subsequent problems using the given material, provided that the input parameters used in generating the original PEGS data file are compatible with the input specifications of the EGS run. EGSnrc retains the stand-alone nature of PEGS, but in EGS5, because some of the variables and functions computed in PEGS, especially those involved in step size optimization, are dependent on the energy range of each given problem, it was necessary to couple EGS5 with its version of PEGS, PEGS5. EGS5 is still capable of reusing existing PEGS5 data files, but since the computation time required to generate PEGS data files is usually trivial relative to that consumed in the transport simulation (except when the Goudsmit–Saunderson multiple scattering distribution is employed), there is almost no penalty for recomputing PEGS5 data files.

Despite this modification of PEGS in EGS5, the format of PEGS input files is the same in EGS5 and EGSnrc, taken from PEGS4. (Note though that EGSnrc-PEGS and PEGS5 output files are not compatible.) Materials are defined and options specified using the FORTRAN NAMELIST I/O extension. Note that many of the features of EGS5 and EGSnrc described in the previous sections as being options must be defined in the PEGS material specifications. As with user code compilation and execution, EGSnrc provides both a graphical interface and a command line facility for constructing PEGS data files. The GUI version precludes the need for users to construct NAMELIST input files, but it does not present ways to access all of the options which are built into PEGS.

Basic Structure of an EGS User Code

At a minimum, user codes for both EGSnrc and EGS5 must consist of a MAIN program and subroutines HOWFAR (for geometry) and AUSGAB (for scoring). EGSnrc also requires a second geometry subroutine, HOWNEAR. MAIN interacts with the EGS code base by manipulating the values of COMMON block variables and by calling EGS subroutine HATCH to read PEGS data and EGS subroutine SHOWER to initiate each particle history. MAIN must also perform initializations specific to the application, including specification of the materials to be used prior to calling HATCH. Additionally, in EGSnrc, MAIN must call an initialization routine egs_init for file initializations, while in EGS5, MAIN must call a new initialization routine BLOCK_SET, which is used to initialize data not defined in BLOCK DATA subprograms because the arrays are too large. Also in EGS5, if the PEGS data files required for the given simulation do not exist, MAIN must call PEGS5 prior to calling HATCH. After all initializations and the call to HATCH, MAIN then controls the simulation by recursively calling SHOWER after first defining the state variables and geometric region of each primary particle. HOWFAR is called by various EGS subroutines and returns the straight-line distance to the nearest problem boundary for a particle's given position and direction, which are passed into HOWFAR through its argument list or through EGS COMMON blocks. HOWFAR must also determine new particle regions after boundary crossings, and control particle history termination. HOWNEAR is required by EGSnrc to define the distance to nearest problem boundary for a particle's given position in any direction. Calls to AUSGAB are embedded in the various physics and tracking subroutines of EGS at a variety of places, but are executed only if requested by the user through the initialization in MAIN of certain variables. Scoring for the problem at hand is performed in AUSGAB by interrogating the values of various EGS COMMON block variables. Additionally, AUSGAB may be used to modify many EGS variables, at the discretion of the user.

Writing EGSnrc User Code

As noted earlier, when users wish to model problems that fall outside of the capability of the sample codes provided, they must provide their own user code. The steps required to construct an EGSnrc user code from scratch are given in the EGSnrc user manual as follows [23]:

1. call egs_init for file initialization
2. User Over Ride Of EGS Macros and Defaults
3. Pre-HATCH Call Initialization
4. HATCH Call
5. Initialization For HOWFAR & HOWNEAR
6. Initialization For AUSGAB
7. Initialization For Variance Reduction
8. Determination Of Incident Particle Parameters
9. SHOWER Call
10. Output Of Results

Macros, MORTRAN

EGSnrc inherits from EGS4 a preprocessing and macroprocessing language known as "MORTRAN," in which user codes are written. The primary advantage of the use of MORTRAN, which is described in the EGS4 manual [2], is that default EGSnrc parameters, function calls, and even full procedures can be readily replaced by user-written MORTRAN "MACROS," tailored to the specific problem being studied.

EGSnrc C++ Library

As noted earlier, in 2009, a set of C++ EGSnrc libraries were developed and made available as "the availability of flexible and general geometry and particle source packages would greatly reduce the amount of work needed to perform simulations that cannot be done with the standard set of user codes" [73]. In the current release of EGSnrc, therefore, the expectation is that most user code will be written in C++, taking advantage of the C++ class library included in the distribution and described in PIRS-898. Thus users may be completely absolved from the need to delve into MORTRAN.

When using the C++ library, users derive their application class from one of the two application classes provided, one for simple applications, and one for modeling advanced problems. Users are still required to implement AUSGAB and construct the portions of MAIN which control problem output, and may also be required to provide other functions when modeling advanced applications. In almost all situations, however, source and geometry modeling can be defined in input files. Additionally, the EGSnrcMP package provides a C interface to EGSnrc, allowing user codes to be written in C or C++.

Step Size Control

One of the biggest drawbacks of EGS4 was the dependence of simulation results on the user selected value of the maximum fractional energy loss over an electron step, ESTEPE. Running EGS4 with optimal efficiency meant determining the largest possible step size (to speed execution) which still yielded the results converged upon at small step sizes. Optimal values of ESTEPE were almost never apparent *a priori*, and so had to be searched for by the user. Because of the development of its new transport mechanics and boundary crossing method, EGSnrc does not exhibit step size dependence for values of ESTEPE <25%. Therefore, while user control of electron step sizes is still available as an option, it is not a requirement of EGSnrc.

Additional Features

Some of the optional variance reduction routines typically used in EGS4 have been internalized and included by default in EGSnrc. These include bremsstrahlung splitting, Russian Roulette of secondary charged particles and range rejection.

The default random number generator in EGSnrc is RANLUX [74], though access to the EGS4 RANMAR [75] generator is retained as an option.

EGSnrc Sample User Codes

All or part of the above exercise of developing user code described above can, of course, be dispensed with if the problem being investigated can be handled with one of the generalized sample user codes provided with the EGSnrc package. In addition to the seven (plus variants) tutorial programs based on the EGS4 tutorials,* designed to demonstrate construction of basic user codes for MAIN, HOWFAR, HOWNEAR, and AUSGAB, several of the standard NRC user codes are distributed with the system. The most important ones, which can be readily applied to cylindrical geometry problems, are DOSRZnrc, FLURZnrc, CAVRZnrc, and SPRRZnrc. DOSRZnrc and FLURZnrc tally absorbed radiation dose and particle fluence, respectively, in cylindrical geometries. CAVRZnrc is tailored to ion chamber simulations, and SPRRZnrc computes stopping power ratios. These programs are described in detail in NRCC report PIRS-702 [76]. In addition, a user program called EXAMIN is provided for examining and plotting the contents of PEGS data files.

A graphical interface, egs_inprz [4], is available to assist in the construction of cylindrical geometry data files for the EGSnrc RZ user codes. C++ versions of two of the tutor programs, tutor2 and tutor7, are included in the distribution and demonstrate all of the steps required to construct EGSnrc user codes with the C++ library.

Writing EGS5 User Codes

The steps required to construct an EGS5 user code are given in the EGS5 user manual as follows [5]:

1. Pre-PEGS initializations
2. PEGS call
3. Pre-HATCH initializations
4. Specification of incident particle parameters
5. HATCH call
6. Initializations for HOWFAR
7. Initializations for AUSGAB
8. SHOWER call
9. Output of results

---

* The program "tutor6" has been substantially rewritten for EGSnrc.

The order in which steps 4, 6, and 7 are performed can be changed by the user, provided that they fall after step 1 and before step 8. In addition, step 2 may be skipped if an existing PEGS data file is to be used.

FORTRAN "include" Extension

Perhaps the biggest system change implemented in EGS5 is the departure from the use of MORTRAN and the conversion of the entire package into FORTRAN. Some of the functionalities of MORTRAN are retained by the use of the FORTRAN "include" extension, which is used to synchronize common blocks throughout all the EGS5 and PEGS5 subroutines (thus emulating the function of MORTRAN macros in specifying array dimensions). The common blocks of EGS5 are contained in a subdirectory of the EGS5 distribution called "include," and common blocks intrinsic to PEGS5 are found in a subdirectory called "pegscommons." The commons are accessed in user codes via symbolic links which are created by egs5run in each user code directory.

Step Size Control

Because of the use of the dual random hinge transport mechanics, EGS5 takes both multiple scattering steps (which are actually implemented in terms of scattering strength (K1) steps) and energy loss steps, meaning two sets of step sizes must be specified. Further, EGS5 permits users to specify step sizes for both types of hinges at both high and low energies, to get around the EGS4 problem of the small fractional energy loss steps required for high energies slowing the simulation without benefit at low energies. Thus, four fractional energy loss step sizes may be set by the user in EGS5.

Since this method of step size selection places even a greater burden on the user for optimizing the performance of the program, an alternative (and much simpler) method of step size control is provided. First, since the only purpose of the energy hinge in the dual-hinge approach is to provide accurate integration of energy-dependent variables over electron pathlengths, energy steps are computed in PEGS as the largest steps which assure energy integration within a specified tolerance (set by default to 1%). Second, new subroutines have been added to PEGS5 which pre-determine multiple scattering steps (in terms of K1) by selecting the longest steps yielding converged values of total tracklength of electrons impinging on cylinders of diameters and thicknesses corresponding to a "characteristic dimension" (CHARD) specified by the user for each medium in a problem. The characteristic dimension CHARD of a medium depends in a logical manner on the shapes of the regions containing the medium. For instance, the characteristic dimension of a semi-infinite slab would be the slab thickness, CHARD for a cylinder would be the minimum of the diameter and the length of the cylinder, and for a voxel geometry, CHARD would be the minimum voxel dimension.

Auxiliary Geometry and Scoring Routines

The EGS5 distribution contains a set of auxiliary subprograms (and files containing their associated common blocks) which can be used for some generic tallying and, more importantly, in modeling Cartesian, cylindrical, and combinatorial geometries (CG). The programs, which can be readily used in constructing subroutine HOWFAR, are accessed via symbolic links in each user code directory, similar to the manner in which the main EGS5 and PEGS5 source files and common block files are used.

To further assist in the construction of complex geometries, a graphical interface program, Cgview, though not part of the EGS5 distribution, is publicly available from the KEK website. Geometries created with Cgview can be stored in file formats which can be read in by the CG subroutines included in the EGS5 auxiliary codes.

EGS5 Sample User Codes

In addition to the EGS5 versions of the EGS4 tutorial codes, the EGS5 package includes five more sophisticated user codes: ucbend, which contains user code for transporting particles in magnetic fields; uccyl, which is a generalized cylindrical geometry program; uc lp, which is a version of uccyl with specialized variance reduction implemented; ucsampl5, described as an example of a "complete" EGS5 user code; and ucsampcg, which illustrates the use of the EGS5 combinatorial geometry package. Because the version of HOWFAR included with ucsampcg is generic, the often challenging task of geometry modeling can be performed in EGS5 rather easily by employing ucsampcg's HOWFAR to do boundary tracking computations in geometries created graphically with Cgview.

An additional user program, uc_examin, the EGS5 version of the original EGS4 EXAMIN program for extracting the contents of PEGS data files, is also provided.

Additional Features

As with EGSnrc, EGS5 has incorporated some standard EGS4 variance techniques internally. These include, bremsstrahlung splitting, splitting of x-rays generated from EII, and range rejection. In addition, sample user code uc–lp demonstrates the use of leading particle biasing, and sample user code uccyl contains an illustration of the manner in which importance sampling can be implemented in EGS5 user codes. Like EGSnrc, EGS5 employs RANLUX [74] as its random number generator.

Applications of EGS in Nuclear Medicine

A small selection of papers and conference proceedings describing the use of EGS4, EGSnrc, and EGS5 in various areas of nuclear medicine are presented here.

Internal Emitter Dosimetry

Sato et al. [77] used EGS5 to simulate the transport of ^{90}Y beta particles in CT-derived voxel phantoms in mice. Monte Carlo results compared favorably with experimental measurements, in which small fluorescent glass dosimeters were used to measure dose from a capsule of ^{90}YCl$_3$ solution implanted in a phantom roughly simulating the size of mice.

Prideaux et al. [78] used EGSnrc to compute patient-specific tumor dosimetry for thyroid cancer being treated with radio-iodine. Multiple registered CT/SPECT scans were used to define the patient anatomy and to determine 3D radionuclide activity distributions in the thyroid. With this data as input, EGSnrc was used to calculate simulated 3D dose rates, which were then used to compute the biological effective dose and uniform equivalent dose for the therapy.

Simulations of SPECT Imaging

Yokoi et al. used EGS4 to develop a 3D Brain SPECT Simulator (3DBSS) [79], from which various radioactivity distribution patterns, such as the cerebral blood flow pattern, the striatum image pattern, and the skull bone image pattern, can be generated. The simulation package uses the Zubal digital brain phantom to define brain geometry, and assumes ^{99m}Tc as the isotope. They found that the FWHM of the spatial resolution of imaging systems can vary by up to a factor of 2 depending on the collimator resolution, even though scatter was found to be independent of the collimator.

Hashimoto et al. used EGS4 to evaluate compensation for detector depth dependency in SPECT reconstruction of simulated images of a myocardial numerical phantom infused with ^{201}Tl [80]. EGS4 was used to generate projection data, which was reconstructed using an iterative maximum likelihood expectation maximization (MLEM) method, with the frequency–distance relation (FDR) method used to correct for depth dependency of the detector response.

Lanconelli et al. [81] used results of EGSnrc simulations to optimize the design of a dedicated SPECT breast imaging system. Their study included models of a breast phantom with various size tumors, three different collimators, and the NaI(Tl) detector. Signal-to-noise ratios and image contrast were computed for various tumor-to-background activity and simulated breast thicknesses for both tumor sizes and all three collimators, and images were reconstructed using the simulated data.

PET Simulations Using EGS

Adam and Karp used EGS4 to investigate the potential effectiveness of additional shielding in mitigating scatter contamination in 3D-PET [82]. In that work, the patient, the proposed shielding, and the detector were modeled as concentric cylinders of varying lengths and thicknesses, and various activity distributions were examined. They concluded that the use of septa is effective at reducing scatter contamination for large objects, and the use of external patient shielding is effective for objects with small diameters. They also pointed out that the degree of the effect is closely related to the value of low energy threshold for detection.

Kitamura and Murayama used EGS4 simulations of a 3D-PET system to investigate the effect of changing the detector ring diameter on count rate performance of GOS-DOI PET scanners [83]. Count rates derived from EGS4 using a standard count rate model were compared to count rate measurements taken from existing scanners to verify the model. Results from additional EGS4 simulations were then used to investigate the performance of the proposed, next-generation detector systems.

Palmer et al. [84] developed a semiempirical model of positron range distributions which were used to compute the range-blurring effect on system resolution for a variety of positron emitters. EGSnrc simulations of positron transport were used to validate the range distributions of the model, which was formulated to facilitate range-blurring corrections during reconstruction.

Zhang et al. [85] developed a whole-body PET image reconstruction platform which employs an ordered subset expectation maximization (OS-EM) algorithm with the system matrix generated by EGSnrc simulations of the scanner. Reconstructions using Monte Carlo-based system matrices produced better quality images (in terms of contrast recovery and contrast noise) relative to reconstructions performed with a standard OS-EM algorithm and with clinical software.

References

1. Ford RL, Nelson WR. The EGS Code System: Computer Programs for the Monte Carlo Simulation of Electromagnetic Cascade Showers (Version 3). Stanford Linear Accelerator Center, Report SLAC-210, 1978.
2. Nelson WR, Hirayama H, Rogers DWO. The EGS4 Code System. Stanford Linear Accelerator Center, Report SLAC-265, 1985.
3. Kawrakow I. Accurate condensed history Monte Carlo simulation of electron transport, Part I: EGSnrc, the new EGS4 version. *Med. Phys.* 2000;27:485–498.
4. Kawrakow I, Rogers DWO. The EGSnrc Code System. National Research Council of Canada, Report PIRS-701, 2002.

5. Hirayama H, Namito Y, Bielajew AF, Wilderman SJ, Nelson WR. The EGS5 Code System. Stanford Linear Accelerator Center, Report SLAC-R-730, 2005.

6. Berger MJ. Monte Carlo calculation of the penetration and diffusion of fast charged particles, In *Methods in Computational Physics*, Vol. I, Adler B, Fernbach S, Rotenberg M Eds. Academic Press, New York, 1963, 135–215.

7. Sempau J, Wilderman SJ, Bielajew AF. DPM, a fast, accurate Monte Carlo code optimized for photon and electron radiotherapy treatment planning dose calculations. *Phys. Med. Biol.* 2000;45:2263–2291.

8. Bielajew AF, Salvat F. Improved electron transport mechanics in the PENELOPE Monte-Carlo model. *Nucl. Instrum. Methods.* 2001;B173:332–343.

9. Rogers DWO. Low energy electron transport with EGS. *Nucl. Instrum. Methods.* 1984;227:535–548.

10. Bielajew AF, Rogers DWO. PRESTA: The Parameter Reduced Electron-Step Transport Algorithm for Electron Monte Carlo Transport. National Research Council of Canada, Report PIRS-0042, 1986.

11. Larsen EW. A theoretical derivation of the condensed history algorithm. *Ann. Nucl. Energy.* 1992;19:701–714.

12. Bielajew AF, Rogers DWO. Effects of a Møller cross section error in the EGS4 code. *Med. Phys. (Abstract).* 1996;23:1153.

13. Namito Y, Hirayama H. Implementation of the electron impact ionization into the EGS4 code. *Nucl. Instrum. Meth. A.* 1999;423:238–246.

14. Namito Y, Hirayama H. LSCAT: Low-Energy Photon-Scattering Expansion for the EGS4 Code (Inclusion of Electron Impact Ionization). High Energy Accelerator Research Organization (KEK) Japan, Report 2000-4, 2000.

15. Casnati E, Tartari A, Baraldi C. An empirical approach to K-shell ionisation cross section by electrons. *J. Phys. B.* 1982;15:155–167.

16. Casnati E, Tartari A, Baraldi C. An empirical approach to K-shell ionisation cross section by electrons. *J. Phys. B.* 1983;16:505.

17. Kolbenstvedt H. Simple theory for K-ionization by relativistic electrons. *J. Appl. Phys.* 1967;38:4785–4787.

18. Middleman LM, Ford RL, Hofstadter R. Measurement of cross sections for x-ray production by high-energy electrons. *Phys. Rev.* 1970;2:1429–1443.

19. Jakoby C, Genz H, Richter A. A semi-empirical formula for the total K-shell ionization cross section by electron impact. *J. Phys. (Paris) Colloq.* 1987;C9:487.

20. Gryziński M. Classical theory of atomic collisions. I. Theory of inelastic collisions. *Phys. Rev. A.* 1965;138:336–358.

21. Gryziński M. Two-particle collisions. II. Coulomb collisions in the laboratory system of coordinates. *Phys. Rev. A.* 1965;138:322–335.

22. Gryziński M. Two-particle collisions. I. General relations for collisions in the laboratory system. *Phys. Rev. A.* 1965;138:305–321.

23. Kawrakow I. Electron impact ionization cross sections for EGSnrc. *Med. Phys. (Abstract).* 2002;29:1230.

24. Koch HW, Motz JW. Bremsstrahlung cross-section formulas and related data. *Rev. Mod. Phys.* 1959;31:920–955.

25. ICRU. Stopping Powers for Electrons and Positrons. International Commission on Radiation Units and Measurements, Report ICRU-37, 1984.

26. Rogers DWO, Duane S, Bielajew AF, Nelson WR. Use of ICRU-37/NBS Radiative Stopping Powers in the EGS4 System. National Research Council of Canada, Report PIRS-0177, 1989.

27. Namito Y, Nelson WR, Seltzer SM, Bielajew AF, Rogers DWO. Low-energy x-ray production studies using the EGS4 code system. *Med. Phys.* 1990;17:557 (abstract).
28. Seltzer SM, Berger MJ. Bremsstrahlung spectra from electron interactions with screened atomic nuclei and orbital electrons. *Nucl. Instrum. Meth. B.* 1985;12:95–134.
29. Seltzer SM, Berger MJ. Bremsstrahlung energy spectra from electrons with kinetic energy from 1 keV to 10 GeV incident on screened nuclei and orbital electrons of neutral atoms with Z = 1–100. *At. Data Nucl. Data Tables.* 1986;35:345–418.
30. Pratt RH, Tseng HK, Lee CM, Kissel L, MacCallum C, Riley M. Bremsstrahlung energy spectra from electrons of kinetic energy 1 keV ≤ T1 ≤ 2000 keV incident on neutral atoms 2 ≤ Z ≤ 92. *At. Data Nucl. Data Tables.* 1977;20:175–209.
31. Tessier F, Kawrakow I. Calculation of the electron-electron bremsstrahlung cross-section in the field of atomic electrons. *Nucl. Instrum. Meth. B.* 2008;266:625–634.
32. Bielajew AF, Mohan R, Chui CS. Improved Bremsstrahlung Photon Angular Sampling in the EGS4 Code System. National Research Council of Canada, Report PIRS-0203, 1989.
33. Heitler W. *The Quantum Theory of Radiation.* Oxford: Clarendon Press;1954.
34. Bethe HA. Theory of passage of swift corpuscular rays through matter. *Ann. Physik.* 1930;5:325.
35. Berger MJ, Seltzer SM. Tables of Energy Losses and Ranges of Electrons and Positrons. National Aeronautics and Space Administration, Report NASA-SP-3012, 1964.
36. Duane S, Bielajew AF, Rogers DWO. Use of ICRU-37/NBS Collision Stopping Powers in the EGS4 System. National Research Council of Canada, Report PIRS-0173, 1989.
37. Moliére GZ. Theorie der Streuung schneller geladener Teilchen. I. Einzelstreuung am abgeschirmten Coulomb-Field. *Z. Naturforsch.* 1947;2a:133–145.
38. Bethe HA. Moliére's theory of multiple scattering. *Phys. Rev.* 1953;89:1256–1266.
39. Kawrakow I, Bielajew AF. On the representation of electron multiple elastic-scattering distributions for Monte Carlo calculations. *Nucl. Instrum. Meth. B.* 1998;134:325–336.
40. Mott NF. The scattering of fast electrons by atomic nuclei. *Proc. R. Soc. Lon.* 1929;A124:425.
41. Riley ME. Relativistic Elastic Electron Scattering from Atoms at Energies Greater than 1 keV. Sandia Laboratories, Report SLA-74-0107, 1974.
42. Desclaux JP. A multiconfiguration relativistic Dirac-Fock program. *Comput. Phys. Commun.* 1975;9:31–45.
43. Goudsmit SA, Saunderson JL. Multiple scattering of electrons. *Phys. Rev.* 1940;57:24–29.
44. Bielajew AF. Plural and multiple small-angle scattering from a screened Rutherford cross section. *Nucl. Instrum. Meth. B.* 1994;86:257–269.
45. Motz JW, Olsen HA, Koch HW. Pair production by photons. *Rev. Mod. Phys.* 1969;41:581–639.
46. Storm E, Israel HI. Photon cross sections from 1 keV to 100 MeV for elements Z = 1 to Z = 100. *Atom. Data Nucl. Data.* 1970;7:565–681.
47. Bielajew AF. Improved Angular Sampling for Pair Production in the EGS4 Code System. National Research Council of Canada, Report PIRS-0287, 1991.

48. Øverbø I, Mork KJ, Olsen A. Pair production by photons: Exact calculations for unscreened atomic field. *Phys. Rev. A.* 1973;8:668–684.
49. Mork KJ. Pair production by photons on electrons. *Phys. Rev.* 1967;160:1065–1071.
50. Hirayama H, Namito Y, Ban S. Effects of Linear Polarisation and Doppler Broadening on the Exposure Buildup Factors of Low Energy Gamma Rays. National Laboratory for High Energy Physics (KEK) Japan, Report KEK 93–186, 1994.
51. Namito Y, Hirayama H. Improvement of low energy photon transport calculation by EGS4–Electron bound effect in Compton scattering. *Presented at the meeting of the Japan Atomic Energy Society,* Osaka, Japan, 1991;401.
52. Waller I, Hartree DR. On the intensity of total scattering of X-rays. *Proc. R. Soc. Lon. Ser.-A.* 1929;124:119–142.
53. Ribberfors R, Berggren KF. Incoherent x-ray-scattering functions and cross sections $(d\sigma/d\omega')_{incoh}$ by means of a pocket calculator. *Phys. Rev. A.* 1982;26: 3325–3333.
54. Namito Y, Ban S, Hirayama H. Implementation of Doppler broadening of Compton-scattered photons into the EGS4 code. *Nucl. Instrum. Meth. A.* 1994;349: 489–494.
55. Namito Y, Ban S, Hirayama H. Implementation of linearly-polarized photon scattering into the EGS4 code. *Nucl. Instrum. Meth. A.* 1993;322:277–283.
56. Brusa D, Stutz G, Riveros JA, Salvat F, Fernandez-Varea JM. Fast sampling algorithm for the simulation of photon Compton scattering. *Nucl. Instrum. Methods.* 1996;A379:167–175.
57. Hubbell JH, Øverbø I. Relativistic atomic form factors and photon coherent scattering cross sections. *J. Phys. Chem. Ref. Data.* 1979;8:69–105.
58. PHOTX. Photon Interaction Cross-Section Library for 100 Elements. Radiation Shielding Information Center, Oak Ridge National Laboratory, Report DLC-136/PHOTX, 1995.
59. Sakamoto Y. Photon cross section data PHOTX for PEGS4 code. *Proceedings of the Third EGS Users' Meeting,* Japan, 1993, KEK, Japan.
60. Namito Y, Hirayama H. Improvement of the cross-section and branching-ratio evaluation in EGS4 in the energy interval which has an absorption-edge. *Proceedings of the Eighth EGS Users' Meeting,* Japan, KEK, Japan, 1999.
61. Hirayama H, Namito Y. Implementation of a General Treatment of Photoelectric-Related Phenomena for Compounds or Mixtures in EGS4. High Energy Accelerator Research Organization (KEK) Japan, Report 2000–3, 2000.
62. Hirayama H, Namito Y. Implementation of a General Treatment of Photoelectric-Related Phenomena for Compounds or Mixtures in EGS4 (Revised version). High Energy Accelerator Research Organization (KEK) Japan, Report 2004–6, 2004.
63. Cullen DE, Perkins ST, Rathkopf JA. The 1989 Livermore Evaluated Photon Data Library (EPDL) Lawrence Livermore National Laboratory, Report UCRL-ID-103424, 1990.
64. Berger MJ, Hubbell JH. XCOM: Photon Cross Sections on a Personal Computer. National Bureau of Standards, Report NBSIR87–3597, 1987.
65. Bielajew AF, Rogers DWO. Photoelectron Angular Distribution in the EGS4 Code System. National Research Council of Canada, Report PIRS-0058, 1986.
66. Sauter F. Über den atomaren Photoeffekt in der K-Schale nach der relativistischen Wellenmechanik Diracs. *Ann. Physik.* 1931;11:454–488.

67. Foote BJ, Smyth VG. The modeling of electron multiple-scattering in EGS4/ PRESTA and its effect on ionization-chamber response. *Nucl. Instrum. Meth. B.* 1995;100:22–30.
68. Kawrakow I. Electron transport: Longitudinal and lateral correlation algorithm. *Nucl. Instrum. Meth. B.* 1996;114:307–326.
69. Kawrakow I, Bielajew AF. On the condensed history technique for electron transport. *Nucl. Instrum. Meth. B.* 1998;142:253–280.
70. Fernandez-Varea JM, Mayol R, Baro J, Salvat F. On the theory and simulation of multiple elastic scattering of electrons. *Nucl. Instrum. Meth. B.* 1993;73:447–473.
71. Bielajew AF, Wilderman SJ. Innovative Electron Transport Methods in EGS5. *Proceedings of the Second International Workshop on EGS4,* Japan, 2000, KEK, Japan.
72. Kawrakow I, Mainegra-Hing E, Rogers DWO. EGSnrcMP: The multi-platform environment for EGSnrc. National Research Council of Canada, Report PIRS-887, 2006.
73. Kawrakow I, Mainegra-Hing E, Tessier F, Walters BRB. The EGSnrc C++ Class Library. National Research Council of Canada, Report PIRS-898 (Rev. A), 2009.
74. James F. RANLUX: A FORTRAN implementation of the high-quality pseudo-random number generator of Lüscher. *Comput. Phys. Commun.* 1994;79:111–114.
75. James F. *A Review of Pseudorandom Number Generators.* CERN-Data Handling Division;1988.
76. Rogers DWO, Kawrakow I, Seuntjens JP, Walters BRB. *NRC User Codes for EGSnrc.* Ottawa, Canada: National Research Council of Canada, 2010.
77. Sato Y, Yamabayashi H, Nakamura T. Estimation of internal dose distribution of Y-90 beta-ray source implanted in a small phantom simulated mice. *Proceedings of the 13th EGS Users' Meeting,* Japan, 2006, KEK, Japan.
78. Prideaux AR, Song H, Hobbs RF et al. Three-dimensional radiobiologic dosimetry: Application of radiobiologic modeling to patient-specific 3-dimensional imaging based internal dosimetry. *J. Nucl. Med.* 2007;48:1008–1016.
79. Yokoi T, Hashimoto T, Shinohara H. Development of EGS-Based 3D Brain SPECT Simulator (3DBSS). *Proceedings of the Third International Workshop on EGS4,* Japan, 2005, KEK, Japan.
80. Hashimoto T, Imae T, Usuda D, Momose T, Shinohara H, Yokoi T. Implementation and Performance Evaluation of Depth-Dependent Correction in SPECT for Myocardial Numerical Phantom: A Simulation Study Using EGS4. Paper presented at *Proceedings of the Third International Workshop on EGS4 2005;* KEK, Japan.
81. Lanconelli N, Campanini R, Iampieri E et al. Optimization of the acquisition parameters for a SPET system dedicated to breast imaging. Paper presented at *Conference Records of the IEEE Nuclear Science and Medical Imaging Conference 2006;* San Diego, CA.
82. Adam L-E, Karp JS. Optimization of PET Scanner Geometry. Paper presented at *the Second International Workshop on EGS4 2000;* KEK, Japan.
83. Kitamura K, Murayama H. Count Rate Performance Simulations for Next Generation 3D PET Scanners. *Proceedings of the Ninth EGS4 Users' Meeting,* Japan, 2001, KEK, Japan.
84. Palmer MR, Zhu XP, Parker JA. Modeling and simulation of positron range effects for high resolution PET imaging. *IEEE Trans. Nucl. Sci.* 2005;52:1391–1395.
85. Zhang L, Staelens S, Holen RV et al. Fast and memory-efficient Monte Carlo-based image reconstruction for whole-body PET. *Med. Phys.* 2010;37:3667–3676.

11

ASIM: An Analytic PET Simulator

Brian F. Elston, Claude Comtat, Robert L. Harrison, and Paul Kinahan

CONTENTS

Simulation plays an important role in the development of positron emission tomography (PET). It is used extensively in the design of tomographs, in the investigation of physical effects that cannot be quantified using experiments, in the development of data correction and image reconstruction algorithms, and in the development of imaging protocols. Simulation studies have led to understanding the effect of scatter on PET images, the optimal design of septa, the effects of noise on PET images, and the sensitivity and resolution limits of PET. Simulation continues to be used in these areas for PET and other image modalities [1–3].

In PET patient scans, the true underlying activity distribution is unknown. We have no way of knowing in a given scan how various factors are confounding the data: statistical noise, biological variability, scattered radiation, patient motion, deadtime in the detectors and electronics, detector resolution and partial volume, septal penetration, and random coincidences. Simulation provides an unparalleled window for examining these factors [4] as they can be designed to isolate a single factor of interest, for instance when researchers perform multiple simulations of the same imaging situation to determine the effect of statistical noise or biological variability.

ASIM [2] is an analytic PET simulation that we have recently released as an open-source package. It provides a computationally efficient way to probe a subset of these problems. ASIM is particularly suited to studies where statistical noise, biological variability, and the effects of image resolution are the dominant variables in studies where large numbers of data sets need to be generated.

Role of Simulation in PET Research

The wide range of problems being addressed has led to a tremendous variety of simulation tools. The most exact simulations are the particle-tracking variety used for high-energy physics simulations, for example, Géant [5], EGS [6], MCNP [7], and Penelope [8,9]. These cover the full range of possible interactions and allow for very detailed descriptions of tomographs and imaged objects. Their drawbacks are that they are slow and can be difficult to set up for medical imaging simulations. Efforts have been made to simplify their use for medical imaging, notably the Geant4 Application for Tomographic Emission (GATE) [10] and PET–EGS [11,12]. With proper setup, the high-energy physics packages can provide very accurate simulations. A complete simulation is still, however, painstakingly slow. A whole-body PET scan can take days, weeks, months, or more to simulate on a modern single-processor computer.

There are also general-use photon-tracking simulations that are tailored specifically for medical imaging, for example, SimSET [1] for PET and SPECT, PETSIM [13] for PET, and SIMIND [4,14] for SPECT. These packages support more limited, tomograph-specific geometries and concentrate on tracking photons within a limited energy range, reducing the types of interactions that must be simulated. These simulations are easier to set up and faster than the high-energy physics packages. The restrictions on geometries make it difficult or impossible to use these packages to simulate nonstandard systems (e.g., "nested ring" scanners [15]) and limitations on the types of particles simulated make these packages less exact than the high-energy applications. However, their accuracy is usually satisfactory for comparative design studies, for characterizing the contaminants in data and assessing the relative importance of various contaminants, and for examining the impact of data correction methods (e.g., scatter correction).

The simplest model is line of response (LOR) coincidence generation (e.g., Sim3D [16]), which directly generates true coincidence events. If the orientation and location of the LOR is randomized, then this is also a form of Monte Carlo simulation, albeit much faster and less accurate than the photon-tracking methods described above.

Scattered radiation has been one of the areas most studied with simulation (e.g., [17–19]) and has produced some single-purpose photon-tracking simulations, for example, those of Holdsworth [20] and Beekman [21]. These packages are less flexible, both in their modeled geometry and modeled physics, than the general packages above. However, for the single purpose of estimating scatter distribution, they are considerably faster than the other packages. This allows for the investigation of patient-specific scatter elimination.

The time required for a photon-tracking simulation varies linearly with the number of decays (typically many billions) simulated. Although photon-tracking simulations are amenable to parallelization and/or importance sampling, this does not provide the acceleration necessary for studies requiring many realistic simulations. These kinds of applications are the motivation for providing even faster simulation methods.

Analytic PET Simulation

ASIM belongs to a different genre of simulation, analytic simulation. These simulations are useful for problems that would require too much computation time using photon-tracking simulations, for instance, studies that require sufficient photons for multiple independent and identically distributed (i.i.d.) image formations. Examples of this type of study are those aimed at understanding and optimizing patient PET imaging. In particular, observer studies of tumor detectability, image noise measurements, and kinetic modeling studies often require hundreds or thousands of realizations of a small set of imaging configurations.

Multiple realizations are more quickly produced using analytic simulations [2,22]. An analytic simulation first directly calculates mean estimates of each sinogram bin based on a ray-tracing model. These mean values can then be used as noiseless data; noisy data sets can be produced by replacing each bin with a noisy (typically Poisson) realization based on the known pdf. Done correctly, this is a form of bootstrapping [23].

The trade-offs that analytic simulators make in comparison with the more accurate photon-tracking simulations are that all effects (noise, scatter, randoms, attenuation, etc.) are calculated rather than naturally arising out of following the fate of individual photons. Most of the important effects such as noise, attenuation, and detector blurring can be accurately calculated, but unfortunately not scattered or random coincidences. Analytic simulations must rely on simplified models or object and tomograph-specific models for these contaminants (for instance, derived from a photon-tracking simulation). However, image reconstructions of data from analytic simulations can give quite accurate approximations of noise and resolution properties. Without photon tracking, scatter and randoms can only be roughly estimated. Thus, we can reliably estimate the noise added by scatter and randoms (at least relative to other sources of noise), but not the bias. In practice the bias from scatter is imperfectly estimated [24–26], thus leaving a residual bias after scatter correction. ASIM does not attempt to estimate this residual due to the challenge of using an appropriate model for the residual bias. Instead, the mean estimated scatter is used first to calculate the added noise due to scatter and then used for scatter correction, leaving no residual bias. The scatter and randoms are either predetermined (e.g., from measurements or photon-tracking simulations) or estimated using the Klein–Nishina equation, using a simplified representation of the object to reduce computation time.

Another simulation approach that some researchers have adopted is directly adding noise to clinical images or image-derived data for kinetic modeling (e.g., [27,28]). These simulations are very fast, but have the disadvantage of being prone to bias, inaccurate noise texture, and other inaccuracies. They are often used when researchers desire a quick way to generate data and to find that there are no public domain or open-source tools appropriate for their task (e.g., [29,30]). However, if the mean of the acquired data is estimated, and the estimated mean values replaced with Poisson-distributed random variables about the means, a new Poisson-distributed random variable is generated. Thus, new realizations can be very quickly generated once the mean values are estimated. A trade-off of image-based analytic simulations is that noise correlation effects are not included. These effects are often a dominant factor in real images.

In summary, the strength of analytic simulations is that while the first noiseless data set can require extensive computation, noisy realizations can be generated very quickly by adding noise. Thus, generating hundreds or thousands of realizations requires little more time than one realization, making analytic simulation for producing many noisy i.i.d. realizations of the same configuration, for example, in the quantitative assessment of image quality (e.g., [31]). The trade-off is that analytic simulation methods cannot, by themselves, accurately simulate such data contaminants as scatter, random coincidences, pulse pileup, or detector blurring. Analytic simulators have been developed for SPECT [29,30] and for PET [2,22,32].

From analytic simulation methods to photon-tracking methods, Table 11.1 summarizes the advantages of the various simulation packages discussed above. In general, flexibility and accuracy come at the cost of speed and ease-of-use. The hybrid packages, GATE and PET–EGS, trade some of the flexibility of the high-energy physics simulations for ease-of-use similar to the general-purpose photon-tracking simulations, while retaining the accuracy advantage and speed disadvantage of their underlying high-energy simulations.

Development and History of ASIM

The analytic simulator (ASIM) was developed under NIH funding (R29-CA74135: PI Kinahan, "Strategies for Clinical Oncology Imaging with 3DPET") starting in 1997 and continuing through today by a

TABLE 11.1

Major Simulation Packages Arranged by Increasing Flexibility and Accuracy/Decreasing Speed

Trade-Off Mode	More Accurate (Slower)					More Precise (Faster)	
	Photon-Tracking					Analytic	
Role	High-energy physics (HEP)	Adapted HEP	General-purpose nuclear medicine	Special-purpose nuclear medicine	True coincidence only	Sinogram based	Image based
Instances	Geant, EGS4, MNCP	GATE	SimSET, SIMIND, PET-EGS, PETSIM	Beekman, Holdsworth	Sim3D, PETTRACE	ASIM	Alpert
Pros	Most accurate and flexible. Carefully validated. Freely available	Accurate and flexible. Freely available	Accurate and faster than HEP-based methods. Freely available.	Faster than general-purpose medical imaging.	Fastest photon-tracking method. Relatively simple. Useful for debugging and testing reconstruction algorithms	Much faster than photon-tracking methods. Can include important physics effects. Some are freely available	Fastest, simplest method
Cons	Very slow and complex to set up and run	Slow, somewhat complex	Less flexible less accurate than HEP-based methods	Only specific interactions (e.g., Compton scatter). Inflexible and less accurate than general-purpose packages. No independent validation	Does not model important effects (e.g., attenuation, scatter, randoms)	Some physics effects (e.g., scatter) must be approximated	Does not model physics or noise correlations in images. No independent validation

Note: A more comprehensive comparison (without analytic simulations) can be found in [3]. References for the packages can be found in [2,6,10,12–14,16,20,21,27].

collaborative work between the SHFJ in Orsay, France, and the University of Washington (R01-CA115870, R01-CA126593, R01-CA42593, and U01-CA148131) in Seattle, USA.

In 1996, Michel Defrise and Christian Michel designed and coded the original version (1.0) of the ASIM analytic simulator with a low-level, but fast, Poisson noise generator. In 1997, Claude Comtat and Paul Kinahan expanded on the project, creating version 3.0, which progressed to version 5.52 by March 2011 with many new additions to the physics model. These versions are coupled to the CTI–PET scanner geometry through the ECAT software library of M. Sibomana and C. Michel. In 2008, Brian Elston and Paul Kinahan created from ASIM 5.39 a version with scanner-independent input–output (I/O). This version has been publicly available as an open-source project since 2010 at http://depts.washington.edu/asimuw/.

Researchers have utilized the ASIM simulation package to study lesion detectability in response to reconstruction techniques [33,34], scan duration [35], and weight-based scanning protocols [36] in PET oncology imaging. These studies utilized multiple independent realizations to quantify numerically the detectability of various sized lesions simulated within an extended MCAT phantom in an emulated ECAT HR+ scanner [34]. More recent studies have examined the detectability of patient response to therapy as a result of tumor change between PET scans using common evaluation techniques (e.g., contrasting PERCIST and EORTC) [37]; and the effects of noise, calibration, reconstruction, and analysis methods on correctly determining change in PET SUVs through simulation studies [38]. The code was also used to simulate dynamic PET studies and evaluate pharmacokinetic measurements [39]. There are examples of images reconstructed from ASIM simulation data at the end of this chapter.

Structure of ASIM

ASIM is a software toolkit designed as a fast analytic simulator for positron emission scanners. The package provides a variety of options, including the simulation of emission data, attenuation correction, randoms, scatter, detector efficiency variations, 2D and 3D mode, detector blurring, normalization, and noise propagation. The scanner-independent input–output (I/O) is supported with the YAFF (Yet Another File Format), which is based on the interfile model of two output files, one an ASCII header and the other a sinogram data block. The extension used for the data block is .yaff, and .yhdr is used for the header file in the YAFF format.

The header file contains parameters relevant to the description of the simulation data block. The sinogram data block is comprised of four-dimensional projection data (radial, azimuthal, axial, and ring difference) in big-endian

4-byte floating point format. The software is coded in ANSI-compliant C, using modern software architecture and coding practices, tested and compiled under UNIX using a Macintosh system. The code is hierarchically organized in a functionally derived C style, so as to be easily modifiable, expandable, and debuggable.

Algorithmic Description

There are three core modules of the ASIM software package:

1. **Simul()**: Generates noiseless PET emission and attenuation sinograms.
2. **Noise()**: Applies Poisson-distributed noise to a sinogram.
3. **Normalize()**: Applies corrections to these data sets.

The three modules are run in series for investigative studies, in conjunction with reconstruction and analysis, by creating a chain of software simulation command calls with common parameters. This process can be automated with shell scripts to create multiple realizations easily as diagramed in Figure 11.1.

As an input to **Simul()**, mathematically described phantoms or objects give both emission and attenuation properties. These objects are typically specified as collections of truncated ellipsoids with constant or linearly varying radioactivity concentration and attenuation. For simulated patient studies, studies with ASIM can use a generic extended MCAT phantom, developed by C. Lartizien, by using the geometric phantom scripting system provided within the software [34], as can be seen in Figure 11.2. Here, a collection of truncated ellipsoids (including variations such as cylinders) are used to represent a generic human form. Of note in the emission image, we can clearly identify the heart, brain, arms, spine, lungs, bladder, stomach, liver, kidneys, and spleen. Identifiable in the attenuation image are the skull, lungs, ribs, arm bones (humerus, ulna, radius), and spine.

For phantoms composed of truncated ellipsoids, the unattenuated mean true coincidences are determined by a closed-form analytic (i.e., fast) calculation of the intersection length of the LOR between two detector elements. This is repeated for each emission object along the LOR and the results are summed. To improve the accuracy of calculation, a subsampling factor can be specified to subdivide the surface areas of the two detectors that are in coincidence. In this case, the calculations are performed for each pair-combination of detector subareas, and the results are averaged. This approach can provide a more accurate estimate of the unattenuated mean true coincidence rate, at the cost of more computation time. Attenuation sinograms are calculated in

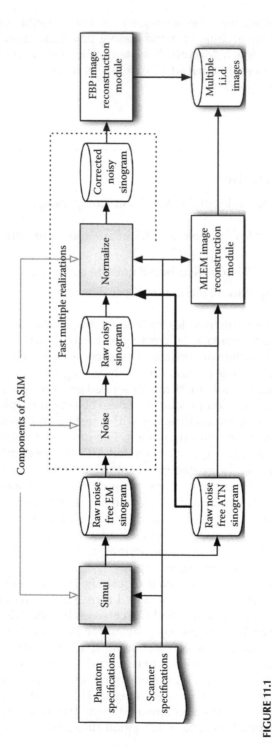

FIGURE 11.1

Typical procedural flow using Simul, Noise, and Normalize to generate multiple realizations for an example study. Here, one (or potentially more) raw noise-free emission (EM) sinogram is created along with a raw noise-free attenuation (ATN) sinogram. An image reconstruction is performed with filtered-backprojection (FBP) or maximum-likelihood expectation-maximization (MLEM) to obtain multiple independent identically distributed (i.i.d.) images. (From ASIM, http://depts.washington.edu/asimuw/. With permission.)

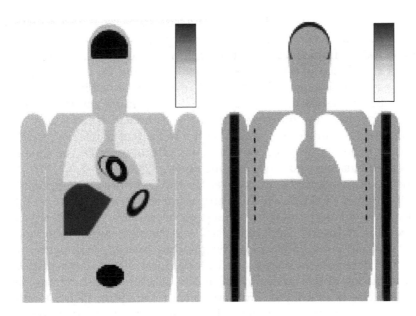

FIGURE 11.2
Central coronal section of the three-dimensional Arsene phantom designed by Carole Lartizien, which is built up from truncated ellipsoids and can be used as an input to the **simul()** component of ASIM. Emission density (left) in *e* (counts/mL) and linear attenuation coefficient in μ (1/mm). Lartizien C et al. *J. Nucl. Med.* 2003;44:276–290.

a similar manner. It is also possible to input a pixelized phantom image volume, which can be useful for brain imaging simulations and other detailed objects not easily represented by truncated ellipsoids.

An advantage of the analytical calculation method is computational speed. A potential further advantage is that the forward projection calculation is different from forward projection methods typically used in iterative image reconstruction algorithms, thus avoiding an artificially close connection between a reconstruction algorithm and the true physics of PET data acquisition.

ASIM includes the option of modeling the spatially dependent resolution in a clinical PET scanner using either simple sinogram smoothing or a more precise detector-specific blurring effect that accurately mimics PET scanner spatial variations in intrinsic detector resolution. By separately generating noise-free sinograms of lesions, composite sinograms can be created with varying contrast and noise levels by adding scaled versions of the sinograms.

The mean scattered or random coincidence rates are calculated using a model where the total activity in each slice is concentrated along the axis of the scanner, taking into account the activity outside the field of view. Similarly, the attenuation objects are replaced with a rotationally symmetric volume of water with the same total attenuation at each axial location as the original attenuation object. This rotational symmetry enables a fast calculation of the mean scattered and random mean coincidence rates, the former using a 1D

Klein–Nishina model-based calculation. Alternatively, the mean scattered and random mean coincidence rates can be prespecified based on measured scanner data or from more accurate (but slower) photon-tracking simulations.

Exact calculation of the scattered coincidence rates for each sinogram element is not feasible as ASIM does not use photon-tracking. However, the rough approximation used is acceptable as the simulation data is later corrected for scatter assuming no bias in the estimated mean scatter. In this model, it is the influence of scatter on the noise that is relevant, thus better accuracy is not required as the dominant effects on noise are true coincidences, attenuation, and random coincidences [2].

The sinogram elements are indexed by the bed position, temporal frame, and location of the LOR, which in turn is parameterized by the radial position, azimuthal angle, mid-slice axial position, and the ring difference.

Once the noise-free sinogram data is generated, multiple independent noisy realizations can be quickly generated by **Noise()**. Attenuation, scatter, randoms, detector efficiencies, and isotope decay effects can be included in the calculation of the Poisson random deviates for each sinogram element. Poisson noise is contributed from up to four sources in PET: the emission, transmission, normalization, and blank scans. Typically, only noise in the emission scan is modeled. In a scan simulation, three types of events can be detected: true, random, and scattered coincidences. As a result, three independent Poisson noise sources are generally considered: emission trues, emission randoms, and emission scatter.

The count levels for the simulations are set either explicitly by specifying the number of true, scattered, and random coincidences for a given bed position (whole-body scans) or temporal frame (dynamic scans), or implicitly by specifying a unique calibration factor for true and random coincidences, plus a scatter fraction. Given the isotope half-life and the emission and transmission scan durations, prompt coincidence count levels are calculated for each bed position and temporal frame.

Once the raw noisy sinograms are generated, they are processed and reconstructed with the same procedures used with measured data to ensure the same correlated noise properties in the reconstructed images. Quantitative corrections that can be applied by **Normalize()** include attenuation correction, efficiency normalization, random and scattered coincidences correction, and arc correction, among others.

Usage Illustration

For an usage example, we will generate two emission sinograms, one for the torso and another for the tumor, as well as an attenuation and normalization scan. We then sum the two emission sinograms, applying a multiplicative factor to one (i.e., the tumor), and add noise to the composite sinogram. This sinogram is then preprocessed prior to image reconstruction by applying attenuation correction and axial normalization. We generated the following example running the simulation software from the command line parameters shown in Table 11.2, in the order of execution denoted by process. We used UNIX on a Macintosh computer, but the software can run with no or minor modifications on most UNIX/Linux and Windows systems.

A **Simul()**-generated emission sinogram was created with arc effect and line of response solid angle scaling (LORSAS) turned off. We modeled a simplified three-ring PET scanner based on the ECAT HR+. We placed the phantom so that the top of the tomograph field-of-view axially was a point 365 mm from the top of the phantom (see the command process 1 in Table 11.2). In Table 11.3, we see the command line parameters used for Simul in the usage illustration here. Note that for this example the file named "301" contains the appropriate definitions for the ECAT HR+ scanner (i.e., a scanner definition file), with the number of rings set to three, and the torso.data input file is a Simul phantom representation of an extended MCAT phantom. The torso data output file was named torso_em.yaff, with the header file name defaulting to torso_em.yhdr as a convention. The simulation run was specified to also output scatter and delayed data, as well as specifying a 2D scan with a ring difference of 1. The tumor emission sinogram was generated using the command seen in Table 11.2, process [2]. The tumor input file is tumor.data, and represents a 3 cm diameter lesion in the chest wall. Other

TABLE 11.2

Command Line Parameters, in the Order of Execution, for Generating the Composite Noisy, and Subsequently Corrected Sinograms, That Were Then Reconstructed into the Images Displayed in Figure 11.3

Process	Command Line	Effect
1.	Simul -i torso.data -o torso_em.yaff -E -k -h -S -D -f 365 -m 301 -d 1	Generate emission sinogram for torso
2.	Simul -i tumor.data -o tumor_em.yaff -E -k -h -S -D -f 365 -m 301 -d 1	Generate emission sinogram for tumor
3.	Simul -i torso.data -o torso_atten.yaff -A -k -h -f 365 -m 301 -d 1	Generate attenuation sinogram for torso
4.	Simul -i torso.data -o torso_norm.yaff -N -k -h -f 365 -m 301 -d 1	Generate normalization sinogram for torso
5.	Noise -i torso_em.yaff -t 6000000 -r 6000000 -R -s 2100000 -S -J 2.0 -I tumor_em.yaff -o torso_tumor_noisy.yaff	Sum torso and tumor sinogram, apply multiplicative factor to tumor sinogram, and apply noise to composite sinogram
6.	Normalize -i torso_tumor_noisy.yaff -a torso_atten.yaff -n torso_norm.yaff -k -o torso_tumor_normalized.yaff	Apply attenuation correction and axial normalization to input sinogram

Source: From ASIM, http://depts.washington.edu/asimuw/. With permission.

TABLE 11.3

Command Line Parameters and Their Description Used with Simul in the Usage Illustration

Parameter	Effect
-E	Generate noise-free emission sinogram
-A	Generate attenuation sinogram
-N	Generate normalization sinogram
-S	Generate scatter sinogram
-D	Generate delayed sinogram
-i	Select filename for input file (geometric phantom or pixelized)
-o	Select filename for output file (and by convention sets header filename)
-k	Simulate arc effect
-h	Turn line of response solid angle scaling (LORSAS) off
-f	Set start of bed position in mm
-m	Select file that the scanner definition is located in
-d	Set the scan dimension to 2D or 3D

Source: From ASIM, http://depts.washington.edu/asimuw/. With permission.

parameters generating the tumor-emission sinogram are identical to that for generating the emission sinogram for the torso.

An attenuation sinogram was generated from the torso with arc effect and LORSAS turned off, using the same three-ring emulated ECAT HR+ scanner, and starting from the same bed position (Table 11.2, process 3). The input file to the attenuation simulation is the torso phantom file, torso.data. The input and output filename conventions are the same as that used for the emission sinogram, including the convention for naming header files. All other parameters in the command perform are the same as those described above.

A normalization scan was created using the same methodology and process as for the attenuation and emission data (Table 11.2, process 4).

Noise was added by specifying the number of true counts, in this case 60 million, with random and scatter correction applied, and counts specified representing a typical clinical noise level for randoms and scatter (Table 11.2, process 5). A description of the command line parameters used with noise can be seen in Table 11.4. The primary input file is torso_em.yaff, with the output data block filename, torso_tumor_noisy.yaff, and the output header filename torso_tumor_noisy.yhdr. Additionally, the noise application has the ability to sum two sinograms and apply a multiplicative factor to one of them. The additional input sinogram is tumor_em.yaff, and the multiplicative factor specified for it is a value of 2.0. In this study, the additional input file was used to model a tumor with varying activity levels within a torso to investigate lesion detectability.

This noisy emission sinogram was then axially normalized and attenuation corrected (Table 11.2, process 6). The input file is attenuation corrected

TABLE 11.4

Command Line Parameters and Their Description Used with Noise in the Usage Illustration

Parameter	Effect
-S	Apply scatter correction
-R	Apply random correction
-i	Select filename for input sinogram file
-o	Select filename for output file (and by convention sets header filename)
-I	Select filename for secondary optional input sinogram file
-J	Set multiplicative factor to apply for additional input sinogram from -I command
-t	Set number of coincidences for true count
-r	Set number of coincidences for random count
-s	Set number of coincidences for scatter count

Source: From ASIM, http://depts.washington.edu/asimuw/. With permission.

using the attenuation sinogram torso_atten.yaff, axially normalized using torso_norm.yaff, with arc correction disabled. In this case, arc correction was disabled as that factor was accounted for in the reconstruction process. In Table 11.5, we see a description of the command line parameters used with Normalize in this example. The input file is torso_tumor_noisy.yaff, the output sinogram file name is torso_tumor_normalized.yaff, with the header output filename defaulting to torso_tumor_normalized.yhdr.

The sinogram data was then reconstructed using the filtered-back projection (FBP) implementation FBP2D in the Software for Tomographic Image Reconstruction (STIR) library, and STIR's Ordered Subset Maximum A Posteriori One-Step Late (OSMAPOSL) iterative algorithm, and subsequently analyzed. This process can quickly be repeated many times, creating multiple realizations, with individual instantiations having unique noise characteristics. Examples of these images can be seen in Figure 11.3.

TABLE 11.5

Command Line Parameters and Their Description Used with Normalize in the Usage Illustration

Parameter	Effect
-i	Select filename for input sinogram file
-o	Select filename for output file (and by convention sets header filename)
-a	Select input attenuation sinogram and applies attenuation correction
-n	Select input normalization sinogram and applies axial normalization
-k	Arc correction

Source: From ASIM, http://depts.washington.edu/asimuw/. With permission.

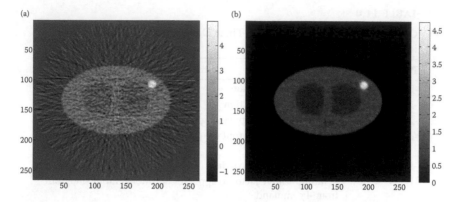

FIGURE 11.3

Filtered backprojection (a) and OSMAPOSL (b) image reconstructions of typical count simulations for tumor SUV = 2. Sinograms simulated using ASIM, and images reconstructed with STIR.

Representative Studies Using ASIM

Study 1: The study "SNR effects in determining change in PET SUVs in response to therapy" [37] conducted an investigation of the relative impact of photon noise and other noise on the ability of PET to detect a response to therapy. ASIM was used to simulate an anthropomorphic phantom with a 3 cm diameter breast tumor in a three-detector ring tomograph loosely emulating the ECAT HR+, shown in Figure 11.3. Reconstruction was performed with FBP using a Hann filter cutoff at .5 of the Nyquist frequency, while the OSMAPOSL reconstruction was run 2 iterations, 12 subsets, with a 3 mm Metz filter. An estimate of the Standard Uptake Value (SUV) was obtained using SUV mean and SUV max. Other (non-photon-counting) noise, such as calibration errors and biological variability, was estimated by adding varying amounts of Gaussian noise to the SUV estimate. Receiver operating characteristics (ROC) curves were created, area under the curve (AUC) was determined, and finally the percentage of realizations that satisfy the European Organization for Research and Treatment of Cancer (EORTC) [40] and PET Response Criteria in Solid Tumors (PERCIST) [41] criteria were calculated for partial metabolic response or progressive metabolic disease.

This investigation concluded that results were much less sensitive to change in scan duration and reconstruction method than to changes in scanner calibration or patient test–retest variability. The PERCIST classification scheme was deemed noticeably less error prone than the EORTC system, misclassifying metabolic response as progression, and vice versa, less often. EORTC was deemed as having criteria that seemed too narrow, or finely grained, given the current typical system measurement errors.

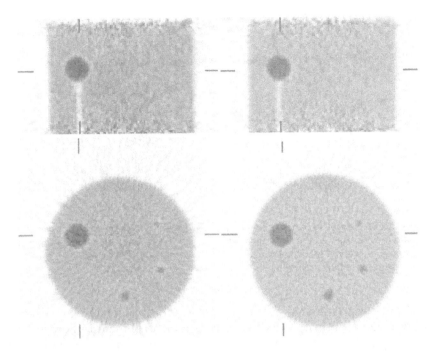

FIGURE 11.4
Transverse orthogonal sections through a 20 cm diameter cylindrical phantom with hot spheres supported by plastic rods. (Left: Measured data from a Siemens/CTI ECAT HR+ scanner. Right: Simulated data for the same acquisition using ASIM. With permission.)

In Figure 11.4, we see a study comparison of simulated and measured data; on the left, measured data from a Siemens/CTI ECAT HR+ scanner acquired in fully 3D mode, and on the right, simulated data for the same acquisition properties using ASIM.

Study 2: An examination of how scan duration affects the detectability of tumor masses was performed in the study "A quantitative approach to a weight-based scanning protocol for PET oncology imaging" [36]. The focus of this study was to investigate the effects of patient attenuation, determined by patient thickness. As attenuation increases, the number of photons detected exponentially decreases, thus affecting tumor detectability in PET imaging. The simulation study was performed using an MCAT phantom, with the addition of 18 lesions ranging in size from 1 to 3 cm in diameter spaced throughout the phantom, seen in Figure 11.5. The simulation was performed using ASIM, corrected for attenuation and detector geometry, and reconstructed using attenuation weighted ordered subset expectation maximization (AWOSEM). Detectability for each lesion's location, size, contrast ratio, and duration was evaluated using a non-prewhitening matched filter (NPWMF).

This study found that target detectability varies nonlinearly with target diameter, as a function of scan duration. Target detectability squared was

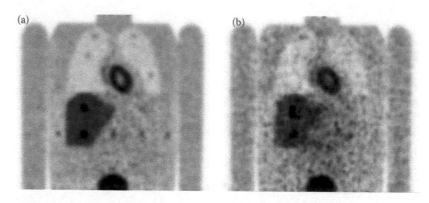

FIGURE 11.5
Coronal slices for image volumes generated by ASIM with 2 cm diameter lesions with a contrast ratio of 3:1 compared to the local background. (a) Noisy realization simulating a 10 min per bed position scan (40 min total scan time). (b) Noisy realization simulating a 1 min per bed position scan (4 min total scan time).

found to be approximately proportional to scan duration and noise equivalent count (NEC) density. Notably, while under current clinical practice scan time is typically not adjusted for body habitus, the research found that image quality within the thorax and abdomen degrades with increased body girth.

Study 3: An investigation of using simulation to predict the progression of tumor growth was conducted in the study "Applying a patient-specific biomathematical model of glioma growth to develop virtual [18F]-FMISO PET images" [42]. The study sought to assess the proliferation and invasion of glioma through mathematical modeling. This research used a pharmacokinetic model to determine the activity of a [18F]-flouromisonidazole (FMISO) tracer within the brain over the time interval of virtual PET imaging. This summed tracer map was then processed by ASIM, using a pixelized input phantom, creating a simulated FMISO–PET image that mimics the noise characteristics and acquisition artifacts seen in clinical FMISO–PET images for the patient.

We see in Figure 11.6 measured and simulated images of MRI and PET images of hypoxic glioma tumor cells. The underlying PET data used for the simulated image is based on a model combining glioma cellular distribution with the pharmacokinetics for FMISO uptake. The model is run forward in time until the glioma volume matches the measured tumor volume (but not necessarily the pattern of distribution) from the patient MRI. Simulated PET images were then generated with ASIM.

This work establishes a direct link between anatomical (MRI) modeling and molecular (PET) imaging and provides a tool for predicting hypoxic distribution on a patient-specific level. It demonstrates capabilities of the proliferation–invasion–hypoxia–necrosis–angiogenesis (PIHNA) model to produce predictive, patient-specific realizations as visualized on clinical MRI.

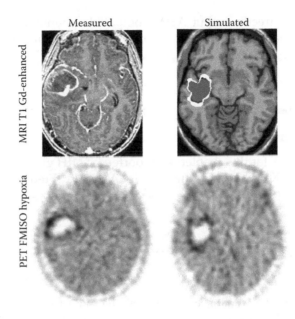

FIGURE 11.6
Measured and simulated images of MRI and PET images of hypoxic glioma tumor cells. (From Gu S et al. *Math. Med. Biol.* 2011.)

Future Directions

We are currently in the deployment phase of the first open-source release of the ASIM software. The source code has been restructured and architected into modular function-based software using the C programming language. Documentation is provided to describe the simulation package and its usage. The source code and documentation are presently available online.

Some researchers have indicated that having a simulation metalanguage interpreter would be valuable. Currently, the inputs and outputs of various simulators, such as ASIM and SimSET, are quite different. Having a common input metalanguage, for example, describing phantoms and scanner geometries, would make it much easier for users to use multiple simulation packages.

Investigators have expressed a desire for a phantom-building application that would display the object graphically and allow for interactive creation of the phantom. ASIM's current system of phantom definition is text based; graphical display would provide the researcher with direct visual feedback when building new phantoms and ease the creation process. The tool would be designed to directly output ASIM-compatible phantom definition files, and possibly phantoms compatible with a limited set of other simulation packages.

Acknowledgments

The research work for this chapter is supported by NIH grants R01-CA115870, R01-CA126593, R01-CA42593 and U01-CA148131, French ANR grant ANR-05-MMSA-014-01.

References

1. Lewellen TK, Harrison RL, and Vannoy S. The SimSET program. In: Ljungberg M, Strand SE, King MA, eds. *Monte Carlo Calculation in Nuclear Medicine: Applications in Diagnostic Imaging*. Bristol and Philadelphia: IOP Publishing 1998: pp. 77–92.
2. Comtat C, Kinahan PE, Defrise M, Michel C, Lartizien C, and Townsend DW. Simulating whole-body PET scanning with rapid analytical methods. *Proceedings of IEEE Nuclear Science Symposium and Medical Imaging Conference*, Seattle, WA, October 24–30, 1999, Vol. 3: pp. 1260–1264.
3. Buvat I, and Castiglioni I. Monte Carlo simulations in SPECT and PET. *J. Nucl. Med.* 2002;46:48–61.
4. Andreo P, and Ljungberg M. General Monte Carlo codes for use in medical radiation physics. In: Ljungberg M, Strand SE, King MA, eds. *Monte Carlo Calculation in Nuclear Medicine: Applications in Diagnostic Imaging*. Bristol and Philadelphia: IOP Publishing 1998: pp. 37–52.
5. Segars WP, Tsui BMW, Frey EC, Johnson GA, and Berr SS. Development of a 4-D digital mouse phantom for molecular imaging research. *Mol. Imaging Biol.* 2004; 6(3):149–159.
6. EGS (Online). 1985. http://www.slac.stanford.edu/.
7. Appelbaum FR. The influence of total dose, fractionation, dose rate, and distribution of total body irradiation on bone marrow transplantation. *Semin. Oncol.* 1993;20:3–10.
8. Arcovito G, Azario L, Cichocki F, Piermattei A, Rossi G, and Galli G. Evaluation of the absorbed dosage in Compton tomography. *Radiol. Med. (Torino).* 1989;77: 540–543.
9. Sempau J, Fernandez-Varea JM, Acosta E, and Salvat F. Experimental benchmarks of the Monte Carlo code PENELOPE. *Nucl. Instrum. Meth.* 2003;207:107–123.
10. GATE: Geant4 Application for Tomographic Emission (Online). 2004. http://www-lphe.epfl.ch/GATE/.

11. Arheden H, Holmqvist C, Thilen U et al. Left-to-right cardiac shunts: Comparison of measurements obtained with MR velocity mapping and with radionuclide angiography. *Radiology*. 1999;211(2):453–458.
12. Castiglioni I, Cremonesi O, Gilardi M et al. Scatter correction techniques in 3D PET: A Monte Carlo evaluation. *IEEE Trans. Nucl. Sci.* 1999;46:2053–2058.
13. Thompson C, Moreno-Cantu J, and Picard Y. PETSIM: Monte Carlo simulation of all sensitivity and resolution parameters of cylindrical positron imaging systems. *Phys. Med. Biol.* 1992;37(3):731–749.
14. Ljungberg M, and Strand SE. A Monte Carlo program simulating scintillation camera imaging. *Comp. Meth. Progr. Biomed.* 1989;29:257–272.
15. Tai YC, Wu H, and Janccek M. Initial study of an asymmetric PET system dedicated to breast cancer imaging. *IEEE Trans. Nucl. Sci.* 2006;53(1):121–126.
16. Daube-Witberspoon. Sim3D (Online). 1994. http://www.slac.stanford.edu/.
17. Bai C, Zeng G, and Gullberg G. A slice-by-slice blurring model and kernel evaluation using the Klein-Nishina formula for 3D scatter compensation in parallel and converging beam SPECT. *Phys. Med. Biol.* 2000;45(5):75–1307.
18. Barret O, Carpenter T, Clark J, Ansorge R, and Fryer T. Monte Carlo simulation and scatter correction of the GE advance PET scanner with SimSET and Geant4. *Phys. Med. Biol.* 2005;50:4823–4840.
19. Haynor DR, Kaplan MS, and Miyaoka RS. Multiwindow scatter correction techniques in single-photon imaging. *Med. Phys.* 1995;22:2015–2024.
20. Holdsworth, H. Investigation of PET image reconstruction and data correction techniques with emphasis on an accelerated Monte Carlo simulation for correction and evaluation of PET. University of California 2002.
21. Beekman FJ, Den Harder JM, Viergever MA, and Van Rijk PP. SPECT scatter modeling in nonuniform attenuating objects. *Phys. Med. Biol.* 1997;42:1133–1142.
22. Furuie SS, Herman GT, Narayan TK et al. A methodology for testing for statistically significant differences between fully 3D PET reconstruction algorithms. *Phys. Med. Biol.* 1994;39:341–354.
23. Haynor DR, and Woods SD. Resampling estimates of precision in emission tomography. *IEEE Trans. Med. Imag.* 1989;8:337–343.
24. Ollinger JM. Model-based scatter correction for fully 3D PET. *Phys. Med. Biol.* 1996;41(1):153.
25. Watson CC. New, faster, image-based scatter correction for 3D PET. *IEEE Trans. Nucl. Sci.* 2000;47(4):1587–1594.
26. Wollenweber SD. Parameterization of a model-based 3-D PET scatter correction. *IEEE Trans. Nucl. Sci.* 2002;49(3):722–727.
27. Alpert NM, Rajendra D, Badgaiyan RD, Elijahu L, and Fischman AJ. A novel method for noninvasive detection of neuromodulatory changes in specific neurotransmitter systems. *NeuroImage*. 2003;19:1049–1060.
28. Marcinkowski A, Layfield D, Tgavalekos N, and Venegas JG. Enhanced parameter estimation from noisy PET data: Part II—Evaluation. *Acad. Radiol.* 2005; 12(11):1448–1456.
29. Lahorte P, Vandenberghe S, and Van Laere K. Assessing the performance of SPM analyses of spect neuroactivation studies. *NeuroImage*. 2000;12:757–764.
30. Ward T, Fleming JS, Hoffmann SM, and Kemp PM. Simulation of realistic abnormal SPECT brain perfusion images: Application in semi-quantitative analysis. *Phys. Med. Biol.* 2005;50:5323–5338.

31. Lartizien C, Kinahan PE, and Comtat C. Volumetric model and human observer comparisons of tumor detection for whole-body PET. *Acad. Radiol.* 2004;11:637–648.
32. Rowe R, and Shubo D. A pseudo-Poisson noise model for simulation of positron emission projection data. *Med. Phys.* 1992;19(4):1113–1119.
33. Janeiro L, Comtat C, Lartizien C et al. Numerical observer studies comparing FORE + AWOSEM, FORE + NECOSEM and NEC based fully 3-D OSEM for 3-D whole-body PET imaging. *IEEE Trans. Nucl. Sci.* 2006;53:1194–1199.
34. Lartizien C, Kinahan PE, Swensson R et al. Evaluating image reconstruction methods for tumour detection in 3-dimensional whole-body PET oncology imaging. *J. Nucl. Med.* 2003;44:276–290.
35. Cheng PM, Kinahan PE, Comtat C, Kim J-S, Lartizien C, and Lewellen TK. Effect of scan duration on lesion detectability in PET oncology imaging. *Proceedings of 2004 IEEE International Symposium on Biomedical Imaging*, Arlington, VA, April 15–18, 2004, Vol. 2: pp.1432–1435.
36. Kinahan PE, Cheng P, Alessio A, and Lewellen T. A quantitative approach to a weight-based scanning protocol for PET oncology imaging. *Conference Records of the IEEE Nuclear Science and Medical Imaging Conference*, Nuclear Science Symposium; Puerto Rico, USA, 2005, pp. 1886–1890.
37. Harrison R, Elston B, Doot R, Lewellen T, Mankoff D, and Kinahan PE. Effects of scan duration, analysis method, and SUV measurement error on assessing change in response to therapy. Abstract presented at the *IEEE Nuclear Science and Medical Imaging Conference*, Knoxville, TN, USA, 2010.
38. Kinahan PE, Harrison R, Elston B, Doot R, Lewellen T, Mankoff D. Effects of noise, calibration, reconstruction and analysis methods on correctly determining change in PET SUVs. *J. Nucl. Med.* 2011;52(Supplement 1):1772.
39. Maroy R, Boisgard R, Comtat C et al. Segmentation of rodent whole-body dynamic PET images: An unsupervised method based on voxel dynamics. *IEEE Trans. Med. Imag.* 2008;27:342–354.
40. EORTC (European Organization for Research and Treatment of Cancer) (Online). 2011. http://www.eortc.be/.
41. PERCIST (PET Response Criteria in Solid Tumors). 2010. http://www.medical.siemens.com.
42. Gu S, Chakraborty G, Champley K et al. Applying a patient-specific biomathematical model of glioma growth to develop virtual [18F]-FMISO PET images. *Math. Med. Biol.* 2012;29(1):31–48.

12

Monte Carlo in SPECT Scatter Correction

Kenneth F. Koral and Yuni K. Dewaraja

CONTENTS

The subjects of this chapter are the use of Monte Carlo simulation for scatter correction in single-photon emission computed tomography (SPECT) and its use for the evaluation of scatter-correction methods. Gamma-ray scattering that involves no energy loss (coherent scattering, alias Rayleigh scattering) is of negligible importance compared to Compton scattering for high gamma ray energies and low atomic number elements [1]. Patient scattering usually involves high gamma ray energies and low atomic number elements and so usually no corrections are made for coherent scattering. Possible exceptions might be Bremsstrahlung imaging with respect to energy and bone interactions with respect to atomic number. Collimator scattering usually involves lead, a high atomic number element, and so it might also be a possible exception.

Compton scattering in the patient or in the collimator of a SPECT camera leads to miss-positioned counts in the SPECT projections. If these counts are reconstructed without correction, they lead to loss of contrast and resolution and incorrect quantification. The extent of the contrast deficit has been demonstrated by a ^{201}Tl cardiac perfusion SPECT study in 10 patients: the percent of the volume of the left-ventricular wall that was characterized as below normal in blood flow increased from 12.4% to 20.2% with a Monte Carlo-based correction for scatter compared to without scatter correction [2]. The incorrect quantification has been illustrated with ^{99m}Tc in phantoms: for a sphere in a cylinder, activity quantification that ignored scatter correction while using a scatter-free camera calibration was 18.2% high when there was no "tissue" background and the sphere was off axis and increased to 41.7% when the activity concentration equaled that in the sphere and the sphere was on axis [3].

Quantification improvements with Monte Carlo-based scatter correction have more recently been shown in other applications such as brain imaging [4]. The general consensus is that the existence of scattered gamma rays calls for a "correction" (alias a "compensation") to achieve the best results. Currently, typical clinical studies [5,6] are indirectly indicating whether the results with correction justify the effort required. The existence of the capability to carry out triple-energy-window (TEW) [7], dual-energy-window, and effective-scatter-source-estimation (ESSE) [8] scatter correction within commercial systems (those from Siemens Medical Solutions, Phillips Healthcare, and GE Healthcare, for example) should increase the number of clinical studies. In addition, a Monte Carlo-based correction has been speeded up to enough of an extent [9,10] that it is poised to enter at least some clinical applications.

Solution by Energy Discrimination?

In the case of radionuclides, such as ^{131}I and ^{67}Ga, that have complex decay schemes that feature the emission of a gamma ray with an energy higher than the energy of the gamma ray being detected for the purpose of imaging, Compton scattering lowers the energy of the scattered higher-energy gamma ray so that its energy signal (alias "detected energy") may fall within the detected-energy acceptance window set for the gamma ray that is being imaged. So, there is clearly a problem that needs a solution involving scatter correction. The problem is known as that due to "down scatter."

In the case of a radionuclide, such as ^{99m}Tc, that has only a single emission, one might assume that scattered gamma rays produce a lower detected energy than unscattered gamma rays, and that, therefore, detected-energy discrimination can keep scattered gamma rays from being included in a projection image. In fact, the energy of a scattered gamma ray increases as the scattering angle approaches zero degrees; at exactly zero degrees, the energy of the scattered gamma ray equals that of the unscattered gamma ray. Also, the differential cross section for gamma-ray scattering is nonzero at zero degrees. In contrast, the exit solid angle for the scattered gamma ray approaches zero at zero degrees. Moreover, the product of the differential cross section and the solid angle also approaches zero at zero degrees. Therefore, it would seem that fortuitously the detected energy of scattered photons might always be less than that of the unscattered gamma rays and the scattered gamma rays could be eliminated by energy discrimination. However, all Anger cameras have finite energy resolution. (With a sodium iodide crystal, the resolution is around 10% at 140 keV.) This finite energy resolution distributes the energy of a gamma ray over a range of detected energies, some lower than the actual energy but some higher. Therefore, the detected energy of a given scattered gamma ray can be even greater than the actual energy of the unscattered gamma ray. Moreover, to accept a reasonable number of unscattered gamma rays, which have a spread of detected energy signals as well, one needs a finite-width detected-energy acceptance window. In using such a window, one accepts the detected energy signal of some scattered gamma rays. So, energy discrimination by itself cannot keep scattered gamma rays from being included in a projection image.

Use of Monte Carlo

A comprehensive review of scatter correction methods, irrespective of their relationship to Monte Carlo simulation, was published recently [11]. A similar review was published in 1994 by some of the same authors [12]. Another review of the same type was published in 2004 [13]. A 2005 book chapter that emphasizes methods newer than those covered in the 1994 review is also available [14]. A 2004 book chapter [15] and an earlier review paper from 1995 [16] cover scatter corrections as well as other corrections in SPECT. The previous edition of this book detailed the use of Monte Carlo in SPECT scatter correction as of 1998 [17]. In this edition, we will emphasize the newest developments in the subject area.

In general, the forte of Monte Carlo simulation is that it allows separate tracking of detected gamma rays that have undergone scattering and those that have not. It can even separate the results for scattered photons according to the number of times they have scattered. Figure 12.1 shows Monte Carlo simulated energy spectra for all the ^{99m}Tc counts from a point source at the center of a 22 cm diameter, 20 cm long cylindrical phantom, for the Compton-scattered counts alone and for the counts from the different orders of that scattering. Two dedicated codes that have been widely used to address problems specific to SPECT imaging, such as scatter, are the SIMIND [18] and SIMSET [19] algorithms. More recently, a general-purpose simulation platform Geant4 Application for Tomographic Emission (GATE) has been developed for accurate simulation of SPECT and PET scanners. This application

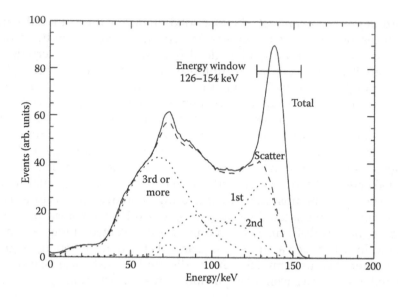

FIGURE 12.1

Energy spectra for a ^{99m}Tc point source in a cylinder of water. Different curves are for all counts, scatter counts, and scatter counts by order of scattering. First-order scatter refers to a single scatter, second order to two successive scatters, and so on. Spectra were generated using the SIMIND Monte Carlo algorithm. (Graph is courtesy of Michael Ljungberg.)

in turn is based on the GEANT4 Monte Carlo package [20–23]. The accuracy of GATE, which has been demonstrated in validation studies with experimental data [24], and recent acceleration [25] is highly promising for future applications.

Scatter Correction within Iterative Reconstruction

One appealing way of correcting for scattered gamma rays is to simply include them in the iterative reconstruction problem as a separate term. The maximum likelihood expectation maximization (ML–EM) algorithm equation below, first proposed by Bowsher et al. in 1996 [26], shows where a term for the scatter intensity can be placed:

$$\lambda_i^{k+1} = \frac{\lambda_i^k}{\sum\limits_{j=1}^{nbin} t_{ji}} \cdot \sum\limits_{j=1}^{nbin} \frac{t_{ji} \cdot p_j}{s_j^k + \sum\limits_{m=1}^{nvox} t_{jm} \cdot \lambda_m^k} \tag{12.1}$$

Here, λ_i^{k+1} is the ith voxel of the image at the $(k+1)$th iteration of the algorithm, λ_i^k is that voxel at the kth iteration, p_j is the jth projection bin, t_{ji} is the element of the transition matrix that relates the jth projection bin to the ith image voxel, and s_j^k is the estimate of the scatter count in the jth projection bin at the kth iteration. Here, the transition matrix gives the probability that a photon is emitted from the ith image voxel, travels to the camera without scatter, and is detected in the jth projection bin. The goal here, as with the other techniques below, is to reconstruct an image that corresponds to photons that have not scattered. Or stated in another way, it is to reconstruct an image where the voxel values equal the number of geometrically collimated detected photons emitted from the source as if they had passed through an "air attenuator," that is, an attenuator with negligible attenuation and negligible Compton scattering. The denominator of the term within the main summation sign is the "forward projection" of the current estimate for the image. The numerator in that term forms the basis of the "backprojection."

The ML–EM algorithm above is usually implemented in a faster approach that depends on subsetting the projection data. The faster method is called ordered subsets expectation maximization (OS–EM). Generally, we will not explicitly call out the division into subsets during our discussions in this chapter, nor usually distinguish whether ML–EM or OS–EM is being used in a study, although it is usually OS–EM that is employed. Another practical approach was championed by Kampius et al. It was dubbed a "dual matrix" approach [27]. In the method, the transition matrix that transforms the image into projections is allowed to be different from the matrix that backprojects the projections into the image. The appropriate equation is

$$
\lambda_i^{k+1} = \frac{\lambda_i^k}{\displaystyle\sum_{j=1}^{nbin} b_{ji}} \cdot \sum_{j=1}^{nbin} \frac{b_{ji} \cdot p_j}{\displaystyle\sum_{m=1}^{nvox} c_{jm} \cdot \lambda_m^k}
\tag{12.2}
$$

Here, most symbols are the same as in Equation 12.1, but two new matrices, b and c, appear. The difference between the forward-projection transition matrix, b, and the backprojection transition matrix, c, is that b includes scattered gamma rays, as well as attenuation and detector response, while c includes only attenuation and detector response. Zeng and Gullberg list five references for papers in which unmatched matrices are used to speed up the scatter-included reconstruction problem [28]. They also state that "... when noise is present, both Landweber and ML–EM algorithms demonstrate first a short convergent trend then diverge from the desired solution, no matter if a valid or an invalid projector/backprojector pair is used. Therefore, choosing a valid backprojector may not be a very critical factor in a practical image reconstruction problem."

Monte Carlo-Based Correction

Monte Carlo simulation can be used in Equation 12.1 to estimate the scatter in the projection, *s*, given some estimate of the true object. One then ideally iterates the reconstruction over and over to approach the best answer, while updating the estimates of *s* and the image at each iteration. However, the estimation of the scatter in the projection by a Monte Carlo calculation is computationally demanding, and so it may not be practical to update it every time the image estimate is updated. In addition, repeatedly updating the estimate of *s* may not result in significant gain because typically the scatter in a projection is smooth, making its estimate relatively insensitive to the small changes in fine image detail from one iteration to another. Therefore, in some implementations of Monte Carlo-based scatter compensation, the estimate of *s* was generated only once [4] or was updated only after several OS–EM iterations [29].

In the original implementation of iterative-reconstruction-based scatter compensation, the scatter was included in the (single) transition matrix, but the computation time and storage requirements were prohibitive because probability values were included for the many relatively-unlikely-but-possible scatter events [30,31]. The method, in which Monte Carlo simulation created the transition matrix values, was called "inverse Monte Carlo." [Incidentally, the previous edition of this book said that inverse Monte Carlo assigned "the scattered counts to their true origin during tomographic reconstruction." That implies a higher-count final image than if scatter counts are subtracted in the projections or simply accounted for in the reconstruction. Such a higher-count final image was not the case; the statement was due to a misconception on the part of one of the present chapter authors (KFK).] The lack of practicality of inverse Monte Carlo, especially at the time it was introduced, is indicated by the fact that one study investigating it [30] needed to employ a phantom that had no variation along the *z* axis (2D) in order to allow the study to be carried out with realistic reconstruction times and realistic storage requirements.

The past decade has seen considerable development of Monte Carlo-based scatter compensation in fully 3D iterative reconstruction [4,29,32,33]. The approach originally proposed by the group at Utrecht University [32] avoids the need for massive transition matrix storage by performing scatter estimation using a Monte Carlo simulator as a forward projector (to reduce noise in the reconstructions, a nonstochastic projector was used for the primary photons). They developed promising acceleration techniques [34] to make the 3D Monte Carlo-based reconstruction practical including convolution-based forced detection, which combines stochastic photon transport in the object with analytic detector modeling. Other approaches to acceleration were reducing the number of photon histories simulated in the early iterations and reusing the photon tracks calculated in a previous iteration. Thus far, 3D iterative reconstruction with Monte Carlo-based scatter correction has been applied to ^{99m}Tc, ^{201}Tl, ^{131}I, ^{111}In, and simultaneous ^{99m}Tc/^{123}I SPECT studies, as detailed below.

^{99m}**Tc:** Evaluations by Beekman et al. [32] for ^{99m}Tc using a simulated "water-filled" cylindrical phantom showed that global errors (NMSE and NME) were strongly reduced with Monte Carlo-based scatter estimation compared to using an approximate object-shape-dependent scatter point spread function (PSF). Contrast-to-noise curves for the two scatter correction methods were close to each other, but images based on the approximate scatter PSF showed clear artifacts that were adequately removed by using the more accurate Monte Carlo-based forward projector.

More recently, evaluations of their approach have been carried out with clinically realistic experimental ^{99m}Tc emission data and attenuation maps from a dual-detector SPECT system [9]. Four phantom configurations mimicking cardiac perfusion imaging were used for this evaluation. The OS–EM reconstructions included one with only attenuation correction (A) and others with correction for attenuation and detector response and no scatter correction (AD), Monte Carlo-based scatter correction (ADS–MC), or TEW-based scatter correction (ADS–TEW). ADS–MC outperforms the other methods in contrast-to-noise ratio, contrast separability of cold defects, and robustness to anatomical defects. A typical reconstructed myocardium of one of the thorax phantoms is shown in Figure 12.2 for each algorithm. The vertical profiles through the inferior defect show that ADS–MC gives about the same defect contrast as ADS–TEW. The main advantage of ADS–MC over ADS–TEW was its superior noise properties as was evident from contrast-to-noise ratio curves. When compared at equal contrast, noise was approximately 14% lower for the ADS–MC reconstruction algorithm than for the other methods. Quantitatively, the contrast achieved with the ADS–MC algorithm was approximately 10–20% higher than with the other methods, when measured at an equal noise level. Although Monte Carlo-based methods are much more time consuming than energy window-based methods, the complete reconstruction time for a typical cardiac acquisition (30 iterations, 8 subsets) was <9 min on a dual processor (2.66 GHz) PC, which is approaching clinical acceptability. The authors point out that because of the linear nature of their algorithm, computation time can be easily reduced by using more processors in parallel.

Cot et al. evaluated the SPECT quantification of the specific uptake of ^{99m}Tc dopamine transporter ligands with Monte Carlo-based scatter correction in fully 3D iterative reconstruction [4]. The scatter estimate was calculated using the SIMSET Monte Carlo algorithm and unlike the work of Beekman et al., acceleration of the simulation was not a focus of their work. The initial input image to the SIMSET simulator was the first iteration of the OS–EM reconstruction without scatter correction. Evaluation studies with a numerical striatal phantom comparing nominal and estimated scattered counts demonstrated that Monte Carlo yielded an accurate estimate of the spatial distribution of scattered counts. Although the SIMSET scatter estimate was

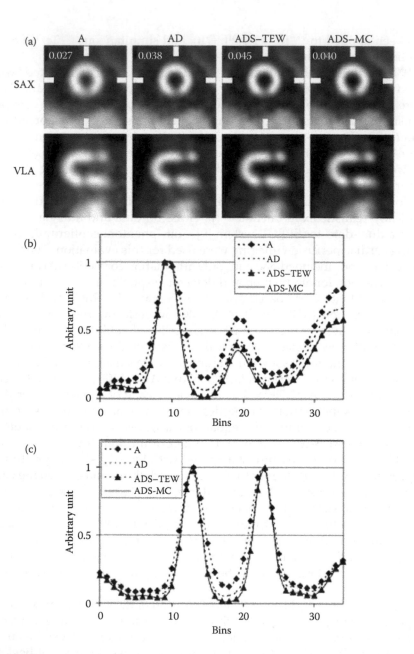

FIGURE 12.2
(a) Top row: short-axis (SAX) views of reconstructions, postfiltered by 3D Gaussian filter
($\sigma = 0.48$ cm). Associated noise level is listed in each image. Bottom row: corresponding vertical
long-axis (VLA) images. (b and c) Vertical profiles through inferior defect (b) and horizontal
profiles (c) in SAX images (each profile was normalized to its maximal pixel value). (Reprinted
by permission of the Society of Nuclear Medicine. Xiao J et al. *J. Nucl. Med.* 2006;47(10):1662–1669,
Figure 4.)

updated multiple times, they showed that the scatter estimate did not change significantly at each update. Hence, the Monte Carlo scatter estimate was generated only once in the rest of their evaluations. Using a numerical striatal phantom, the absolute quantification method that included Monte Carlo-based scatter correction and 3D iterative reconstruction reached 96% and 97% of the nominal specific uptake ratio after 50 iterations for the putamen and the caudate, respectively. The corresponding limits achieved using only primary photons were 98% and 99%. Thus, the quantitative values with Monte Carlo-based scatter correction were almost as good as the ideal results. With all other corrections except scatter correction, the corresponding limits were 89% and 91%. The simulation time to generate the scatter estimate (120 projections) was 12 h on a (1.4 GHz) Linux workstation. Their simulations for ^{99m}Tc did not include scattering in the collimator/detector system because this component is small for low-energy radionuclides, but it should be considered for higher-energy radionuclides.

^{201}**Tl**: Clinically realistic evaluations for ^{201}Tl cardiac imaging similar to those discussed above for ^{99m}Tc cardiac imaging demonstrated that the improvements that can be achieved with Monte Carlo scatter estimation are even larger for ^{201}Tl than for ^{99m}Tc [10]. In experimental phantom measurements, the contrasts achieved with fully 3D Monte Carlo-based scatter correction performed using dual matrix OS–EM were 10–24% higher than those achieved with a method that included only attenuation and detector response during the OS–EM reconstruction. For a typical contrast level, the reconstruction with the Monte Carlo-based scatter correction exhibited noise levels about 27% lower than the other method.

^{131}**I**: Monte Carlo-based scatter compensation with 3D OS–EM reconstruction has also been implemented and evaluated for quantitative ^{131}I SPECT [29]. In ^{131}I SPECT, both object scatter and collimator scatter is highly significant because of the downscatter from the multiple gamma ray emissions at higher energies (637 keV, 722 keV) than the main photopeak energy at 364 keV. Hence, the extensively validated SIMIND Monte Carlo program, which includes accurate physical modeling of the collimator [35], was used to estimate the ^{131}I scatter projections. Simulations included modeling of collimator scatter and penetration and the multiple gamma ray emissions of ^{131}I. As with the implementation of Beekman et al. to reduce noise, an analytical projector was employed for the primary photons. In the 3D OS–EM reconstruction, a TEW scatter estimate was used in the initial iterations and Monte Carlo scatter estimates were used in the latter iterations. Phantom simulation studies demonstrated that after two updates of the Monte Carlo scatter estimate, there was good agreement between the estimated scatter and the true scatter and that further updates did not result in significant improvements. Hence, in the evaluation studies, the scatter estimate was updated only twice (at iteration 20 and iteration 40) with the estimate being held constant in the iterations in between the updates.

In clinically realistic simulations using the voxel-man phantom to mimic tumor and normal organ imaging following ^{131}I radioimmunotherapy, 3D OS–EM reconstruction with Monte Carlo scatter estimation was compared to 3D OS–EM reconstruction with TEW scatter estimation. Figure 12.3 compares profiles across a typical scatter projection. It demonstrates that TEW overestimates the scatter while there is excellent agreement between the second Monte Carlo update (at iteration 40) and the true scatter. Figure 12.4 shows one slice of the defined voxel phantom maps used as input to the Monte Carlo simulator and the corresponding 3D OS–EM reconstructed images (at iteration 41). The degradation in contrast when there is no scatter correction is evident. Both the TEW and Monte Carlo-scatter-corrected images look very similar to the image corresponding to the primary photons. Quantitative evaluations of the results from the voxel-man phantom and an elliptical phantom with spherical "tumors" were carried out using normalized mean square error and bias. Although results with Monte Carlo scatter estimation were superior to results with TEW estimation in almost all cases, the improvements were not large (see Table 12.1). However, more recent comparisons [36] using experimental data have demonstrated shortcomings of the TEW correction under high count rates typical for imaging following

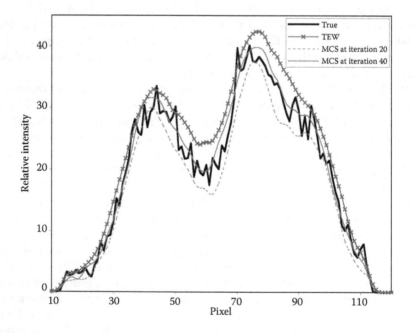

FIGURE 12.3
Profile comparison for the voxel-man phantom. Image is a typical scatter projection. (From Dewaraja YK, Ljungberg M, Fessler JA. 3-D Monte Carlo-based scatter compensation in quantitative I-131 SPECT reconstruction. *IEEE Trans. Nucl. Sci.* 2006;53(1):181–188, Figure 7. Reprinted with permission of the IEEE.)

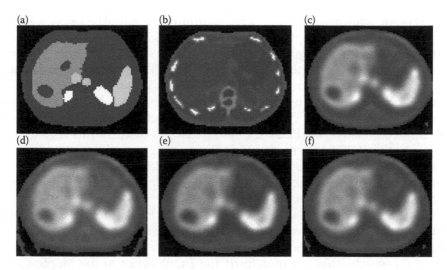

FIGURE 12.4

Slice of voxel-man phantom (a) activity map, (b) attenuation map, (c) reconstructed image corresponding to primary photons only, (d) image reconstructed without scatter correction, (e) image reconstructed with TEW, and (f) image reconstructed with Monte Carlo scatter estimation. (From Dewaraja YK, Ljungberg M, Fessler JA. 3-D Monte Carlo-based scatter compensation in quantitative I-131 SPECT reconstruction. *IEEE Trans. Nucl. Sci.* 2006;53(1):181–188, Figure 8. Reprinted with permission of the IEEE.)

TABLE 12.1

Results from Monte Carlo Simulations for Two Phantoms

	TEW		Monte Carlo	
	%NMSE	%Bias	%NMSE	%Bias
Elliptical Phantom				
Total image	9.9	−5.1	9.5	−1.1
Large hot sphere	10.3	−8.6	9.1	−7.6
Small hot sphere	15.2	−24.3	13.7	−23.1
Warm sphere	8.6	−0.2	9.1	7.6
Voxel Man Phantom				
Total image	15.9	−4.1	15.4	2.3
Liver	6.6	−11.5	5.5	−2.4
Lung	6.8	−3.0	6.1	1.2
Kidney	13.0	−11.4	11.7	−9.0
Spleen	10.5	−11.8	8.4	−7.2

Source: Excerpted from Dewaraja YK, Ljungberg M, Fessler JA. 3-D Monte Carlo-based scatter compensation in quantitative I-131 SPECT reconstruction. *IEEE Trans. Nucl. Sci.* 2006;53(1):181–188. © 2006, IEEE.

Note: Percent normalized mean square error (%NMSE) and percent bias (%bias) for selected regions in images reconstructed using "within-reconstruction" scatter correction that employed a constant TEW scatter estimate or an updatable Monte Carlo scatter estimate.

the therapy administration of ^{131}I radioimmunotherapy. The cause was the observed change in the ^{131}I energy spectra at high count rates, which was attributed to pulse-pile up effects. Under identical scatter conditions, the proportion of photons detected in the narrow TEW scatter windows increased significantly with count rate. Quantification was improved when the TEW correction was replaced by the Monte Carlo correction, which utilizes only the main window data. These experimental phantom measurements were performed on a hybrid SPECT/CT system. In this case, the mis-registration between the SPECT activity map and the x-ray CT-based density map that define the input object to the Monte Carlo simulator is minimized. Hence, the Monte Carlo-based scatter correction is well suited for data acquired with such hybrid systems.

Clinical application of the 3D OS–EM reconstruction with Monte Carlo scatter estimation was demonstrated by a ^{131}I radioimmunotherapy patient study where the patient's measured SPECT activity map and the coregistered CT-based density map defined the input object to the Monte Carlo projector [29]. Although in the patient study, the truth was not known, the scatter estimate approached convergence after 2–3 updates as in the above-discussed phantom studies. The SIMIND simulation time to generate the low-noise scatter estimate with 10^8 photons/projection was 20 h on a DEC alpha workstation, but in this implementation the speed of the simulator was less important than in implementations where the estimate is updated at each iteration.

[111]In: Compensation for scatter is particularly important for [111]In (with gamma emissions at 171 and 245 keV) because of "downscatter," that is, 245 keV photons that scatter such that they lose sufficient energy to be detected within the 171 keV energy window. A fast Monte Carlo simulation algorithm has been developed and included within a joint-ordered subsets expectation maximization approach (denoted MC–JOS–EM) to compensate simultaneously for scatter and cross talk in ^{111}In imaging [33]. The MC–JOS–EM approach extends the Monte Carlo-reconstruction methods developed by the Utrecht group [32,34] by (1) using the "Delta scattering" method to determine photon interaction points, (2) simulating scatter maps for many energy windows simultaneously, and (3) using prestored PSFs that account for all collimator and detector effects. Evaluation studies using a torso phantom were carried out to compare activity quantification accuracy of MC–JOS–EM with that of the TEW approach where the estimated scatter was also included additively within JOS–EM (TEW–JOS–EM). In the Monte Carlo approach, scatter was first corrected by means of the TEW method and the Monte Carlo scatter estimate was calculated (only once) after five iterations of TEW–JOS–EM. The same Monte Carlo scatter estimate was used during the forward projection step of each subsequent iteration. The Monte Carlo method yielded activity estimates for tumor (sphere insert), liver, and background regions in the torso phantom that were more accurate and precise than those obtained with TEW compensation (Table 12.2).

TABLE 12.2

Average Relative Bias and Average Relative Standard Deviation (STD) of ^{111}In Activity Concentration Estimates in the Sphere Insert, Liver, and Background of the Torso Phantom after 40 Iterations

	Sphere Insert		Liver		Background	
	TEW–JOS–EM	MC–JOS–EM	TEW–JOS–EM	MC–JOS–EM	TEW–JOS–EM	MC–JOS–EM
Bias	−15.8%	−6.9%	−10.6%	−5.3%	−13.9%	−2.4%
STD	27.4%	16.1%	4.3%	3.2%	2.5%	1.8%

Source: Excerpted from Ouyang J, El Fakhri G, Moore SC. *Med. Phys.* 2008;35(5):2029–2040.

^{99m}Tc/^{123}I: Simultaneous ^{99m}Tc/^{123}I SPECT allows for the assessment of two physiological functions under the same conditions. However, separation of ^{99m}Tc and ^{123}I images is difficult because their primary emission energies are close (140 and 159 keV, respectively). In addition, ^{123}I has higher-energy emissions between 440 and 625 keV (2.4% of total) and between 625 and 784 keV (0.15% of total), so downscattered ^{123}I photons can be detected in both windows; primary photons of each radionuclide can also be detected in the other window (cross talk). MC–JOS–EM that was discussed above has been investigated for cross talk and scatter compensation in ^{99m}Tc/^{123}I SPECT. In the first of these studies, Monte Carlo-simulated brain data was used for simultaneous brain profusion (^{99m}Tc-HMPAO) and neurotransmission (^{123}I altropane) SPECT [37], while in the second, experimentally measured data from a cardiac torso phantom was used for simultaneous ^{99m}Tc-Sestamibi/^{123}I-BMIPP cardiac SPECT [38]. In these studies, first, the ^{99m}Tc-only and the ^{123}I-only data sets were reconstructed using standard OS–EM and these reconstructed images were used by the Monte Carlo algorithm as the starting images to obtain scatter and cross talk estimates. The OS–EM reconstructions were continued using the Monte Carlo-generated scatter and cross talk estimates during each forward projection process. In both the brain study and the cardiac study, relative bias and relative standard deviation achieved with MC–JOS–EM was significantly better than with standard OS–EM without scatter correction, and with standard OS–EM using a dual-energy-window approach for scatter correction.

Model-Based Correction

In the scatter correction literature, the word "modeling" is not always used rigorously. It sometimes means "calculating" or "calculating approximately." At other times, it means substituting a simpler object, a "model," in place of the true object. Several "model-based" corrections have been derived for ^{99m}Tc and modified for other radionuclides. Model-based corrections have been used with ^{111}In, ^{123}I, simultaneous ^{99m}Tc/^{123}I, and ^{90}Y imaging.

^{99m}Tc, ^{201}Tl: Early efforts to use a model to approximately calculate scattering for an object were those for ^{99m}Tc related to a "slab phantom." The scatter estimate was derived independently by two groups: Frey et al. [39,40] and Beekman et al. [41,42]. The approach required calculating camera scatter response functions for a point source within a slab phantom that had a plane dividing all space into a region filled with a uniformly attenuating medium and a region filled with an air attenuator. The point source was placed within the uniformly attenuating medium at a given distance from the dividing plane and a gamma camera was located on the other side of the plane, parallel to it, and at a specified distance from it. For a given camera setup, the only parameters that governed the scatter response function were the two distances. Without expansion, the model applied well only to a uniform scatter medium. However, extension of the method to account for a nonuniformly attenuating medium was investigated [40].

The correction based on the slab-phantom model has been superseded by one in which the scatter estimate comes from the projection through an attenuating medium of an effective scatter source. Hence, the method is called ESSE [8]. The effective source activity distribution is related to an integral over all space of the true source activity distribution, properly weighted. The implementation of the calculation of the effective scatter source activity distribution involves three convolutions. To use the method, two kernels for those convolutions need to be calculated. To do so, Frey and Tsui employed a modification of the SIMIND Monte Carlo code. The phantom used was a large, but finite, water-filled slab. To validate their new method for both ^{99m}Tc and ^{201}Tl, the authors calculated projections for small cubic sources within a 32 cm by 22 cm elliptical-cross-section phantom that was 40 cm long and compared them to those obtained by Monte Carlo. They found good agreement for both radionuclides. Moreover, scatter to primary ratios were in agreement, and for ^{201}Tl were much superior to those obtained using the slab-based method (in one case, a –1.3% difference with ESSE while the difference was 100% with the slab-based method).

^{111}In: He et al. used Monte Carlo to simulate the data for a comparison of several methods of estimating residence time in an ^{111}In radioimmunotherapy dosimetry and treatment planning application [43]. The quantitative SPECT (QSPECT) method employed attenuation, detector response, partial volume, and ESSE scatter correction. "A modified version of the SIMSET MCS ... code combined with an angular response function (ARF) based simulation of the interactions in the collimator and detector was used to simulate the projection data." The code is documented in a publication by Song et al. [44]. He et al. found that the error in the residence time estimate was only –0.9%, –0.2%, –1.2%, –3.8%, –0.7%, and 1.9% for the heart, lungs, liver, kidney, spleen, and marrow, respectively, when they employed a time series of QSPECT images. The other two methods were a time sequence of planar images (which are frequently used by themselves in the clinic), and a time sequence of planar images that had the resultant activity sequence normalized by one QSPECT image. The all-planar method had much lower accuracy than with the time series of QSPECT images. The normalized method had lower accuracy than with the time series of QSPECT images.

These researchers also carried out a clinical comparison of the all-planar method versus the method employing a time series with the resultant activity sequence normalized by one QSPECT image [6]. Here, the true answer was not known, and so no evaluation of residence time accuracy was attempted. However, for the dose-limiting organ, which was the liver, the average difference between the results from the two methods was −18%. Our conclusion is that using a SPECT normalization in which the SPECT includes a scatter correction has a definite impact in an important clinical application when compared to the frequently used clinical practice.

^{123}**I and** ^{99m}**Tc/**^{123}**I:** The original ESSE has been adapted for estimating scatter and downscatter for ^{123}I [45]. That estimate has then been used during reconstruction, essentially employing Equation 12.1. The extension from the ^{99m}Tc approach to one for simultaneous ^{99m}Tc/^{123}I imaging required including "photon interactions with the collimator-detector system." These were carried out by computing "collimator-detector response functions (CDRFs)." The computation was done using the Monte Carlo N-Particle (MCNP) code. The functions include the effects of backscatter from camera elements behind the crystal [46]. The application of the ^{123}I imaging alone was quantification in the brain. Here, the scatter estimate was validated by Monte Carlo simulation. The application of the simultaneous ^{99m}Tc/^{123}I was again brain imaging. It was shown "for both experimental and simulation studies, the model-based method provided cross talk estimates that were in good agreement with the true cross talk" [46].

^{90}**Y:** For ^{90}Y Bremsstrahlung imaging, the extension from the ^{99m}Tc approach again involved including the collimator-detector response. For the ^{90}Y study, both the ESSE kernels and the collimator-detector response kernels were computed via Monte Carlo simulation using the SIMIND code. The application was quantitative imaging for patients being treated with high-doses of ^{90}Y ibritumomab [47]. Absorbed doses for the liver, spleen, left and right kidney, and left and right lung were calculated from ^{90}Y SPECT for three patients. The absolute value of the difference between the intratherapy ^{90}Y dose and the dose predicted from pretherapy ^{111}In SPECT as a percentage of the predicted dose averaged 11.8% when the left and right lung dose values were excluded. The difference for the lungs was higher by a factor of about four.

Correction by Subtraction

Correction by explicitly subtracting away those counts that are from scattered gamma rays has been applied to many isotopes, as detailed below.

Note that when a method yields an explicit scatter estimate, that estimate could also be used in the within-reconstruction methods already covered and indeed has been with at least one method (TEW).

^{99m}Tc: The first scatter correction scheme to be applied to SPECT was a subtraction method called the dual-energy-window method. It was introduced by Jaszczak et al. in 1984 [48]. It has a key parameter, k, defined as the ratio of the number of scattered counts within a lower-energy scatter window divided by the number of scattered counts within the main photopeak window. To carry out the correction, this parameter multiplies an image reconstructed from the scatter-window projections before it is subtracted from an image reconstructed from the photopeak-window projections (a variation involved subtracting projections). Choosing the value for k is somewhat subjective. Monte Carlo (as well as experiment) was used to make and examine the choice [49,50]. The k value to quantify a hot sphere that is immersed in a warm-cylinder "background" was found to vary strongly with the activity level of the background compared to the activity level of the sphere [3]. To adapt the dual-energy-window method to the task of focal quantification in cases where there was background activity, Luo and Koral introduced an iterative, background-adaptive version. Using Monte Carlo tests, they reported rapid convergence and good preliminary results [51] but the approach has not been implemented and, in general, the dual-energy-window method is no longer used.

In testing another method called spectral fitting, simulated energy spectra within small spatial regions in a projection were simulated using Monte Carlo [52]. The test object was a sphere within a cylinder. The authors then fitted these spectra with a model that assumed the scattered spectrum could be represented by a polynomial. The fitting essentially separated the counts into a scattered total and an unscattered total. The authors used a definition of scatter fraction (SF), as the number of scattered events within a window divided by the number of unscattered counts within that same window. This value, with a standard 20% photopeak window, was calculated from the "spectral fitting" result and compared to the true value from the simulation to test the effectiveness of the method [52]. The authors found that the average difference in SF, averaged over all evaluated points, was 16.2%. Later, a regularized, deconvolution-fitting approach was tested in the same way [53]. The deconvolution-fitting algorithm required knowledge of the camera energy-response function (i.e., the detected-energy spread function for any true gamma-ray energy). Guidance about the regularization parameter, lambda, required for the deconvolution-fitting algorithm was obtained through extensive Monte Carlo testing [53]. However, this method also has not been pursued further.

Currently, a scatter-estimation method widely employed is a subtraction technique called the TEW method. As noted earlier, it has been made available for clinical use by commercial vendors. The method was developed chronologically after the two methods described above; it employs the usual photopeak and adds two windows, one on either side of the photopeak [7]. These windows usually have a width narrower than the window for the photopeak and, ideally, record only scattered counts. A straight line connects the count levels in the two narrow windows to estimate the scatter in the main window. Ogawa et al. emphasized the practicality of their correction method—the computation was simple. Originally, the scatter estimate was subtracted from the total counts in the projection but, as mentioned previously, it can also be used during reconstruction. In developing their method, Ogawa et al. generated tomographic data sets for simple geometric phantoms by Monte Carlo simulation and used images reconstructed from the unscattered photons as their "gold standard." For their "cold-spot" phantom, the visual diameter of the spot was right with TEW correction whereas it was too small without. However, a profile across the image did show slight overcorrection at the bottom of the cold spot [7].

For another method, Monte Carlo simulation was used to generate required "scatter line spread functions" [54]. These functions, computed at given rotation angles with the point source at a given spatial location, are one dimensional, giving the number of scattered counts versus distance. The distance axis can be thought of as in a plane that (1) contains the point source and (2) is perpendicular to the axis of rotation. The object itself is three dimensional and the projection image has the normal two dimensions. The function gets to be one dimensional by integration along the camera face in the direction parallel to the axis of rotation. Knowledge of the functions for the object is required, or some approximate forms are used. Given the functions, Ljungberg and Strand first reconstructed an uncorrected image from the original projection data. This image was used to provide the weights for their convolution. For each pixel, the scatter line spread function whose point-source location was closest to the pixel was employed. The results from the multiple convolutions were estimates of the scatter component. These estimates were subtracted from the original projections and then a corrected image reconstructed. These manipulations can be rationalized as an iterative procedure that, in the implementation, was stopped after only one iteration. A special attenuation correction was combined with the scatter correction. In the reference cited above, Monte Carlo results from a heart phantom were compared to similar results from the phantom containing "gaseous ^{99m}Tc," that is, ^{99m}Tc that neither scatters nor attenuates. The 80 mm diameter ventricle was on the axis of a much larger cylinder or was displaced off axis as much as 100 mm (for the largest, 300 mm diameter cylinder). Reconstructed values for the heart wall were typically improved by correction for both attenuation and scatter from 30% of expected without corrections to 95% of expected with. For the 250 mm diameter, intermediate-sized cylinder, the reconstructed value averaged over five displacements was exactly 100% of expected (the range appeared to be no more than 98–102%) [54].

^{201}Tl **and** ^{111}In: The same authors repeated the tests of their scatter-attenuation-correction method on their heart phantom filled with "^{201}Tl," a pseudoisotope with a single gamma emission at exactly 75 keV [54] and filled with ^{111}In, concentrating only on the emission at 247 keV and neglecting the lower-energy emission at 172 keV, which should be a viable experimental procedure. Resulting reconstructed values for pseudoisotope ^{201}Tl, again averaged over five displacements of the ventricle from the symmetry axis of the enclosing cylinder, were slightly too small for the smallest and largest cylinder sizes, being 88% instead of 100%. For the intermediate-size cylinder, the average value was closer to right at 94% of expected [54]. The resulting values for ^{111}In had almost the same accuracy as did the ^{99m}Tc results quoted above. As far as the authors of this chapter know, the method has not been implemented or tested further.

^{123}I: The higher-energy emissions of ^{123}I make a scatter correction more important than if they were not present, especially in the case where one does not use a medium-energy collimator but rather a low-energy-high-resolution or a low-energy-all-purpose collimator [55]. This fact presumably motivated Takeda et al. to employ subtraction-based TEW scatter correction in a clinical study in which they used a high-resolution parallel-hole collimator [56]. They imaged normal and ischemic hearts during exercise after an injection of an ^{123}I-labeled structurally modified fatty acid, 15-(p-iodophenyl)3-R,S-methyl-pentadecanoic acid (BMIPP). They set the main window at 159 keV ± 12% plus two 3% scatter windows. They calculated a metric from the total counts in the left ventricle, but the details of the study are too clinically complex to be discussed in this chapter. The point of the study inclusion herein is that scatter correction is being employed in clinically relevant research.

Comparison of Correction Methods

Quantification

One Monte Carlo simulation study by Ljungberg and Strand [57] that focused on activity quantification compared four correction methods that use subtraction. These were the dual-energy-window method with $k = 0.5$, the dual-photopeak-window method [58], the TEW method (in two forms but mostly in the form that assumes the counts in the upper scatter window are zero), and the Ljungberg and Strand method that employs scatter line spread functions. According to the authors, the results "indicate that the differences in performance between different types of scatter correction techniques are minimal for ^{99m}Tc brain perfusion imaging. Thus, a user may select a correction method that is easy to implement on a particular system." They also conclude, however, that it is "important to perform a test of scatter correction methods for source distributions which closely match the clinical application to which they will be applied."

A second Monte Carlo simulation comparing methods was concerned with ^{99m}Tc heart imaging [59]. The authors compared three methods using subtraction, the dual-energy-window method, the TEW method, and a factor-analysis method [60,61]. They found that "scatter correction significantly improved contrast and absolute quantitation but did not have noticeable effects on BEM uniformity or on spatial resolution, and reduced the SNR." (Here, BEM stands for bulls eye map and SNR is signal-to-noise ratio.)

A third Monte Carlo study compared "two subtraction based compensation techniques... with image reconstruction using iterative reconstruction including an accurate scatter model" [62]. The subtraction methods were the dual-energy-window method and a method based on convolution of an image with a scatter response function [63]. The scatter response function is defined as the spatial variation of scattered gamma rays about an origin corresponding to a given strength of unscattered gamma rays at that origin. The convolution method assumes that the scatter response function is spatially invariant. Actually, this function is now known to be "highly dependent on the object shape and source position" according to Frey et al. [62]. Among the results from the study are that (1) even judging by the entire image rather than a limited region of interest, the optimum "k" in the dual-energy-window method varies for the case of hot rods compared to that of cold rods, and (2) iterative reconstruction with an accurate scatter model produces the best results by most of the figures of merit employed.

A fourth Monte Carlo study compared nine correction techniques that are based on two or more windows [61]. The test data was generated for a single, top-view projection of a cylindrical ^{99m}Tc-water-solution phantom containing four small cylinders of activity concentration 2, 4, 6, and 8 relative to the background concentration of the cylinder. Two of the nine methods compared have been referenced above: the dual-energy-window method and the TEW method. One of the nine involves no correction at all and was presented merely to characterize the correction problem. The others are the dual-photopeak-energy-window method [64,65], the channel ratio method [66], the method of Logan and McFarland [67], the technique of Bourguignon et al. [68], FADS or the constrained-factor-analysis-of-dynamic-structures method [69], and a variation by Buvat et al. [70] on the factor-analysis method referred to in a paragraph above. The comparison used relative quantification, absolute quantification, and SNR. A pixel-by-pixel analysis of the accuracy of the estimate of the number of unscattered photons gives values of around 10% when the counting statistics are good and worse (15–45% for the dual-energy-window method, for example) and varied among methods when the counting statistics are poor. Here, good counting statistics means the number of unscattered counts is >130 and poor counting statistics means the number of unscattered counts is between 130 down to 30 counts. The factor-analysis approach tended to outperform the other methods. However, the authors acknowledged the need for the "more sophisticated acquisition mode (30 energy windows)" that is still not yet available on the majority of commercial cameras even through list-mode acquisition.

Detection

In a lesion-detectability study that employed ^{67}Ga because of its use in diagnosis or staging of non-Hodgkin's lymphoma, five scatter compensation methods, including two hypothetical ones, were compared to no scatter compensation [71]. The compensation methods were perfect scatter rejection, ideal compensation, TEW scatter estimation, ESSE, and a post-reconstruction scatter subtraction. Compensation for the middle three in this list was carried out during reconstruction. The post-reconstruction method used the TEW scatter estimate to reconstruct an image that was subtracted from the image reconstructed from the photopeak data. Using the area under a human-observer localization receiver operating characteristic curve, A_{LROC}, as the assessment metric, the TEW method ($A_{LROC} = 0.75$) and the ESSE method ($A_{LROC} = 0.73$) were significantly better than no compensation ($A_{LROC} = 0.67$). For comparison purposes, perfect scatter rejection had an A_{LROC} equal to 0.84. Post-reconstruction scatter subtraction had better detection accuracy than no compensation but it was not statistically significant. Surprisingly to us, ideal scatter compensation only achieved that level of improvement as well. The main conclusion is that in this application, scatter compensation had a definite advantage.

Preclinical Studies

Since the last publication of this book, studies have been carried out to determine if scatter correction is needed for the preclinical imaging of small animals (rats and mice), and to test several correction schemes.

One experimental ^{99m}Tc study [72] found that the results for small animals were similar to those encountered for humans in, for example, the study by Cot et al.: quantitative accuracy got better as attenuation correction was added, and then still better as scatter correction was included. The scatter estimate was produced by the classical TEW method or the simplified version of it wherein the small number of counts in the upper window are replaced with a count of exactly zero. The estimate was incorporated within the reconstruction essentially by employing Equation 12.1. The authors imaged a ^{99m}Tc phantom and used either a single pinhole collimator for the SPECT or a collimator that had three pinholes, each of which was focused on a different location along the axis of rotation. Table 12.3 gives the results for their one-pinhole and three-pinhole SPECT. An improvement in the average percent error for the estimate of the activity concentration within nine 1.5 cm

TABLE 12.3

Average Percent Error in the Estimate of Activity Concentration after Four
Types of Reconstruction of Data Acquired with Either One-Pinhole SPECT
or Three-Pinhole SPECT

	% Error	
	One Pinhole	Three Pinhole
No corrections	−30.5 ± 17.9	−24.0 ± 13.3
Only attenuation correction	12.7 ± 6.3	16.6 ± 4.5
Attenuation correction and simplified TEW scatter correction	−5.1 ± 7.6	−0.3 ± 7.5
Attenuation correction and classical TEW scatter correction	1.2 ± 6.0	4.1 ± 5.6

Note: Also given is the standard deviation for that error.

vials containing a range of activity concentrations is seen when scatter is
taken into account. In these results, the classical TEW and the simplified
TEW give comparable results, which are perhaps not too surprising.

Another experimental study employed a small cylindrical phantom
(4.5 cm in diameter and 4.0 cm in length) also to test quantification of the
activity concentration of ^{99m}Tc [73]. The researchers used six iterations of a
pinhole OS–EM algorithm with 16 subsets and the TEW method to estimate
the scatter which was taken account of during reconstruction by the method
indicated in Equation 12.1. Their results are shown in Figure 12.5. It can be
seen that the activity-concentration-versus-distance profile with scatter correction included follows the true level while with attenuation correction but
without scatter correction, the result is about 10% high. The authors used
a point-source measurement to calibrate their results. Even though there is
little scatter in the point source itself, it was important to carry out the TEW
scatter correction for this measurement (there was a 4.35% difference with
the correction versus without). The authors propose that the main reason for
the difference is nonnegligible collimator scatter.

Hwang et al. [74] used Monte Carlo to simulate a 4 mm diameter radioactive
sphere within a rat-sized cylinder of water. The simulations "modeled physical
effects including Compton scatter, coherent (Rayleigh) scatter and photoelectric absorption." They were performed using GATE. The authors found that
when imaging ^{125}I (simulated source energy = 27.5 keV), the scatter-to-primary
ratio increased with cylinder diameter and exceeded 30% when the cylinder
diameter reached 5 cm. This "... can cause overestimation of the radioactivity concentration when reconstructing data with attenuation correction." The
highest value for the ratio in the case of ^{99m}Tc equaled about 10%.

The conclusion is that although one might expect the opposite to be true
due to the small sizes involved, scatter correction is important for quantitative accuracy in preclinical imaging.

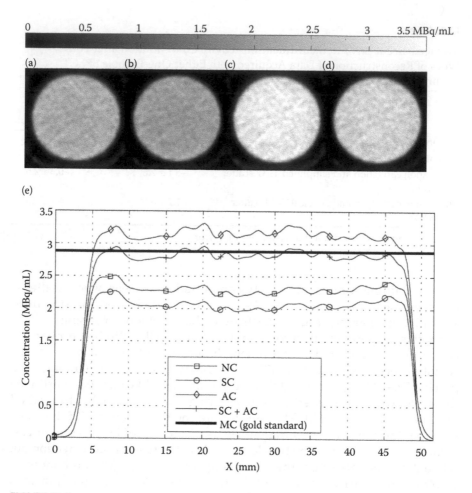

FIGURE 12.5

Averages of 10 transverse SPECT slices of the cylindrical phantom. The gray scales of a–d are the same, from black (0 MBq/mL) to white (3.5 MBq/mL). (a) Without corrections (NC). (b) With scatter correction (SC). (c) With attenuation correction (AC). (d) With scatter and attenuation correction (SC + AC). (e) Line profiles through center of phantom. The line MC indicates the concentration measured with a dose calibrator, as a gold standard. (From *Eur. J. Nucl. Med. Mol. Imag.*, Absolute quantitative total-body small-animal SPECT with focusing pinholes, 37(11), 2010, 2127–2135, Wu et al., Figure 3. Reprinted with permission of Springer Science + Business Media.)

Alternatives to Monte Carlo

An alternative to Monte Carlo for calculating a scatter estimation is a "a theoretically based method which uses analytical calculations" [75]. The method "uses no free parameters and computes the distribution of primary, first-order, and second-order Compton scattered photons." The calculation is based on the Klein–Nishina differential cross section for Compton scattering. Readers are referred to the reference for the details on how the calculation is carried out. At the time of their measurement in 1998, the authors pointed out that their method was faster than Monte Carlo for a point source, required comparable time when the image volume consisted of 800 voxels, and was considerably slower for larger image volumes.

Two experimental alternatives to Monte Carlo simulation are available for estimating the amount of scatter. For a point source, one simply measures first with the point source in the scattering medium and then again at the exact same location in air (i.e., the scattering medium has been removed). For more complicated objects, however, there is no way to avoid self-scatter and self-attenuation, assuming the isotope is not available in a gaseous form. The second experimental approach is to use a high-purity germanium detector. This detector has much higher energy resolution than NaI(T1) and so scattered and unscattered photons are almost completely separated by their detected energies. The complications of measuring scatter fractions with this detector are detailed in a study by Zasadny et al. [76]. Since the method does have its own complications, Monte Carlo is still the method of choice for the development and testing of scatter-correction schemes.

Additions to Monte Carlo

Although it might be possible for Monte Carlo to simulate them, there are complexities in the detection problem which are probably too camera specific for Monte Carlo to be appropriate. The example that comes to mind is local shift of energy spectra with rotation angle [77]. If the effect of such shifts is to be taken into account, it would probably be best to add on a post-simulation, non-Monte Carlo module which would require and utilize input from previous experimental measurements on the specific camera.

Acknowledgments

One of the authors (KFK) would like to acknowledge helpful discussions with Dr. Eric C. Frey and Dr. Jeffrey E. Fessler in regard to certain aspects of this chapter. However, opinions in this chapter are those of the authors and they are responsible for any errors or omissions in the presentation. One of the authors (YKD) would like to acknowledge the support of grant 2R01 EB001994 from the National Institute of Health, United States Department of Health and Human Services.

References

1. Evans RD. *The Atomic Nucleus*. New York: McGraw-Hill 1955.
2. Floyd JL, Mann RB, Shaw A. Changes in quantitative SPECT thallium-201 results associated with the use of energy-weighted acquisition. *J. Nucl. Med.* 1991;32(5):805–807.
3. Luo JQ, Koral KF, Ljungberg M, Floyd CE, Jr., Jaszczak RJ. A Monte Carlo investigation of dual-energy-window scatter correction for volume-of-interest quantification in 99Tcm SPECT. *Phys. Med. Biol.* 1995;40(1):181–199.
4. Cot A, Falcon C, Crespo C et al. Absolute quantification in dopaminergic neurotransmission SPECT using a Monte Carlo-based scatter correction and fully 3-dimensional reconstruction. *J. Nucl. Med.* 2005;46(9):1497–1504.
5. Dewaraja YK, Schipper MJ, Roberson PL et al. 131I-tositumomab radioimmunotherapy: Initial tumor dose-response results using 3-dimensional dosimetry including radiobiologic modeling. *J. Nucl. Med.* 2010;51(7):1155–1162.
6. He B, Wahl RL, Sgouros G et al. Comparison of organ residence time estimation methods for radioimmunotherapy dosimetry and treatment planning-patient studies. *Med. Phys.* 2009;36(5):1595–1601.
7. Ogawa K, Harata Y, Ichihara T, Kubo A, Hashimoto S. A practical method for position-dependent compton-scatter correction in single photon-emission CT. *IEEE Trans. Med. Imag.* 1991;10(3):408–412.
8. Frey EC, Tsui BMW. A new method for modeling the spatially-variant, object-dependent scatter response function in SPECT. *Conference Records of the IEEE Nuclear Science and Medical Imaging Conference*, Albuquerque, NM, November 9–15, 1997;2: pp. 1082–1086.
9. Xiao J, de Wit TC, Staelens SG, Beekman FJ. Evaluation of 3D Monte Carlo-based scatter correction for 99mTc cardiac perfusion SPECT. *J. Nucl. Med.* 2006;47(10):1662–1669.
10. Xiao J, de Wit TC, Zbijewski W, Staelens SG, Beekman FJ. Evaluation of 3D Monte Carlo-based scatter correction for 201Tl cardiac perfusion SPECT. *J. Nucl. Med.* 2007;48(4):637–644.
11. Hutton BF, Buvat I, Beekman FJ. Review and current status of SPECT scatter correction. *Phys. Med. Biol.* 2011;56(14):R85–112.
12. Buvat I, Benali H, Todd-Pokropek A, Di Paola R. Scatter correction in scintigraphy: The state of the art. *Eur. J. Nucl. Med.* 1994;21(7):675–694.
13. Zaidi H, Koral KF. Scatter modelling and compensation in emission tomography. *Eur. J. Nucl. Med.* 2004;31(5):761–782.
14. Zaidi H, Koral KF. Scatter correction strategies in emission computed tomography. In: Zaidi H, ed. *Quantitative Analysis of Nuclear Medicine Images*. New York: Springer Science and Business Media 2005: pp. 205–235.
15. King MA, Glick SJ, Pretorius PH et al. Attenuation, scatter, and spatial resolution compensation in SPECT. In: Wernick MN, Aarsvold JN, eds. *Emission Tomography. The Fundamentals of PET and SPECT*. San Diego: Elsevier 2004.
16. Rosenthal MS, Cullom J, Hawkins W, Moore SC, Tsui BM, Yester M. Quantitative SPECT imaging: A review and recommendations by the Focus Committee of the Society of Nuclear Medicine Computer and Instrumentation Council. *J. Nucl. Med.* 1995;36(8):1489–1513.

17. Koral KF. Monte Carlo in SPECT scatter correction. In: Ljungberg M, Strand S-E, King MA, eds. *Monte Carlo Calculations in Nuclear Medicine: Applications in Diagnostic Imaging.* Bristol; Philadelphia: Institute of Physics Pub. 1998:xxi, p. 309.

18. Ljungberg M, Strand SE. A Monte Carlo program for the simulation of scintillation camera characteristics. *Comput. Methods Programs Biomed.* 1989; 29(4):257–272.

19. Harrison RL, Haynor DR, Gillispie SB, Vannoy SD, Kaplan MS, Lewellen TK. A public-domain simulation system for emission tomography—Photon tracking through heterogeneous attenuation using importance sampling. *J. Nucl. Med.* 1993;34(5):P60.

20. Agostinelli S, Allison J, Amako K, et al. Geant4—A simulation toolkit. *Nuclear Instruments & Methods in Physics Research Section A-Accelerators Spectrometers Detectors and Associated Equipment.* 2003;506(3):250–303.

21. Allison J, Amako K, Apostolakis J et al. GEANT4 developments and applications. *IEEE Trans. Nucl. Sci.* 2006;53(1):270–278.

22. Santin G, Strul D, Lazaro D et al. GATE: A GEANT4-based simulation platform for PET and SPECT integrating movement and time management. *IEEE Trans. Nucl. Sci.* 2003;50(5):1516–1521.

23. Strul D, Santin G, Lazaro D, Breton V, Morel C. GATE (Geant4 Application for Tomographic Emission): A PET/SPECT general-purpose simulation platform. *Nuclear Physics B-Proceedings Supplements.* 2003;125:75–79.

24. Staelens S, Strul D, Santin G et al. Monte Carlo simulations of a scintillation camera using GATE: Validation and application modelling. *Phys. Med. Biol.* 2003;48(18):3021–3042.

25. De Beenhouwer J, Staelens S, Vandenberghe S, Lemahieu I. Acceleration of GATE SPECT simulations. *Med. Phys.* 2008;35(4):1476–1485.

26. Bowsher JE, Johnson VE, Turkington TG, Jaszczak RJ, Floyd CE, Coleman RE. Bayesian reconstruction and use of anatomical a Priori information for emission tomography. *IEEE Trans. Med. Imag.* 1996;15(5):673–686.

27. Kamphuis C, Beekman FJ, van Rijk PP, Viergever MA. Dual matrix ordered subsets reconstruction for accelerated 3D scatter compensation in single-photon emission tomography. *Eur. J. Nucl. Med.* 1998;25(1):8–18.

28. Zeng GSL, Gullberg GT. Unmatched projector/backprojector pairs in an iterative reconstruction algorithm. *IEEE Trans. Med. Imag.* 2000;19(5):548–555.

29. Dewaraja YK, Ljungberg M, Fessler JA. 3-D Monte Carlo-based scatter compensation in quantitative I-131 SPECT reconstruction. *IEEE Trans. Nucl. Sci.* 2006;53(1):181–188.

30. Bowsher JE, Floyd CE, Jr. Treatment of Compton scattering in maximum-likelihood, expectation-maximization reconstructions of SPECT images. *J. Nucl. Med.* 1991;32(6):1285–1291.

31. Floyd CE, Jr., Jaszczak RJ, Greer KL, Coleman RE. Inverse Monte Carlo as a unified reconstruction algorithm for ECT. *J. Nucl. Med.* 1986;27(10):1577–1585.

32. Beekman FJ, de Jong HW, van Geloven S. Efficient fully 3-D iterative SPECT reconstruction with Monte Carlo-based scatter compensation. *IEEE Trans. Med. Imag.* 2002;21(8):867–877.

33. Ouyang J, El Fakhri G, Moore SC. Improved activity estimation with MC-JOSEM versus TEW-JOSEM in ^{111}In SPECT. *Med. Phys.* 2008;35(5):2029–2040.

34. de Jong HWAM, Slijpen ETP, Beekman FJ. Acceleration of Monte Carlo SPECT simulation using convolution-based forced detection. *IEEE Trans. Nucl. Sci.* 2001;48(1):58–64.
35. Ljungberg M, Larsson A, Johansson L. A new collimator simulation in SIMIND based on the delta-scattering technique. *IEEE Trans. Nucl. Sci.* 2005;52(5):1370–1375.
36. Dewaraja YK, Ljungberg M, Koral KF. Effects of dead time and pile up on quantitative SPECT for I-131 dosimetric studies. *J. Nucl. Med.* 2008;49:47P.
37. Ouyang J, El Fakhri G, Moore SC. Fast Monte Carlo based joint iterative reconstruction for simultaneous 99mTc/123I SPECT imaging. *Med. Phys.* 2007;34(8):3263–3272.
38. Ouyang J, Zhu X, Trott CM, El Fakhri G. Quantitative simultaneous 99mTc/123I cardiac SPECT using MC-JOSEM. *Med. Phys.* 2009;36(2):602–611.
39. Frey EC, Ju ZW, Tsui BMW. A fast projector-backprojector pair modeling the asymmetric, spatially varying scatter response function for scatter compensation in SPECT imaging. *IEEE Trans. Nucl. Sci.* 1993;40(4):1192–1197.
40. Frey EC, Tsui BMW. A practical method for incorporating scatter in a projector-backprojector for accurate scatter compensation in SPECT. *IEEE Trans. Nucl. Sci.* 1993;40(4):1107–1116.
41. Beekman FJ, Eijkman EGJ, Viergever MA, Borm GF, Slijpen ETP. Object shape dependent PSF model for SPECT imaging. *IEEE Trans. Nucl. Sci.* 1993;40(1):31–39.
42. Beekman FJ, Frey EC, Kamphuis C, Tsui BMW, Viergever MA. A new phantom for fast determination of the scatter response of a gamma-camera. *Conference Records of the IEEE Nuclear Science and Medical Imaging Conference*, San Francisco, CA, October 30–November 6, 1993: pp. 1847–1851.
43. He B, Wahl RL, Du Y et al. Comparison of residence time estimation methods for radioimmunotherapy dosimetry and treatment planning—Monte Carlo simulation studies. *IEEE Trans. Med. Imag.* 2008;27(4):521–530.
44. Song X, Segars WP, Du Y, Tsui BMW, Frey EC. Fast modelling of the collimator-detector response in Monte Carlo simulation of SPECT imaging using the angular response function. *Phys. Med. Biol.* 21 2005;50(8):1791–1804.
45. Du Y, Tsui BM, Frey EC. Model-based compensation for quantitative 123I brain SPECT imaging. *Phys. Med. Biol.* 2006;51(5):1269–1282.
46. Du Y, Tsui BM, Frey EC. Model-based crosstalk compensation for simultaneous 99mTc/123I dual-isotope brain SPECT imaging. *Med. Phys.* 2007;34(9):3530–3543.
47. Minarik D, Sjogreen-Gleisner K, Linden O et al. 90Y Bremsstrahlung imaging for absorbed-dose assessment in high-dose radioimmunotherapy. *J Nucl Med.* 2010;51(12):1974–1978.
48. Jaszczak RJ, Greer KL, Floyd CE, Harris CC, Coleman RE. Improved SPECT quantification using compensation for scattered photons. *J. Nucl. Med.* 1984;25:893–900.
49. Floyd CE, Jaszczak RJ, Harris CC, Greer KL, Coleman RE. Monte Carlo evaluation of Compton scatter subtraction in single photon emission computed tomography. *Med. Phys.* 1985;12(6):776–778.
50. Ljungberg M, Msaki P, Strand SE. Comparison of dual-window and convolution scatter correction techniques using the Monte-Carlo method. *Phys. Med. Biol.* 1990;35(8):1099–1110.
51. Luo JQ, Koral KF. Background-adaptive dual-energy-window correction for volume-of-interest quantification in 99m-Tc SPECT. *Nucl. Instr. Meth Phys. Res.* 1994;A353:340–343.

52. Koral KF, Wang X, Zasadny KR et al. Testing of local gamma-ray scatter fractions determined by spectral fitting. *Phys. Med. Biol.* 1991;36(2):177–190.
53. Wang X, Koral KF. A regularized deconvolution-fitting method for Compton-sctter correction in SPECT. *IEEE Trans. Med. Imag.* 1992;11:351–360.
54. Ljungberg M, Strand SE. Scatter and attenuation correction in SPECT using density maps and Monte Carlo simulated scatter functions. *J. Nucl. Med.* 1990;31(9):1560–1567.
55. Gilland DR, Jaszczak RJ, Turkington TG, Greer KL, Coleman RE. Volume and activity quantitation with iodine-123 SPECT. *J. Nucl. Med.* 1994;35(10):1707–1713.
56. Takeda K, Saito K, Makino K et al. Iodine-123-BMIPP myocardial washout and cardiac work during exercise in normal and ischemic hearts. *J. Nucl. Med.* 1997;38(4):559–563.
57. Ljungberg M, King MA, Hademenos GJ, Strand SE. Comparison of four scatter correction methods using Monte Carlo simulated source distributions. *J. Nucl. Med.* 1994;35(1):143–151.
58. King MA, Hademenos GJ, Glick SJ. A dual-photopeak window method for scatter correction. *J. Nucl. Med.* 1992;33(4):605–612.
59. El Fakhri G, Buvat I, Benali H, Todd-Pokropek A, Di Paola R. Relative impact of scatter, collimator response, attenuation, and finite spatial resolution corrections in cardiac SPECT. *J. Nucl. Med.* 2000;41(8):1400–1408.
60. Bazin JP, Di Paola R, Gibaud B, Rougier P, Tubiana M. Factor analysis of dynamic scintigraphic data as a modelling method: An application to the detection of metastases. In: Di Paola R, Kahn E, eds. *Information Processing in Medical Imaging*. Paris: INSERM 1980: pp. 345–366.
61. Buvat I, Rodriguez-Villafuerte M, Todd-Pokropek A, Benali H, Di Paola R. Comparative assessment of nine scatter correction methods based on spectral analysis using Monte Carlo simulations. *J. Nucl. Med.* 1995;36(8):1476–1488.
62. Frey EC, Tsui BMW, Ljungberg M. A comparison of scatter compensation methods in SPECT—Subtraction-based techniques versus iterative reconstruction with accurate modeling of the scatter response. *Conference Records of the IEEE Nuclear Science and Medical Imaging Conference*, Orlando, FL, October 25–31, 1992: pp. 1035–1037.
63. Axelsson B, Msaki P, Israelsson A. Subtraction of Compton-scattered photons in single-photon emission computerized tomography. *J. Nucl. Med.* 1984;25(4):490–494.
64. Hademenos GJ, Ljungberg M, King MA, Glick SJ. A Monte-Carlo investigation of the dual photopeak window scatter correction method. *Conference Records of the IEEE Nuclear Science and Medical Imaging Conference*, Sante Fe, NM, November 2–9, 1991: pp. 1814–1821.
65. Hademenos GJ, Ljungberg M, King MA, Glick SJ. A Monte-Carlo investigation of the dual photopeak window scatter correction method. *IEEE Trans. Nucl. Sci.* 1993;40(2):179–185.
66. Pretorius PH, van Rensburg AJ, van Aswegen A, Lotter MG, Serfontein DE, Herbst CP. The channel ratio method of scatter correction for radionuclide image quantitation. *J. Nucl. Med.* 1993;34(2):330–335.
67. Logan KW, Mcfarland WD. Single photon scatter compensation by photopeak energy-distribution analysis. *IEEE Trans. Med. Imag.* 1992;11(2):161–164.
68. Bourguignon MH, Wartski M, Amokrane N et al. Le spectre du rayonnement diffuse dams la fenetre du photopic analyse et proposition d'une methode de correction. *Medecine Nucleaire.* 1993;17:53–58.

69. Mas J, Hannequin P, Ben Younes R, Bellaton B, Bidet R. Scatter correction in planar imaging and SPECT by constrained factor analysis of dynamic structures (FADS). *Phys. Med. Biol.* 1990;35(11):1451–1465.
70. Buvat I, Benali H, Frouin F, Bazin JP, Dipaola R. Target apex-seeking in factor-analysis of medical image sequences. *Phys. Med. Biol.* 1993;38(1):123–138.
71. Farncombe TH, Gifford HC, Narayanan MV, Pretorius PH, Frey EC, King MA. Assessment of scatter compensation strategies for (67)Ga SPECT using numerical observers and human LROC studies. *J. Nucl. Med.* 2004;45(5):802–812.
72. Vanhove C, Defrise M, Bossuyt A, Lahoutte T. Improved quantification in single-pinhole and multiple-pinhole SPECT using micro-CT information. *Eur. J. Nucl. Med. Mol. Imag.* 2009;36(7):1049–1063.
73. Wu C, van der Have F, Vastenhouw B, Dierckx RA, Paans AM, Beekman FJ. Absolute quantitative total-body small-animal SPECT with focusing pinholes. *Eur. J. Nucl. Med. Mol. Imag.* 2010;37(11):2127–2135.
74. Hwang AB, Franc BL, Gullberg GT, Hasegawa BH. Assessment of the sources of error affecting the quantitative accuracy of SPECT imaging in small animals. *Phys. Med. Biol.* 2008;53(9):2233–2252.
75. Wells RG, Celler A, Harrop R. Analytical calculation of photon distributions in SPECT projections. *IEEE Trans. Nucl. Sci.* 1998;45(6):3202–3214.
76. Zasadny KR, Koral KF, Floyd CE, Jaszczak RJ. Measurement of compton-scattering in phantoms by germanium detector. *IEEE Trans. Nucl. Sci.* 1990;37(2):642–646.
77. Koral KF, Luo JQ, Ahmad W, Buchbinder S, Ficaro EP. Changes in local energy spectra with SPECT rotation for two anger cameras. *IEEE Trans. Nucl. Sci.* 1995;42:1114–1119.

13

Monte Carlo in Quantitative 3D PET: Scatter

Robert L. Harrison and Robert S. Miyaoka

CONTENTS

Most clinical and preclinical PET scans are now done in 3D mode (septaless). Many modern tomographs do not even provide the option of 2D scanning. 3D PET offers a factor of 6–7 in true coincidence sensitivity over 2D PET systems for typical clinical scanners, and potentially higher sensitivity improvements if longer axial field-of-view (FOV) is built [1,2]. In the clinic, higher coincidence sensitivity translates into better statistical quality of collected data or a reduction in scanning time or administered activity of studies while maintaining similar counting statistics. For preclinical studies, ultra-high-resolution small animal imaging systems with <1.6 mm axial slice spacing are already in use [3–6]. These scanners require the sensitivity of 3D data acquisition [7].

3D PET has been particularly successful for detection studies, as seen in the increasing use and acceptance of whole-body scans to detect metastases. Its application to quantitative tasks has proved more difficult with challenges caused by both noise and bias. Many factors contribute to these, including the Poisson noise inherent in the sinograms; correction for geometric factors and normalization, random events, attenuation, and scatter coincidences; image reconstruction; partial volume effects; dose calibration noise and errors; tomograph drift and intertomograph variations; and the variability seen in test–retest experiments, one component of which is biological variability [8–23].

In this chapter, we focus on scatter coincidences and corrections for them. Scatter correction remains one of the major sources of both noise and bias in the image reconstruction, especially for imaging of the torso. Its contribution to image noise is similar to that of randoms correction, with randoms being a bigger problem at high coincidence rates and scatter at low coincidence rates. Attenuation and scatter correction are major contributors of bias in the image reconstruction in the torso. Attenuation corrections are plagued by patient motion, partial volume, and problems translating the CT measurements (the most common way of measuring the attenuation map) to 511 keV [24]; however, the correction method itself is clear, the errors are in the measurement process. In contrast, there is no clear measurement process for scatter contamination in PET. Monte Carlo simulation is an essential tool for the study of scatter coincidences.

System characteristics are usually defined by global measurements (e.g., noise equivalent count rate [25]) and are measured using homogeneous and symmetric phantoms [26]. Here, the Monte Carlo method provides accurate information about true, scatter, and random events for the simulated imaging protocols and tomograph geometries. Furthermore, the method is also useful for performing detailed studies of the characteristics of true, scatter, and random events for complex imaging environments. Since accurate methods exist to estimate random events [20,27], the Monte Carlo method has not been required to study these effects. In contrast, when imaging complex, inhomogeneous objects it is difficult to distinguish scattered events from true events. The development of efficient Monte Carlo simulation packages allows the study of scatter for realistic activity distributions and attenuation maps to formulating and evaluating scatter correction techniques.

In 2D PET, if an unscattered photon from a coincidence passed through the septa, it was quite likely that its annihilation partner, being close to antiparallel, would also; if the partner photon scattered, however, the change in direction made this much less likely (Figure 13.1). Thus the septa that defined the slices also helped to reduce the scatter fraction, F_s, in the data [28]

$$F_s = \frac{S}{T + S} \tag{13.1}$$

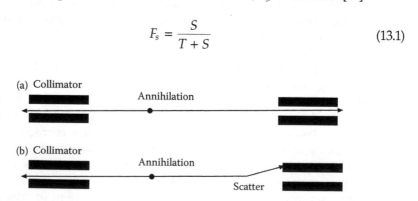

FIGURE 13.1
Reduction of scatter fraction with collimation. Here, we show cross sections through a collimator with (a) an annihilation with both photons unscattered and (b) an annihilation with one photon scattered. With the unscattered photons (a), if one photon passes through the collimator, the fact that the two photons are collinear makes it quite likely that the second photon will also pass. However, when one of the photons scatters (b), the loss of collinearity makes this less likely.

TABLE 13.1

Scatter Fraction versus Lower Energy Threshold for a
37 × 48 cm Chest Phantom

LET (keV)	Scatter Fraction (%)
250	60
300	57
350	52
400	48
450	41

Source: Results taken from Badawi R et al. *Phys. Med. Biol.*
1996;41:1755.

where S is the number of scatter coincidences and T is the number of true coincidences [26]. Typically scatter fractions were <20%; as a result, relatively large biases in the scatter correction resulted in still-acceptable image bias. However, without septa the scatter fraction increases dramatically, reaching over 50% in torso imaging (meaning there are more scatter coincidences than true coincidences—see Table 13.1 [29]). Thus biases in 3D PET scatter correction have a large impact on the accuracy of reconstructed images.

Characteristics of Scatter in 3D PET

The first step in developing scatter corrections for 3D PET is to understand the characteristics of scatter. To get an idea of the distribution of scatter, we performed several simulations using SimSET. The tomograph geometry we simulated was loosely based on the General Electric Discovery STE PET/CT scanner: we represented the detector as a 3-cm-thick cylinder of either bismuth germanium oxide (BGO) or lutetium oxyorthosilicate (LSO) and included collimator endplates, but did not split the detector into blocks or include the patient port covers, detector housing, and so on. This way we can explore the characteristics of scatter in the object without complicating factors like the variation of sensitivity and energy resolution within the detector blocks. These simulations reproduce earlier work reported in the literature, but by using more modern tomograph geometries and assumptions—for instance, the endplates on most modern tomographs are not as deep. While we only include data from our simulations, we have tried to include references to previous work. The simulations can thus be seen as validation of this previous work and a demonstration of the insensitivity of many characteristics of scatter to tomograph geometry.

Almost all scatter in the patient is Compton scatter: for 511 keV photons in water, 99.8% of scatters are Compton; in bone 99.5%; even in lead and tungsten, over 85% of 511 keV photon scatters are Compton. As shown in Figure 13.2, photons lose energy and are often deflected through large angles when they Compton scatter (in contrast to coherent scatter, where they do not lose energy and are usually deflected <5°). In Figure 13.2a, we show the probability distribution for Compton scatter angles at 511 keV; Figure 13.2b shows the relationship between scatter angle and photon energy. We could eliminate almost all scatter coincidences from the object and the great majority from the collimator using an energy window. Unfortunately, the detectors used for PET do not have good enough energy resolution to do so, with energy resolutions ranging from 7% to 20% full-width half maximum (FWHM): over 40% of all scatter is <30°, leaving the photon energy over 450 keV.

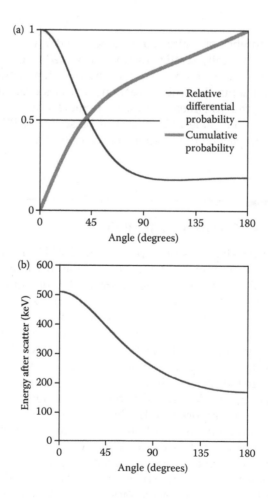

FIGURE 13.2
(a) The differential and cumulative probability of Compton scattering by the deflection angle.
(b) The energy of a 511 keV photon after it Compton scatters through a given angle.

In Figure 13.3, we show the energy spectrums for true coincidences, single-scattered coincidences (one unscattered photon, one single-scattered photon), and multiple-scattered coincidences (both photons scattered or one photon scattered more than once) emitted from a 20-cm-diameter, 70-cm-long flood source in water. Coincidences consist of two photons: for the

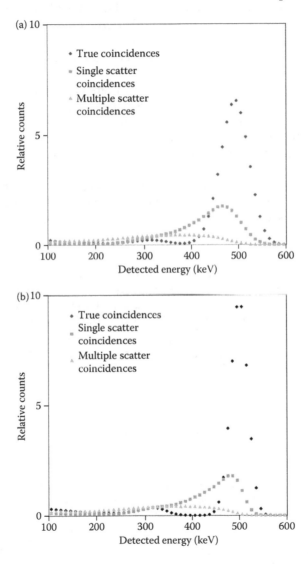

FIGURE 13.3
Energy spectrum from a 20-cm-diameter, 70-cm-long flood source in water for true coincidences, single-scatter coincidences, and multiple-scatter coincidences. The spectrum uses the lower detected energy of the two photons in a coincidence. (a) Simulation of a tomograph with a cylindrical BGO detector, 17% FWHM energy resolution. (b) Simulation of a tomograph with a cylindrical LSO detector, 10% FWHM energy resolution.

energy spectrums we assigned each coincidence the lower detected energy
of its photon pair. We simulated 17% energy resolution for the BGO detectors
and 10% for the LSO. We see that in both cases the peak of the single-scatter
energy spectrum lies well within the peak area of the true-photon spectrum.
To avoid losing too many true coincidences, the lower energy threshold
needs to be set somewhere near the lower edge of the photopeak; the threshold is usually set between 350 and 480 keV, depending on energy resolution.
These thresholds also accept a lot of scatter. 3D PET scatter fractions for clinical scanners, measured in a 20-cm-diameter cylinder, have been reported in
the range of 27–52% [30–35].

The majority of the scatters are single scatters (Tables 13.2 and 13.3, [36]).
The percentage of multiple scatters gets larger as the phantom gets larger and
as the lower energy threshold is relaxed. For typical lower energy thresholds
and objects, the single scatter portion of the scatter ranges from 60% to 90%.

In addition to the energy resolution and threshold, the scatter fraction
depends on the tomograph geometry: detector ring diameter, the axial FOV,
and the depth and positioning of the endplates [2,28,36,37]. While the function of collimating events and eliminating some of the scatter events from
inside the FOV is lost in 3D PET, the endplates do block some of the scattered

TABLE 13.2

Percent of Scatter Coincidences Which Are Single Scatter (One
Unscattered Photon, One Single-Scattered Photon) in Simulations of a
20-cm-Diameter Cylinder, a 35-cm-Diameter Cylinder, and the Zubal
Phantom in a Tomograph with a Cylindrical BGO Detector

Lower Energy Threshold (keV)	20-cm-Diameter Cylinder (%)	35-cm-Diameter Cylinder (%)	Zubal Phantom (%)
300	69	54	58
350	75	61	66
400	81	71	75
450	88	81	83

TABLE 13.3

Percent of Scatter Coincidences Which Are Single Scatter (One
Unscattered Photon, One Single-Scattered Photon) in Simulations of a
20-cm-Diameter Cylinder, a 35-cm-Diameter Cylinder, and the Zubal
Phantom in a Tomograph with a Cylindrical LSO Detector

Lower Energy Threshold (keV)	20-cm-Diameter Cylinder (%)	35-cm-Diameter Cylinder (%)	Zubal Phantom (%)
300	68	53	58
350	75	61	66
400	81	71	75
450	89	82	84

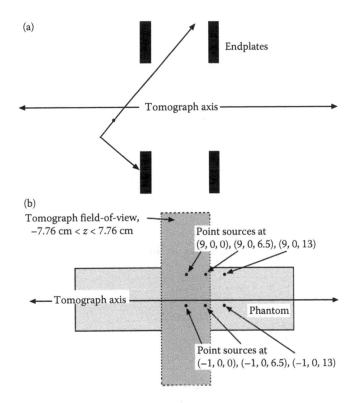

FIGURE 13.4
3D PET scanners are quite open to scatter from outside the field-of-view. (a) The scanner end-plates do, however, provide some protection. (b) Experimental setup for quantifying scatter from outside the field-of-view. Point sources were simulated at six different positions in a 20-cm-diameter, 70-cm-long water cylinder and sinograms of the scatter acquired for direct slices at $z = 0$ cm and $z = 6.5$ cm.

photons from outside the FOV (Figure 13.4a). Despite this, 3D PET scanners are very sensitive to scatter from outside the FOV. We placed point sources at a variety of positions in our 20 cm phantom (Figure 13.4b) to demonstrate sensitivity: three points close to the axis of the phantom, one near tomograph center at (−1, 0, 0), one about a centimeter inside the FOV at (−1, 0, 6.5), and one a little outside the FOV at (−1, 0, 13); another three points were placed at the same axial locations, but near the edge of the phantom, at $x = 9$ cm rather than $x = −1$ cm. Figure 13.5 shows profiles through sinograms of the point source simulations for slices centered at $z = 0$ cm and $z = 6.5$ cm.

The scatter profiles are very smooth. However, the tomograph is quite sensitive to activity outside the FOV, indeed, in the sinogram slice at $z = 6.5$ cm the sensitivity to sources at $z = 13$ cm—over 5 cm outside the FOV—is only slightly less than the sensitivity to sources at $z = 0$ cm, and the shape of the response looks very similar. Even at the tomograph center, the sources outside the FOV contribute substantially to the scatter response [38–41]: with the

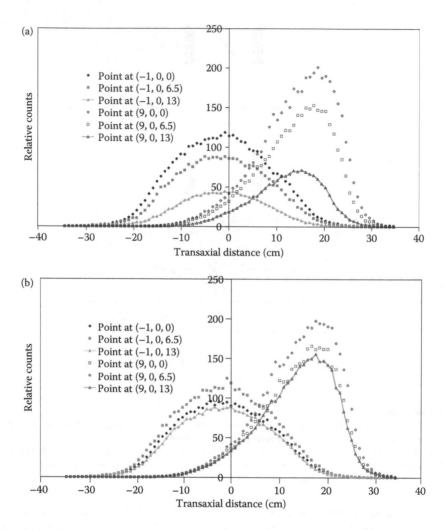

FIGURE 13.5
Profiles through sinograms of scatter coincidences from point sources. (a) Profiles through a direct slice centered at $z = 0$ cm axially for a BGO tomograph with 17% FWHM energy resolution. (b) Profiles through a direct slice centered at $z = 6.5$ cm axially for a BGO tomograph with 17% FWHM energy resolution. (c) Profiles through a direct slice centered at $z = 0$ cm axially for an LSO tomograph with 10% FWHM energy resolution. (d) Profiles through a direct slice centered at $z = 6.5$ cm axially for an LSO tomograph with 10% FWHM energy resolution.

425 keV threshold, the points outside the FOV contribute 13–15% as much scatter as the points at $z = 0$ cm; at 375 keV this increases to 31–35%. In other simulations, we found that this percentage is dependent mainly on the lower energy threshold, not on the crystal material or energy resolution, increasing to ~45% if the threshold is dropped to 300 keV. Another interesting point is that the peaks of the responses to the sources at $x = 9$ cm are not aligned with

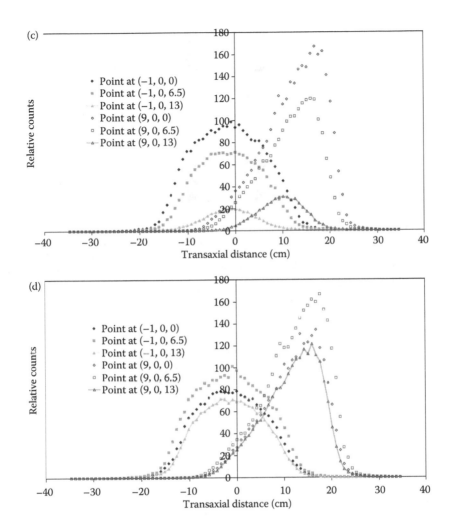

FIGURE 13.5 Continued.

the sources [42], but shifted to between $x = 15$ and 22 cm, moving further outward for the lower energy thresholds.

The scatter response is also sensitive to the attenuation outside the FOV: photons from annihilations in the FOV may contribute after leaving the FOV and scattering off of matter outside the FOV as shown in Figure 13.6 [40]. To test the relative importance of the attenuation and the activity outside the FOV, we simulated a series of sections of (a) the aforementioned cylindrical phantom and (b) the Zubal phantom with uniform activity in all the soft tissues except the lungs and bones, with the heart axially centered in the FOV.

In the first simulation of the series, we truncated both the attenuation and activity to the FOV (Figure 13.7a). In the second, we still restricted the

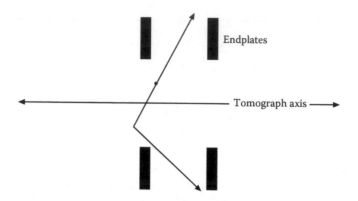

FIGURE 13.6
Matter outside the field-of-view increases detected scatter from activity inside the field-of-view: here, one photon from an annihilation inside the field-of-view passes unscattered between the endplates; the other photon passes through the endplates after scattering off matter outside the field-of-view.

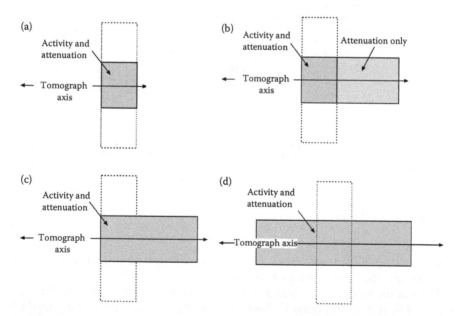

FIGURE 13.7
Phantoms for exploring the contribution of matter and activity outside the field-of-view to detected scatter. (Both the uniform 20-cm-diameter cylinder and the Zubal phantom with uniform activity in the soft tissue except the lungs are used for these experiments.) (a) The phantom is truncated to the field-of-view. (b) The activity is truncated to the field-of-view, the attenuation is modeled in the field-of-view, and to the right of the field-of-view. (c) The activity and attenuation are simulated in the field-of-view and to the right of the field-of-view. (d) Activity and attenuation are simulated in the entire phantom.

activity to the FOV, but we added the attenuation from one side of the phantom to the simulation (Figure 13.7b). In the third, we included both the activity and attenuation on this side of the phantom (Figure 13.7c). In the final simulation of the series, we included the entire cylindrical or Zubal phantom (Figure 13.7d). We did the simulations using BGO crystal with a lower energy threshold of 375 keV and with LSO crystal with a lower energy threshold of 425 keV. We found that the out-of-field activity and attenuation increased the scatter by about 81% for the BGO simulations and about 62% for the LSO simulations (Table 13.4). We also computed the percentage of multiple scattered coincidences to total scattered coincidences. This ranged from 13% to 24% for the FOV-only phantoms and from 14% to 31% for the full phantoms, with the lowest values coming from the cylindrical phantom in the LSO tomograph, and the highest values from the Zubal phantom in the BGO tomograph. The difference between the phantoms restricted to the FOV and the phantoms with attenuation added to one side is small but statistically significant, ranging between 4% and 9%. Adding in the activity from one side of the phantom, so that there is both activity and attenuation outside the FOV on one side, on the axial side near the added activity/attenuation the scatter response is nearly as high as when the complete phantom is used; on the opposite side, the response is nearly as low as when the phantom is restricted to the FOV (Figure 13.8). This results in significant axial asymmetry in the scatter response. With the inclusion of the whole phantom, the axial scatter response becomes roughly symmetric for these phantoms.

TABLE 13.4

Percent Additional Scatter Coincidences Caused by Activity and Attenuation Outside the Field-of-View (FOV) for a 20-cm-Diameter Water Cylinder with Uniform Activity and for the Zubal Phantom

Detector Crystal	Phantom	Full Phantom (%)	Activity and Attenuation in FOV, Plus in One Axial Direction (%)	Activity and Attenuation in FOV, Plus Attenuation in One Axial Direction (%)
BGO	20-cm-diameter cylinder	87	46	6
BGO	Zubal	88	49	9
LSO	20-cm-diameter cylinder	61	32	4
LSO	Zubal	63	35	6

Note: The full 20-cm-diameter phantom is 70 cm long and centered axially in the FOV; the full Zubal phantom extends from the top of the head to the thighs and is imaged with the heart centered axially in the FOV. Results are given for a BGO detector with a 375 keV lower energy threshold and an LSO tomograph with a 425 keV lower energy threshold. The percentages given are the additional scatter coincidences when compared to the same phantom with all attenuation and activity outside the FOV truncated; results are also given for the phantom in the FOV plus the attenuation or the activity and attenuation on just one side of the FOV (for the Zubal, the leg side).

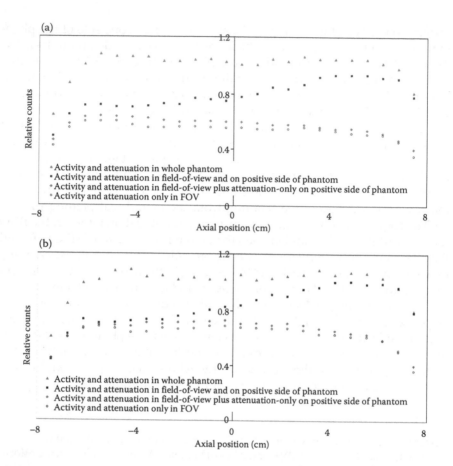

FIGURE 13.8
Contribution to detected scatter from matter and activity outside the field-of-view. (a) For a uniform 20-cm-diameter cylinder in a BGO tomograph with 17% FWHM energy resolution. (b) For a uniform 20-cm-diameter cylinder in an LSO tomograph with 10% FWHM energy resolution. (c) For the Zubal phantom with uniform activity in the soft tissue in a BGO tomograph with 17% FWHM energy resolution. (d) For the Zubal phantom with uniform activity in the soft tissues in an LSO tomograph with 10% FWHM energy resolution.

To see how activity outside the FOV might affect the scatter response of a clinical scan, we repeated the simulation of the Zubal phantom except with the activity in the liver, which is almost entirely outside the FOV, 10 times the activity in the other soft tissues. (This is a higher activity concentration than is usually seen in the liver, but we are using the liver to represent any unexpectedly high activity concentration outside the FOV.) We looked at horizontal and vertical profiles through a sinogram of the scatter coincidences in the central 3 cm axial of the tomograph (Figure 13.9). The extra activity in the liver contributes substantially to the scatter

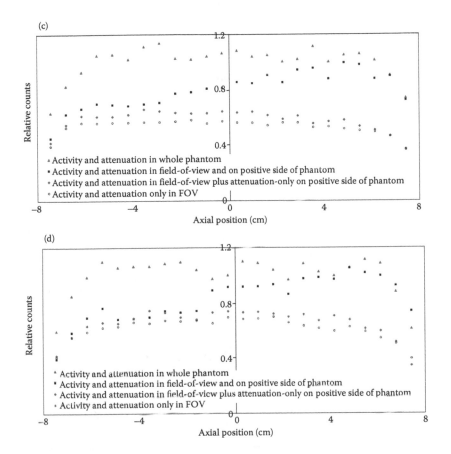

FIGURE 13.8 Continued.

response—adding approximately 45% to the uniform activity response for the BGO tomograph and 35% for the LSO tomograph. The extra liver activity also contributes significant transaxial asymmetry to the scatter profiles.

So we have seen that scatter is a significant source of contamination in 3D PET. It can be substantially reduced by using collimation (2D PET), but only with a large loss of sensitivity. It can be somewhat reduced but not eliminated by using higher low-energy thresholds with detectors with better energy resolution. The scatter response is very smooth, with the majority of the scatter being single-scattered and from activity inside the FOV. Nevertheless, both multiple-scatter coincidences and scatter coincidences from activity outside the FOV are a significant portion of the scatter. Sources outside the FOV can add significant axial and transaxial asymmetries to the scatter response. These are the main characteristics that confront the designer of a scatter correction algorithm.

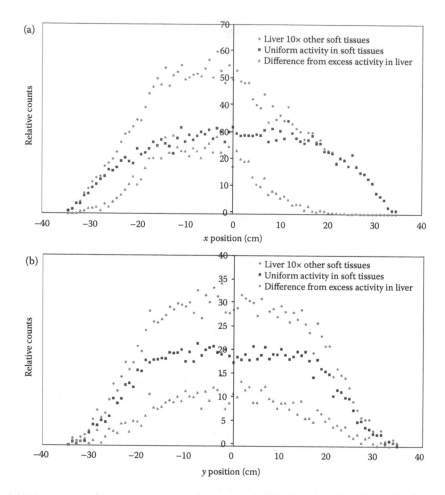

FIGURE 13.9

Horizontal and vertical profiles through an axially centered sinogram of scatter coincidences for the Zubal phantom with liver activity 10 times that in the other soft tissues and with uniform activity in all the other soft tissues. (a) Horizontal profile in a BGO tomograph with 17% FWHM energy resolution. (b) Vertical profile in a BGO tomograph with 17% FWHM energy resolution. (c) Horizontal profile in an LSO tomograph with 10% FWHM energy resolution. (d) Vertical profile in an LSO tomograph with 10% FWHM energy resolution.

Scatter Corrections for 3D PET

Scatter correction methods, proposed for 3D PET, can be divided into seven groups: (1) convolution subtraction; (2) tail fitting; (3) energy based; (4) analytic; (5) methods involving temporary collimation; (6) Monte Carlo; and (7) incorporation of the scatter response into the system matrix for iterative

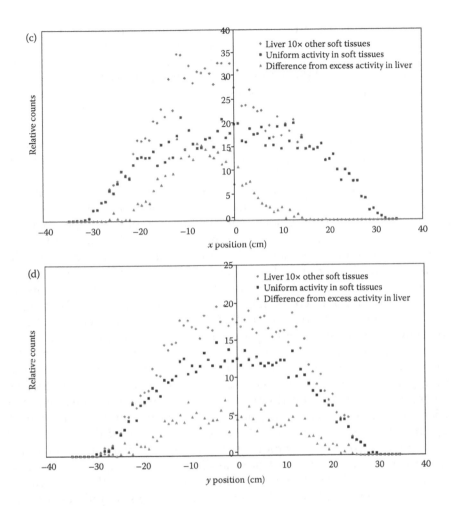

FIGURE 13.9 Continued.

reconstruction. We discuss each of these groups in relation to the scatter characteristics given earlier.

The convolution subtraction or "deconvolution" technique [43,44] was widely used for 2D PET and has also been extended to 3D PET [45–48]. The method requires the empirical determination of the system's point source scatter response function for a standard attenuation object. The scatter distribution is assumed to depend only on the point source position on the selected projection plane and is usually modeled as a 2D Gaussian or a polynomial. This fast method explicitly accounts for the activity distribution. However, the depth-independent scatter response function is characterized for a specific object, which may differ significantly from the object being imaged. This method was quite successful for 2D PET, where the scatter correction is small, but has not been widely adopted for 3D PET, where the errors in scatter estimation result in considerably more bias. The method does not account for activity outside the FOV, also a limitation for 3D PET. There may be a place for this method in small animal imaging, however, as the scatter fractions are lower and the entire animal is often in the FOV.

The second technique fits Gaussian [49,50] or polynomial [51] functions to the lines of response (LOR) outside the object and works well for head-sized objects and with some success in a NaI(Tl)-based PET scanner for torso scans [51]. It has been less successful when applied to torso scans acquired using BGO PET systems [52] but the method is easy to implement; it can account for scatter from activity outside the FOV; and it does not require *a priori* information about the activity and attenuation distributions. Its main weakness is its estimation of the magnitude and shape of scatter from the scatter tails. Information in the tails may not be sufficient to account for complexity in the central part of the distribution—see, for instance, how little difference the liver makes in the tails of the distributions in Figure 13.9a and c compared to the difference it makes in the center of the torso. Furthermore, for large patients, there may be far too few lines not intersecting the patient for a reliable estimation.

Energy-based scatter corrections have been studied for SPECT imaging [53]. With the introduction of 3D PET, two dual-energy window scatter correction techniques were proposed. In the first technique, data from a lower energy window (200–380 keV) is used to estimate the scatter in the photopeak window (380–850 keV, [54]). In the second technique, data from an upper energy window (550–850 keV), that is assumed to be scatter free, is used to estimate the contribution of trues in a wide photopeak window (250–850 keV, [55]). Both techniques depend on an empirical factor relating the ratio between scatter and true events in each of the windows. A shortcoming of the first technique is that the spatial distribution of scatter in the lower window differs from the distribution of scatter in the photopeak window [56,57]. In Figure 13.10a and b, we show this for different energy windows: a photopeak window of 375–650 keV and scatter window of 300–375 keV for a BGO tomograph, and a photopeak window of 425–650 keV and scatter window of 375–425 keV for an LSO tomograph. We look at two options for the scatter estimate, coincidences with at least one photon in the scatter window or coincidences with both photons in the scatter window. The curves have been normalized to have the same area-under-the-curve so that we can compare the shapes more easily. Neither scatter window option gives curves with a shape comparable to the scatter in the photopeak window. The first option has shoulders outside the object. This is caused by the fact that the majority of these scatters are single scatters through a relatively large angle—375 keV, for instance, corresponds to a 50° scatter angle—so that when the scattered photon is combined with an unscattered photon, the line-of-response is shifted outside the object. The second option is considerably flatter than the actual scatter response: most of the coincidences here have both photons scattered, so there is very little spatial information.

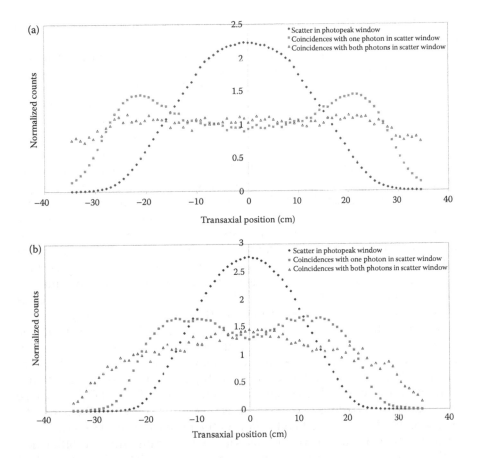

FIGURE 13.10
Comparison of the distribution of scatter in the photopeak window to the distribution of coincidences detected with one photon in a lower energy window or coincidences with both photons detected in a lower energy window for a 20-cm-diameter water cylinder with uniform activity. The curves have been normalized to have the same area-under-the-curve. (a) For a BGO tomograph with photopeak window 375–650 keV, the lower energy window is 300–375 keV. (b) For an LSO tomograph with photopeak window 425–650 keV, the lower energy window is 375–425 keV.

The shortcomings with the upper energy window method are the limited statistics in the upper window (note how few photons are seen above 550 keV in Figure 13.3a and b) and possible problems with pulse pile up and sensitivity to fluctuations in the detector/electronics. A triple-energy window technique was also proposed [58]. This technique worked marginally better than the dual-energy window method when applied on data from an NaI(Tl)-based PET system. Holospectral (multiple-energy window) techniques were also investigated [58]. More recently, Popescu et al. [59] proposed a technique using list mode data including the detected energies of both photons that looked promising in simulations. To our knowledge, this technique has not been tested experimentally. Energy-based techniques do not make any assumptions about the activity or attenuation distribution. Therefore, they could potentially correct for scatter from activity outside the FOV and for heterogeneous objects. They do require extra data to be collected and additional data processing.

Analytic methods comprise the fourth class of scatter correction techniques [42,60–62]. Given the emission data and an accurate attenuation map, Ollinger [61] originally proposed computing the single scatter distribution using the Klein–Nishina equation, a process that has come to be called single scatter simulation. As the distribution of multiple scattered events and scatter from outside the FOV is very different from single scattered events (Figures 13.8a through d and 13.9a through d, [38–41,63]), their contributions are usually extrapolated from the emission data on lines-of-response that do not intersect the patient. Variations of this method are used in most clinical systems, including an adaptation to time-of-flight PET [64]. Similar methods that also compute multiple scatters analytically have been proposed and are becoming ever more feasible with improvements in computer power [65,66]. These methods do not add noise to the data. Their limitations are the estimation of multiple scatters (in the case of single-scatter simulation) and scatter from attenuation and activity outside the FOV from the tails of the emission data. As this estimation relies on lines-of-response that do not intersect the patient, they are very noisy for large patients, as there are very few such lines; also, they cannot reliably estimate scatter from attenuation and activity outside the FOV, as variations outside the FOV can cause relatively large variations in the central section of the scatter profile without a correspondingly large effect on the tails of the profile (Figure 13.9a–d).

The fifth type of scatter correction uses temporary or intermittent collimation to distinguish scatter and true coincidence response [67,68]. The first method uses an auxiliary scan with 2D collimation, then uses the difference between normalized versions of the 2D and 3D scans to estimate the scatter in the 3D data. The second method uses a beam-stopper, a block of lead or tungsten, to block the true coincidences along the lines-of-response, leaving only scatter. The requirement of an additional scan or additional hardware is a significant drawback for these methods. Neither is used in current clinical practice.

The Monte Carlo method of estimating scatter has long served as a "gold standard" when developing and testing scatter correction techniques. Using Monte Carlo, one can test a scatter estimation technique on multiple distributions and compare the scatter estimates to the scatter distributions from the Monte Carlo; one cannot do this experimentally as the scatter distributions cannot be separated from the true coincidence distributions (Figure 13.3a and b). Monte Carlo estimation has also been proposed as a scatter correction technique (e.g., [69]), but has been seen as too computationally intensive for routine use. However, with computing power becoming increasingly affordable, particularly in the form of graphical processing units, and with acceleration techniques for Monte Carlo simulation [70–74], this method looks increasingly feasible. Like the analytic methods, this technique requires an accurate attenuation map and can only account for scatter from activity in the FOV.

The last method in our list is not a technique for estimating scatter, but rather a method of incorporating a scatter estimate into the reconstruction. Iterative image reconstruction algorithms depend on a system matrix, the elements of which describe the likelihood of detecting a decay from a given location (i.e., voxel) on a given line-of-response. To reduce computation time and reduce storage requirements, this huge matrix is usually made into a sparse matrix by considering only true coincidences; given the broad point spread function of scatter coincidences, the matrix is no longer sparse when they are included. As a result, the scatter estimate has usually been subtracted before iterative reconstruction, but this makes it such that the data are no longer Poisson distributed and negates one of the strengths of iterative reconstruction. Increasingly affordable computing power and storage are making the inclusion of scatter (and randoms) in the system matrix an active area of research [75].

Conclusion

This chapter focused on how Monte Carlo simulation has been used to study scatter associated with 3D PET imaging. Because of the finite energy resolution of PET detector systems, it is difficult to discriminate between true and scatter events. Most experimental investigations of scatter have been limited to homogeneous and symmetric phantoms. The development of efficient Monte Carlo simulation packages has allowed the study of scatter for realistic activity distributions and attenuation maps and according to a variety of physical parameters. Monte Carlo investigations have been used to study the spatial characteristics of scatter versus energy. In addition, the Monte Carlo method is uniquely able to provide information about multiple versus single scatter events. Knowledge gained from Monte Carlo investigations has contributed to the formulation and evaluation of new scatter correction techniques. The increasing affordability and power of computing resources has allowed for increasingly accurate estimates of scatter to be incorporated into image reconstructions for clinical scanners.

Acknowledgments

The work by the authors presented in this chapter was supported by PHS grants CA42593 and CA126593.

References

1. Badawi R, Marsden P, Cronin B, Sutcliffe J, Maisey M. Optimization of noise-equivalent count rates in 3D PET. *Phys. Med. Biol.* 1996;41:1755.
2. Eriksson L, Townsend D, Conti M et al. An investigation of sensitivity limits in PET scanners. *Nucl. Instrum. Methods Phys. Res. Section A: Accel. Spectrometers, Detectors Assoc. Equip.* 2007;580(2):836–842.
3. Bao Q, Newport D, Chen M, Stout D, Chatziioannou A. Performance evaluation of the Inveon dedicated PET preclinical tomograph based on the NEMA NU-4 standards. *J. Nucl. Med.* 2009;50(3):401–408.
4. Bioscan. NanoPET/CT: In-Vivo Preclinical Imager. 2011. http://www.bioscan.com/molecular-imaging/nanopet-ct.
5. Tai Y, Ruangma A, Rowland D et al. Performance evaluation of the microPET focus: A third-generation microPET scanner dedicated to animal imaging. *J. Nucl. Med.* 2005;46(3):455–463.
6. Wang Y, Seidel J, Tsui B, Vaquero J, Pomper M. Performance evaluation of the GE Healthcare eXplore VISTA dual-ring small-animal PET scanner. *J. Nucl. Med.* 2006;47(11):1891–1900.
7. Surti S, Karp J, Perkins A, Freifelder R, Muehllehner G. Design evaluation of A-PET: A high sensitivity animal PET camera. *IEEE Trans. Nucl. Sci.* 2003;50(5):1357–1363.
8. Boellaard R, O'Doherty M, Weber W et al. FDG PET and PET/CT: EANM procedure guidelines for tumour PET imaging: Version 1.0. *Eur. J. Nucl. Med. Mol. Imag.* 2010;37(1):181–200.
9. Doot RK, Scheuermann JS, Christian PE, Karp JS, Kinahan PE. Instrumentation factors affecting variance and bias of quantifying tracer uptake with PET/CT. *Med. Phys.* 2010;37(11):6035–6046.
10. Wahl R, Jacene H, Kasamon Y, Lodge M. From RECIST to PERCIST: Evolving considerations for PET response criteria in solid tumors. *J. Nucl. Med.* 2009;50(Suppl_1):122S.
11. Kinahan P, Doot R, Wanner-Roybal M et al. PET/CT assessment of response to therapy: Tumor change measurement, truth data, and error. *Transl. Oncol.* 2009; 2(4):223.
12. Scheuermann J, Saffer J, Karp J, Levering A, Siegel B. Qualification of PET scanners for use in multicenter cancer clinical trials: The American College of Radiology imaging network experience. *J. Nucl. Med.* 2009;50(7):1187.
13. Soret M, Bacharach S, Buvat I. Partial-volume effect in PET tumor imaging. *J. Nucl. Med.* 2007;48(6):932.
14. Hoffman E, Huang S, Phelps M. Quantitation in positron emission computed tomography: 1. Effect of object size. *J. Comput. Assist. Tomogr.* 1979;3(3):299.
15. Hoffman E, Huang S, Phelps M, Kuhl D. Quantitation in positron emission computed tomography: 4. Effect of accidental coincidences. *J. Comput. Assist. Tomogr.* 1981;5(3):391.
16. Hoffman E, Huang S, Plummer D, Phelps M. Quantitation in positron emission computed tomography: 6. Effect of nonuniform resolution. *J. Comput. Assist. Tomogr.* 1982;6(5):987.

17. Huang S, Hoffman E, Phelps M, Kuhl D. Quantitation in positron emission computed tomography: 2. Effect of inaccurate attenuation correction. *J. Comput. Assist. Tomogr.* 1979;3(6):804–814.
18. Huang S, Hoffman E, Phelps M, Kuhl D. Quantitation in positron emission computed tomography: 3. Effect of sampling. *J. Comput. Assist. Tomogr.* 1980; 4(6):819.
19. Mazziotta J, Phelps M, Plummer D, Kuhl D. Quantitation in positron emission computed tomography: 5. Physical-anatomical effects. *J. Comput. Assist. Tomogr.* 1981;5(5):734.
20. Casey M, Hoffman E. Quantitation in positron emission computed tomography: 7. A technique to reduce noise in accidental coincidence measurements and coincidence efficiency calibration. *J. Comput. Assist. Tomogr.* 1986;10(5):845.
21. Fahey F, Kinahan P, Doot R, Kocak M, Thurston H, Poussaint T. Variability in PET quantitation within a multicenter consortium. *Med. Phys.* 2010;37:3660.
22. Velasquez L, Boellaard R, Kollia G et al. Repeatability of 18F-FDG PET in a multicenter phase I study of patients with advanced gastrointestinal malignancies. *J. Nucl. Med.* 2009;50(10):1646.
23. Weber W, Ziegler S, Thodtmann R, Hanauske A, Schwaiger M. Reproducibility of metabolic measurements in malignant tumors using FDG PET. *J. Nucl. Med.* 1999;40(11):1771.
24. Alessio A, Kinahan P, Cheng P, Vesselle H, Karp J. PET/CT scanner instrumentation, challenges, and solutions. *Radiol. Clin. North Am.* 2004;42(6):1017–1032.
25. Strother S, Casey M, Hoffman E. Measuring PET scanner sensitivity: Relating count rates to image signal-to-noise ratios using noise equivalent counts. *IEEE Trans. Nucl. Sci.* 1990;37(2):783–788.
26. Daube-Witherspoon M, Karp J, Casey M et al. PET performance measurements using the NEMA NU 2–2001 standard. *J. Nucl. Med.* 2002;43(10):1398–1409.
27. Knoll G. *Radiation Detection and Measurement.* Vol 2: Wiley, New York; 1989.
28. Thompson C. The effect of collimation on scatter fraction in multi-slice PET. *IEEE Trans. Nucl. Sci.* 1988;35(1):598–602.
29. MacDonald L, Schmitz R, Alessio A et al. Measured count-rate performance of the Discovery STE PET/CT scanner in 2D, 3D and partial collimation acquisition modes. *Phys. Med. Biol.* 2008;53:3723.
30. Herzog H, Tellmann L, Hocke C, Pietrzyk U, Casey M, Kuwert T. NEMA NU2–2001 guided performance evaluation of four Siemens ECAT PET scanners. *IEEE Trans. Nucl. Sci.* 2004;51(5):2662–2669.
31. Jong H, Velden F, Kloet R, Buijs F, Boellaard R, Lammertsma A. Performance evaluation of the ECAT HRRT: An LSO-LYSO double layer high resolution, high sensitivity scanner. *Phys. Med. Biol.* 2007;52:1505.
32. Kemp B, Kim C, Williams J, Ganin A, Lowe V. NEMA NU 2–2001 performance measurements of an LYSO-based PET/CT system in 2D and 3D acquisition modes. *J. Nucl. Med.* 2006;47(12):1960.
33. Jakoby B, Bercier Y, Watson C, Bendriem B, Townsend D. Performance characteristics of a new LSO PET/CT scanner with extended axial field-of-view and PSF reconstruction. *IEEE Trans. Nucl. Sci.* 2009;56(3):633–639.

34. Surti S, Kuhn A, Werner M, Perkins A, Kolthammer J, Karp J. Performance of Philips Gemini TF PET/CT scanner with special consideration for its time-of-flight imaging capabilities. *J. Nucl. Med.* 2007;48(3):471.
35. Teräs M, Tolvanen T, Johansson J, Williams J, Knuuti J. Performance of the new generation of whole-body PET/CT scanners: Discovery STE and Discovery VCT. *Eur. J. Nucl. Med. Mol. Imag.* 2007;34(10):1683–1692.
36. Lupton L, Keller N. Performance study of single-slice positron emission tomography scanners by Monte Carlo techniques. *IEEE Trans. Med. Imag.* 1983;2(4):154–168.
37. Badawi R, Kohlmyer S, Harrison R, Vannoy S, Lewellen T. The effect of camera geometry on singles flux, scatter fraction and trues and randoms sensitivity for cylindrical 3D PET—A simulation study. *IEEE Trans. Nucl. Sci.* 2000;47(3):1228–1232.
38. Harrison R, Haynor D, Vannoy S, Kaplan M, Lewellen T. Continuing investigations of scatter in PET: The effects of collimation and heterogeneous attenuation. *Conference Records of the IEEE Nuclear Science and Medical Imaging Conference.* 1993:1149–1153.
39. Harrison R, Vannoy S, Kohlmyer S, Sossi V, Lewellen T. The effect of scatter on quantitation in positron volume imaging of the thorax. *Conference Records of the IEEE Nuclear Science and Medical Imaging Conference.* 1994;3:1335–1338.
40. Laymon C, Harrison R, Kohlmyer S, Miyaoka R, Lewellen T. Characterization of single and multiple scatter from matter and activity distributions outside the FOV in 3-D PET. *IEEE Trans. Nucl. Sci.* 2004;51(1):10–15.
41. Sossi V, Barney J, Harrison R, Ruth T. Effect of scatter from radioactivity outside of the field of view in 3D PET. *IEEE Trans. Nucl. Sci.* 1995;42(4):1157–1161.
42. Barney J, Rogers J, Harrop R, Hoverath H. Object shape dependent scatter simulations for PET. *IEEE Trans. Nucl. Sci.* 1991;38(2):719–725.
43. Bergström M, Eriksson L, Bohm C, Blomqvist G, Litton J. Correction for scattered radiation in a ring detector positron camera by integral transformation of the projections. *J. Comput. Assist. Tomogr.* 1983;7(1):42.
44. King P, Hubner K, Gibbs W, Holloway E. Noise identification and removal in positron imaging systems. *IEEE Trans. Nucl. Sci.* 1981;28(1):148–151.
45. Bailey D, Meikle S. A convolution-subtraction scatter correction method for 3D PET. *Phys. Med. Biol.* 1994;39:411.
46. Lercher M, Wienhard K. Scatter correction in 3-D PET. *IEEE Trans. Med. Imag.* 1994;13(4):649–657.
47. McKee B, Gurvey A, Harvey P, Howse D. A deconvolution scatter correction for a 3-D PET system. *IEEE Trans. Med. Imag.* 1992;11(4):560–569.
48. Shao L, Karp J. Cross-plane scattering correction-point source deconvolution in PET. *IEEE Trans. Med. Imag.* 1991;10(3):234–239.
49. Cherry S, Huang S. Effects of scatter on model parameter estimates in 3D PET studies of the human brain. *IEEE Trans. Nucl. Sci.* 1995;42(4):1174–1179.
50. Stearns C. Scatter correction method for 3D PET using 2D fitted Gaussian functions. *J. Nucl. Med.* 1995;36:105.
51. Karp J, Muehllehner G, Mankoff D et al. Continuous-slice PENN-PET: A positron tomograph with volume imaging capability. *J. Nucl. Med.* 1990;31(5):617.
52. Lewellen T, Kohlmyer S, Miyaoka R, Kaplan M, Stearns C, Schubert S. Investigation of the performance of the General Electric ADVANCE positron emission tomograph in 3D mode. *IEEE Trans. Nucl. Sci.* 1996;43(4):2199–2206.

53. Luo J, Koral K, Ljungberg M, CE Jr F, Jaszczak R. A Monte Carlo investigation of dual-energy-window scatter correction for volume-of-interest quantification in 99Tcm SPECT. *Phys. Med. Biol.* 1995;40:181.
54. Grootoonk S, Spinks T, Jones T, Michel C, Bol A. Correction for scatter using a dual energy window technique with a tomograph operated without septa. Paper presented at: *Conference Records of the IEEE Nuclear Science and Medical Imaging Conference.* 1991.
55. Bendriem B, Trebossen R, Frouin V, Syrota A. A PET scatter correction using simultaneous acquisitions with low and high lower energy thresholds. Paper presented at: *Conference Records of the IEEE Nuclear Science and Medical Imaging Conference.* 1993.
56. Harrison R, Haynor D, Lewellen T. Dual energy window scatter corrections for positron emission tomography. Paper presented at: *Conference Records of the IEEE Nuclear Science and Medical Imaging Conference.* 1991.
57. Harrison R, Haynor D, Lewellen T. Limitations of energy-based scatter correction for quantitative PET. Paper presented at: *Conference Records of the IEEE Nuclear Science and Medical Imaging Conference.* 1992.
58. Shao L, Freifelder R, Karp J. Triple energy window scatter correction technique in PET. *IEEE Trans. Med. Imag.* 1994;13(4):641–648.
59. Popescu L, Lewitt R, Matej S, Karp J. PET energy-based scatter estimation and image reconstruction with energy-dependent corrections. *Phys. Med. Biol.* 2006; 51:2919.
60. Hiltz L, McKee B. Scatter correction for three-dimensional PET based on an analytic model dependent on source and attenuating object. *Phys. Med. Biol.* 1994;39:2059.
61. Ollinger J. Model-based scatter correction for fully 3D PET. *Phys. Med. Biol.* 1996;41:153.
62. Watson C, Newport D, Casey M. A single scatter simulation technique for scatter correction in 3D PET. *Three-Dim. Imag. Recon. Rad. Nucl. Med.* 1996;4: 255–268.
63. Adam L, Karp J, Brix G. Investigation of scattered radiation in 3D whole-body positron emission tomography using Monte Carlo simulations. *Phys. Med. Biol.* 1999;44:2879.
64. Watson C. Extension of single scatter simulation to scatter correction of time-of-flight PET. *IEEE Trans. Nucl. Sci.* 2007;54(5):1679–1686.
65. Markiewicz P, Tamal M, Julyan P, Hastings D, Reader A. High accuracy multiple scatter modelling for 3D whole body PET. *Phys. Med. Biol.* 2007;52:829.
66. Moehrs S, Defrise M, Belcari N et al. Multi-ray-based system matrix generation for 3D PET reconstruction. *Phys. Med. Biol.* 2008;53:6925.
67. Cherry S, Meikle S, Hoffman E. Correction and characterization of scattered events in three-dimensional PET using scanners with retractable septa. *J. Nucl. Med.* 1993;34(4):671.
68. Chuang K, Wu J, Jan M, Chen S, Hsu C. Novel scatter correction for three-dimensional positron emission tomography by use of a beam stopper device. *Nucl. Instrum. Methods Phys. Res. Section A: Accel. Spectrometers, Detectors Assoc. Equip.* 2005;551(2–3):540–552.
69. Levin C, Dahlbom M, Hoffman E. A Monte Carlo correction for the effect of Compton scattering in 3-D PET brain imaging. *IEEE Trans. Nucl. Sci.* 2002;42(4):1181–1185.
70. De Beenhouwer J, Staelens S, Vandenberghe S, Lemahieu I. Acceleration of GATE SPECT simulations. *Med. Phys.* 2008;35:1476.

71. De Jong H, Slijpen E, Beekman F. Acceleration of Monte Carlo SPECT simulation using convolution-based forced detection. *IEEE Trans. Nucl. Sci.* 2001;48(1):58–64.

72. Haynor D, Harrison R, Lewellen T. The use of importance sampling techniques to improve the efficiency of photon tracking in emission tomography simulations. *Med. Phys.* 1991;18:990.

73. Holdsworth C, Levin C, Janecek M, Dahlbom M, Hoffman E. Performance analysis of an improved 3-D PET Monte Carlo simulation and scatter correction. *IEEE Trans. Nucl. Sci.* 2002;49(1):83–89.

74. Rehfeld N, Stute S, Apostolakis J, Soret M, Buvat I. Introducing improved voxel navigation and fictitious interaction tracking in GATE for enhanced efficiency. *Phys. Med. Biol.* 2009;54:2163.

75. Kadrmas D. LOR-OSEM: Statistical PET reconstruction from raw line-of-response histograms. *Phys. Med. Biol.* 2004;49:4731.

14

Design of Molecular Imaging Systems Assisted by the Monte Carlo Methods

Irène Buvat

CONTENTS

Designing new detectors for imaging purpose is a long, time-consuming, and expensive task, during which the geometry of the detector, the physics involved in the detector components, and the resulting performance have to be thoroughly investigated. Other aspects such as cost and manufacturing constraints also have to be accounted for. At some point, a prototype is usually built to determine the validity of the detector concept and perform some initial characterization of the performance. In the process of new detector development, Monte Carlo simulations can play an important role to better understand the detector response, to test various options without building the corresponding components, and to identify the most promising setup without having to perform costly experiments. This chapter illustrates the use of Monte Carlo simulations for designing or optimizing SPECT (the sections "Monte Carlo Simulations to Optimize the Collimator Design in SPECT" and "Monte Carlo Simulations to Optimize the Detector Design in SPECT") and PET (the section "Monte Carlo Simulations to Optimize the Detector Design in PET") molecular imaging systems. It shows the broad variety of investigations in which Monte Carlo simulations can guide the design of original imaging systems. The advantages and limitations of the use of Monte Carlo simulations in that context are also discussed (the section "Discussion").

Monte Carlo Simulations to Optimize the Collimator Design in SPECT

In SPECT, simulations are widely used for the design and optimization of collimators. Although analytical formulae are available for predicting detection efficiency and spatial resolution in most conventional collimation geometries (parallel hole, fan-beam, cone-beam, pinhole), effects such as septal or edge penetration and collimator scattering can only be accurately predicted using Monte Carlo simulations. Monte Carlo simulations are thus frequently taken advantage of to investigate and optimize collimator characteristics (e.g., diameter of the hole, thickness of the collimator, septal thickness) when analytical calculations cannot be reliable enough. As examples, simulations have been found to be especially useful in the following contexts.

Slit–slat collimation (e.g., [1,2]). Slit–slat collimation consists of a slit in the direction parallel to the axis of rotation of the gamma camera combined with a set of parallel septa perpendicular to the direction of the slit. This combination of fan-beam and parallel-beam collimation improves detection efficiency compared to the multi-pinhole collimation. Slit–slat collimation can also achieve better transaxial resolution than fan-beam collimation. Detailed understanding of the detection sensitivity and spatial resolution properties of slit–slat collimation is needed to compare its performance to other collimator types and to identify the applications that could benefit from such collimation. In such investigations, failure to account for scatter and penetration effects can produce misleading results [2]. Monte Carlo simulations offer a more complete picture of the behavior of different collimator characteristics than analytical calculations [2].

Rotating slat collimation. Rotating slat collimators have been studied through Monte Carlo simulations (e.g., [3–5]). The rotating slat collimator consists of a set of parallel plates oriented perpendicular to the detector. The complete sampling of an object is obtained by spinning the collimator around its own axis. Monte Carlo simulations have been performed to optimize the slat height in a small imaging SPECT device [5] and to characterize the detection sensitivity and spatial resolution performance as a function of the collimator parameters. Monte Carlo simulations have also been used to check the validity of analytical formulae derived for predicting the sensitivity and spatial resolution over the field of view [4,6] and to test various reconstruction methods suited to that type of collimation geometry [3].

Multi-pinhole collimation. Monte Carlo simulations are very helpful to investigate various multi-pinhole collimation designs. Multi-pinhole tomography is an extension of conventional single-pinhole tomography to more than one pinhole per collimator. The pinholes are arranged in such a way that each pinhole only views a certain part of the object, and all pinholes together cover the entire field of view. In multi-pinhole imaging, projections through different pinholes may partially overlap on the detector, resulting in the so-called multiplexing. This multiplexing yields a high increase in system sensitivity, but the price to pay is an ill-conditioning of the system matrix involved in tomographic reconstruction, which makes the reconstruction problem harder to solve. Multi-pinhole collimator can vary by the number and geometrical arrangement of the pinhole and by the thickness of the plate in which the pinholes are drilled. Monte Carlo simulations are extremely helpful to investigate the impact of these parameters on the overall imaging system performances (e.g., [7–10]).

In unconventional collimation setting, such as those mentioned above or when using dual-planar cone-beam collimation, consisting in two cone beam collimators with focal points in different axial planes, Monte Carlo simulations can be used to study the impact of the radius of rotation on spatial resolution, detection efficiency and signal-to-noise ratios [11].

In unconventional as well as in conventional collimation setup, Monte Carlo simulations are usually needed to predict the gamma camera response when scanning radionuclides including high-emission energies, such as I123 or I131. Indeed, for these radionuclides, scatter and septal penetration cannot be neglected [10]. A great advantage of Monte Carlo simulations in collimator studies is that it is possible to determine the respective magnitude of scatter in the collimator, of septal or edge penetration, of lead fluorescence, and of "trues counts" entering the detector through a hole of the collimator without scattering. Monte Carlo simulations are the only way to precisely assess the magnitude of these different components. Scatter events can be separated as a function of the scatter order and location (multiple scattering being usually more penalizing for image quality than first-order scattering). Scatter in the crystal (as a function of the crystal thickness), in the light guide, in photomultiplier tubes, and in shielding can be distinguished so that important information regarding the degradation of image quality introduced by different detector components can be derived. The variation of the scatter components as a function of the energy windows can be studied [10,12] so as to help optimize the acquisition protocols.

Monte Carlo Simulations to Optimize the Detector Design in SPECT

Beyond the optimization of the collimation design in SPECT system, Monte Carlo simulations are often performed to better understand the detector response. When considering scintillation detectors, accurate modeling of the interaction of the gamma rays with the scintillator, of the scintillation process, of the optical transport of the photons produced by the scintillator, and of the photodetector response should ideally be used. For solid-state detectors, a model of the global response of the detector is often used instead of a detailed modeling of the physics in the detectors, although such models do exist as described below.

Scintillation Detectors

Modeling of the interactions of the gamma rays with the scintillator is a core component of any Monte Carlo simulator of particle interactions with matter, be it dedicated to emission tomography or not.

Modeling of the scintillation process and of the transport of the optical photons produced by the scintillator is not always available in Monte Carlo simulation codes used in emission tomography. Indeed, such modeling is often unpractical as a single photon interacting within the crystal can produce up to ~10^4 scintillation photons. The detailed simulation of the scintillation process and optical photon transportation is thus extremely computationally intensive and is often replaced by a "global response," including the response of the photodetectors collecting the optical photons. This is, for instance, the case in the SimSET and SIMIND simulation codes. Conversely, codes such as Geant4 [13], GATE [14,15] or DETECT2000 [16], and Trace Pro (http://www.lambdares.com/software_products/tracepro/) make it possible to simulate the scintillation process and associated transportation of optical photons in details. For instance, the optical finishing of the crystal surface (paint, reflective, polished, fine ground) can be accounted for.

On the basis of such tools, the detection efficiency, energy resolution, spatial resolution, and uniformity as a function of the thickness of scintillator can be investigated for different continuous crystals (such as CsI(Tl) and NaI(Tl)) and for scintillator arrays [17], usually for small detection modules, so that the simulations remain computationally tractable. Collection efficiency can be studied as a function of the surface treatments and of different schemes of optical coupling between the scintillation crystal and the light photon detector [18,19].

The response of the photon detector can be modeled with different degrees of complexity. For instance, a photomultiplier entrance window can be characterized by its geometry and a refraction index [19] or as a polished quartz window associated with experimentally derived quantum efficiency [20]. Similarly, the position sensitive photomultiplier tube (PSPMT) response can be modeled by means of its photocathode quantum efficiency. The impact of the PSPMT anode structure can be characterized [17]. Readout using avalanche photodiode (APD) or position-sensitive avalanche photodiode (PSAPD) response can be simulated so that the imaging performance of gamma camera based on PSAPD and discrete APD can be compared [21].

Although a complete simulation of a SPECT acquisition including detailed modeling of the scintillation process would be unpractical, such detailed simulation limited to small detection modules is extremely valuable to better understand the response of the detector and then determine how detectors could be assembled into complete imaging systems. Simulated data can also be used to test various positioning algorithms based on the signal delivered by the detector and to study potential spatial distortions at the edge of the detector [18].

Semiconductor Gamma Cameras

Semiconductor detectors offer a promising alternative to scintillation detectors in emission tomography. In a semiconductor detector, photons produce electron–hole pairs that are directly involved in the electronic signal formation, unlike in the scintillation detectors for which the electronic signal is derived indirectly after the creation and propagation of the optical photons in the scintillator and collection by a photodetector. As a result, the energy resolution is improved compared to scintillation-based detectors, although this is at the expense of time resolution (due to the fluctuation of charge carrier drifting time). The overall advantage of semiconductor detector over conventional scintillation detector is therefore an active topic of research.

Gamma cameras involving solid-state detectors instead of scintillation crystals have been modeled using Monte Carlo simulations to study the impact of various parameters related to the collimator and to the detector on the imaging performance. For instance, CdZnTe detectors have been modeled using GATE [14,15] to determine how the overpixellization of the detector affects the compromise between contrast recovery and noise and whether the measurement of the depth of interaction (DOI) within the CZT detector improves the imaging performance [22,23]. In [24], the spatial resolution and sensitivity of 25 collimators fitted with a CdZnTe detector have been studied both analytically and using Monte Carlo simulations. In these studies, however, a global model of the semiconductor detector is used, and there is no Monte Carlo simulation of the electron and hole transportation and trapping. The simulation approach rather consists in characterizing the energy response of a single-pixelated module. Yet, a detailed analytical simulation of semiconductor detector response has been reported [25]. Alternatively, a model for Monte Carlo simulation of the charge collection processes in semiconductor CZT detectors has also been described based on the Geant4 toolkit [26]. This model accounts for the electrical properties of the detectors, the transport properties of the material, the trapping induced by impurities, and the γ-ray–matter interaction processes. Similar to the extensive simulation of the scintillation process in scintillators, a sophisticated simulation of semiconductor detector is only practical when considering one or only a few detectors. In this situation, it can provide a valuable insight into the behavior of the detector as a function of its intrinsic components.

Monte Carlo Simulations to Optimize the Detector Design in PET

Similar to SPECT, when building a PET detector, many parameters need to be optimized such as the dimensions and type of crystals and photomultiplier tubes (PMTs), geometry and field of view (FOV), sampling, electronics, light guide, and shielding. Given that the manufacturing of a PET imaging system is often even costlier than that of a SPECT scanner, Monte Carlo simulations play a key role in the design of PET systems.

Scintillation Detectors in PET

Monte Carlo simulations are widely used to investigate and optimize the performance of PET scanners, as a function of a number of detector features. For instance, they have been used to explore the influence of the crystal material and crystal lengths on the NEC curves which characterize the global PET imaging performance based on the statistical quality of the data [27]. In small-animal PET, Monte Carlo simulations have been used to study different dual layer phoswhich detectors involving LSO and LuYAP and determine which length of the crystal layers should be used to achieve a high and uniform radial spatial resolution for scanners with different ring diameters [28]. The performance obtained for different geometrical configurations of the scanner and various crystal lengths, crystal materials, energy windows, and ring arrangements can be characterized for small-animal imaging system and also for human PET scanners (e.g., [29,30]). Monte Carlo simulations are especially useful when planning to modify an existing PET scanner to determine the impact of the planned modifications on the system performance. This is, for instance, the case when an existing PET system is going to be used for a specific purpose, for instance, for being combined with another imaging device such as an MR scanner [30] or an optical imaging system [29]. For such applications, a model of the existing system can be designed and validated through a careful comparison of simulated and measured performance indices (sensitivity, NEC curves, spatial resolution), before modifying some model features (for instance, scanner diameter, shielding) to determine the impact of these changes on the performance indices. The influence of the presence of the other detector components on the PET detector performance, for instance, due to additional scattering material, can also be investigated [31].

Even the impact of mechanical structures of the detector (i.e., the frames supporting the detector modules) that are sources of scattering can be investigated using Monte Carlo simulations [32], by analyzing how sensitivity and spatial resolution are altered due to the presence of these structures. In particular, strength of the simulations is that they make it feasible to untangle the origin of the detected events: scatter coincidences can be precisely identi-fied and the material in which scatter occurred can be retrieved. Similarly, coincidences involving photons that have undergone multiple interactions into several detector units can be picked out. The impact of these events on sensitivity and spatial resolution can be precisely quantified.

Monte Carlo simulations can be used to determine the complete PET scan-ner response when only the response of one or few block detectors have been characterized [33–35].

Comprehensive modeling of the scintillation process and optical photon transportation has been performed in PET to get insight into the impact of the scintillation crystal and scintillator–photodetector coupling onto the resulting imaging performance. For instance, Monte Carlo simulations have been used to systematically optimize the detector design with regard to the scintillation crystal, optical diffuser, surface treatment, layout of large-area APDs, and signal-to-noise ratio of the APD devices [36,37]. Monte Carlo simulations have also been performed to optimize the design of the light guide and select the best configuration for prototyping [38]. The dependence of light output and time resolution on crystal length, light yield, decay time, surface treatment, and photon interaction depth can also be thoroughly investigated through simulations to help in the design of PET detector that combine time-of-flight and depth-of-interaction measurements [39]. As in SPECT, when performing extensive simulation of the scintillation process, only one or few detection modules are modeled as tracking optical photons for a complete PET scans would be totally impractical given the zillions of optical photons involved.

Semiconductor Detectors in PET

Semiconductor detectors are also studied in the context of dedicated PET imaging. For instance, Monte Carlo simulations have been performed to predict the performance of a dedicated breast PET system, consisting of two opposite panels of CdZnTe detectors [40]. Using simulations, the sen-sitivity, the count rate performance, the spatial resolution, and the spatial resolution uniformity could be determined, as well as the lesion visu-alization capabilities. In that study, the DOI resolution needed to detect small lesions (1.5 mm in diameter) was also studied. Various architecture designs involving pixelated CZT detectors were also studied in the context of small-animal PET, varying the module arrangement, the detector thick-ness, and the pixel pitch [41]. Similarly, the performance of system archi-tectures based on strip CdTe detectors have been investigated using Monte Carlo simulations [42]. As for the SPECT imaging systems, the response of the individual semiconductor detectors is usually modeled using macro-parameters, such as energy resolution, time resolution, and quantum effi-ciency. These parameters can be derived from a detailed simulation of the individual detectors [25,26].

Discussion

As demonstrated by the numerous examples listed in the previous sections, Monte Carlo simulations can be extremely helpful to study the impact of a very large number of detector components or detector parameters on the overall performance of a detection system. Simulations can thus avoid costly experiments and allow for the investigation of an almost unlimited range of detector design. However, the usefulness of the results derived from simulations highly relies on how well the numerical models reproduce the real response of the detector. Validation of the models is thus a crucial step to ensure physically meaningful and reliable results.

The validation task is of foremost importance and excellent agreement between simulated and experimental data has been reported at different levels of the imaging system response, for instance, modeling of the optical photons in scintillating detectors (e.g., [20]), modeling the count rate curves as a function of the source activity in PET (e.g., [43]), or modeling prototypes for scintigraphic imaging (e.g., [44]). Many published results demonstrate the high reliability of Monte Carlo simulations to predict the imaging system response, after they have been properly parameterized. However, it should be emphasized that a number of parameters in the simulations usually have to be fitted to experimental data to properly reproduce measured data. For instance, when performing Monte Carlo simulations of the optical response of a scintillator, the smoothness or roughness of each surface surrounding the crystal must be carefully tuned, as well as the reflectivity of the crystal surfaces. When the detailed scintillation process is not modeled, the amount of spatial blurring associated with the readout of the scintillator signal has to be adjusted to experimental data [27]. Similarly, dead times parameters have to be determined experimentally so as to reproduce the count rate variations with the source activity [27]. In other words, simulations are never fully predictive and always involve some parameters that can be set only from experiments.

The reliability of the simulation model of an imaging system should be first validated by comparing the simulated and experimental value of a number of figures of merit characterizing the response of the imaging system. This step is also necessary to tune the free parameters that are unknown theoretically and that need to be properly set to closely mimic the behavior of the imaging system. The relevant figures of merit depend on the system under consideration. In PET, for instance, there is a large number of figure of merit that can be compared between simulated and real data, such as single and coincidence count rates, system sensitivity, scatter fraction, energy resolution, and spatial resolution. A list of these figures of merit for non-time-of-flight PET systems has been suggested in Ref. [45]. Slightly different figures of merit are appropriate for SPECT systems, but also including energy and spatial resolutions, detector sensitivity, or scatter fractions.

The validation step is often tedious but should definitely not be neglected before relying on a simulation study to predict the response of an imaging system. Ideally, a numerical model of an existing system should be first designed and validated against real experiments, before changing some parameters to see what difference it makes on the imaging performance (e.g., [28,46]). Alternatively, validation of the simulation model can be performed by considering a single detector module (e.g., [41]). Once the response of a module is properly reproduced by the numerical model, the performance of a complete imaging system for different geometrical arrangements of the modules can be investigated.

In the real world, SPECT and PET scanners always suffer from some imperfections that are difficult to simulate accurately. For instance, in PET, block decoding effects and mechanical misalignment of the detector blocks are difficult to simulate and change with production techniques and production errors. As a result, it should be kept in mind that simulation results might sometime be too optimistic regarding the actual performance of the imaging system. Still, simulations remain extremely useful to better understand the best one can get with a system under ideal conditions and to see how the performance of a system can be improved by modifying some components, or the geometry of the system, or even the data processing (for instance, the impact of the readout electronic [47]). Simulations are often the only approach to precisely untangle effects that are impossible to separate using real experiments.

Conclusion

Monte Carlo simulations can play a key role in the long process of designing original detectors or improving existing imaging systems. Using simulations, an unlimited range of detector designs, varying by their geometry and detector components, can be investigated at almost no cost (except for the cost of running the simulation software on an appropriate computer). The advantages and limitations of any design can be precisely characterized. Even more important, simulations can give access to pieces of information that cannot be precisely measured in realistic conditions in experimental data, such as the scatter fraction in PET or SPECT detectors for instance. Yet, it should be emphasized that simulations should be relied on only after the numerical models involved in the simulations have been carefully validated against experimental data. These experimental data can be available from existing systems close to the prototype under investigation, or from components close to the ones present in the prototype under investigation. This validation step, although tedious, is mandatory to properly interpret simulation results when designing new detectors. It is also an important step to deeply understand the response of a complex imaging system such as a SPECT or a PET scanner and to get some clear insight into all parameters affecting this response. As a result, in addition to assisting detector design, simulations of imaging systems certainly have a role to play to better understand, explain, and teach the respective impact of the different components of an imaging system on the overall imaging performance it can deliver.

References

1. Metzler SD, Accorsi R, Novak JR, Ayan AS, Jaszczak RJ. On-axis sensitivity and resolution of a slit-slat collimator. *J. Nucl. Med.* 2006;47:1884–1890.
2. Novak JR, Ayan AS, Accorsi R, Metzler SD. Verification of the sensitivity and resolution dependence on the incidence angle for slit-slat collimation. *Phys. Med. Biol.* 2008;53:953–966.
3. Van Holen R, Vandenberghe S, Staelens S, Lemahieu I. Comparing planar image quality of rotating slat and parallel hole collimation: Influence of system modeling. *Phys. Med. Biol.* 2008;53:1989–2002.
4. Vandenberghe S, Van Holen R, Staelens S, Lemahieu I. System characteristics of SPECT with a slat collimated strip detector. *Phys. Med. Biol.* 2006; 51:391–405.
5. Boisson F, Bekaert V, El Bitar Z, Wurtz J, Steibel J, Brasse D. Characterization of a rotating slat collimator system dedicated to small animal imaging. *Phys. Med. Biol.* 2011;56:1471–1485.
6. Staelens S, Koole M, Vandenberghe S, D'Asseler Y, Lemahieu I, Van de Walle R. The geometric transfer function for a slat collimator mounted on a strip detector. *IEEE Trans. Nucl. Sci.* 2005;52:708–713.
7. Cao Z, Bal G, Accorsi R, Acton PD. Optimal number of pinholes in multi-pinhole SPECT for mouse brain imaging—A simulation study. *Phys. Med. Biol.* 2005;50:4609–4624.
8. Schramm NU, Ebel G, Engeland U, Schurrat T, Béhé M, Behr TM. High-resolution SPECT using multipinhole collimation. *IEEE Trans. Nucl. Sci.* 2003;52:408–413.
9. Shokouhi S, Metzler SD, Wilson DW, Peterson TE. Multi-pinhole collimator design for small-object imaging with SiliSPECT: A high-resolution SPECT. *Phys. Med. Biol.* 2009;54:207–225.
10. Staelens S, Vunckx K, De Beenhouwer J, Beekman F, D'Asseler Y, Nuyts J, Lemahieu I. GATE simulations for optimization of pinhole imaging. *Nucl. Instrum. Meth. Phys. Res. A* 2006;469:359–363.
11. Lalush DS, DiMeo AJ. A Monte Carlo investigation of dual-planar circular-orbit cone-beam SPECT. *Phys. Med. Biol.* 2002;47:4357–4370.
12. Staelens S, Santin G, Vandenberghe S et al. Transmission imaging with a moving point source: Influence of crystal thickness and collimator type. *IEEE Trans. Nucl. Sci.* 2005;52:166–173.

13. Agostinelli S, Allison J, Amako K et al. GEANT4—A simulation toolkit. *Nucl. Instrum. Meth. Phys. Res. A.* 2003;506:250–303.
14. Jan S, Benoit D, Becheva E et al. GATE V6: A major enhancement of the GATE simulation platform enabling modelling of CT and radiotherapy. *Phys. Med. Biol.* 2011;56:881–901.
15. Jan S, Santin G, Strul D et al. GATE: A simulation toolkit for PET and SPECT. *Phys. Med. Biol.* 2004;49:4543–4561.
16. Cayouette F, Laurendeau D, Moisan C. DETECT2000: An improved Monte-Carlo simulator for the computer aided design of photon sensing devices. *Proc. SPIE* 2003;4833:69–76.
17. Garibaldi F, Cisbani E, Cusanno F et al. Optimization of compact gamma cameras for breast imaging. *Nucl. Instrum. Meth. Phys. Res. A.* 2001;471:222–228.
18. Netter E, Pinot L, Ménard L, Duval MA, Janvier B, Lefebvre F, Siebert R, Charon Y. Designing the scintillation module of a pixelated mini gamma camera: The spatial spreading behaviour of light. *IEEE Nucl. Sci. Symp. Conf. Record* 2009: 3300–3302.
19. Vittori F, de Notaristefani F, Malatesta T, Puertolas D. A study on light collection of small scintillating crystals. *Nucl. Instr. Meth. Phys. Res. A.* 2000;452:245–251.
20. Lo Meo SL, Bennati P, Cinti MN et al. A Geant4 simulation code for simulating optical photons in SPECT scintillation detectors. *J. Inst.* 2009:P07002.
21. Despres P, Funk T, Shah KS, Hasegawa BH. Monte Carlo simulations of compact gamma cameras based on avalanche photodiodes. *Phys. Med. Biol.* 2007;52:3057–3074.
22. Robert C, Montemont G, Rebuffel V, Buvat I, Guerin L, Verger L. Simulation-based evaluation and optimization of a new CdZnTe gamma-camera architecture (HiSens). *Phys. Med. Biol.* 2010;55:2709–2726.
23. Robert C, Montemont G, Rebuffel V, Verger L, Buvat I. Optimization of a parallel hole collimator/CdZnTe gamma-camera architecture for scintimammography. *Med. Phys.* 2011;38:1806–1819.
24. Weinmann AL, Hruska CB, O'Connor MK. Design of optimal collimation for dedicated molecular breast imaging systems. *Med. Phys.* 2009;36:845–856.
25. Montemont G, Gentet MC, Monnet O, Rustique J, Verger L. Simulation and design of orthogonal capacitive strip CdZnTe detectors. *IEEE Trans. Nucl. Sci.* 2007;54:854–859.
26. Benoit M, Hamel LA. Simulation of charge collection processes in semiconductor CdZnTe γ-ray detectors. *Nucl. Instrum. Meth. Phys. Res. A.* 2009;606:508–516.
27. Michel C, Eriksson L, Rothfuss H, Bendriem B, Lazaro D, Buvat I. Influence of crystal material on the performance of the HiRez 3D PET scanner: A Monte-Carlo study. *IEEE Nucl. Sci .Symp. Conf. Record* 2006:2528–2531.
28. Chung YH, Choi Y, Cho GS, Choe YS, Lee KH, Kim BT. Optimization of dual layer phoswich detector consisting of LSO and LuYAP for small animal PET. *IEEE Trans. Nucl. Sci.* 2005;52:217–221.
29. Rannou F, Kohli V, Prout D, Chatziioannou A. Investigation of OPET performance using GATE, a Geant4-based simulation software. *IEEE Trans. Nucl. Sci.* 2004;51:2713–2717.
30. Vandenberghe S, Keereman V, Staelens S, Schultz V, Marsden P. Effect of geometrical constraints on PET performance in whole body simultaneous PET-MR. *IEEE Nucl. Sci. Symp. Conf. Record* 2009: 3808–3811.

31. Keereman V, Vandenberghe S, De Beenhouwer J, Van Holen R, Staelens S, Schultz V, Solf T. Effect of geometrical constraints on PET performance in whole body simultaneous PET-MR. *IEEE Nucl. Sci. Symp. Conf. Record* 2009: 3804–3807

32. Spanoudaki VC, Lau FW, Vandenbroucke A, Levin CS. Physical effects of mechanical design parameters on photon sensitivity and spatial resolution performance of a breast-dedicated PET system. *Med. Phys.* 2010;37(11):5838–5849.

33. Ramirez RA, Liu S, Liu J et al. High-resolution L(Y)SO detectors using PMT quadrant-sharing for human & animal PET cameras. *IEEE Trans. Nucl. Sci.* 2008;55:862–869.

34. Rey M, Vieira JM, Mosset JB, Moulin Sallanon M, Millet P, Loude JF, Morel C. Measured and simulated specifications of Lausanne ClearPET scanner demonstrator. *IEEE Nucl. Sci. Symp. Conf. Record* 2005;4:2070–2073.

35. Van de Laan DJ, Maas MC, de Jong HWAM, Schaart DR, Bruyndonckx P, Lemaître C, van Eijk CWE. Simulated performance of a small-animal PET scanner based on monolithic scintillation detectors. *Nucl. Instrum. Meth. Phys. Res.* 2007;A 571:227–230.

36. Peng H, Olcott PD, Spanoudaki V, Levin CS. Investigation of a clinical PET detector module design that employs large-area avalanche photodetectors. *Phys. Med. Biol.* 2011;56:3603–3627.

37. Van de Laan DJ, Maas MC, de Jong HWAM, Schaart DR, Bruyndonckx P, Lemaître C, van Eijk CWE. Simulated performance of a small-animal PET scanner based on monolithic scintillation detectors *Nucl. Instrum. Meth. Phys. Res.* 2007, A 571 227–230

38. Song TY, Wu H, Komarov S, Siegel SB, Tai YC. A sub-millimeter resolution PET detector module using a multi-pixel photon counter array. *Phys. Med. Biol.* 2010;55:2573–2587.

39. Spanoudaki VC, Levin CS. Investigating the temporal resolution limits of scintillation detection from pixellated elements: Comparison between experiment and simulation. *Phys. Med. Biol.* 2011;56:735–756.

40. Peng H, Levin CS. Design study of a high-resolution breast-dedicated PET system built from cadmium zinc telluride detectors. *Phys. Med. Biol.* 2010;55:2761–2788.

41. Visvikis D, Lefevre T, Lamare F, Kontaxakis G, Santos A, Darambaravan D. Monte Carlo based performance assessment of different animal PET architectures using pixellated CZT detectors. *Nucl. Instrum. Meth. Phys. Res. A.* 2006;569:225–229.

42. Descourt P, Mathy F, Maitrejean S et al. Performance assessment of a variable field of view and geometry PET animal scanner based on CdTe strip detector blocks. *IEEE Nucl. Sci. Symp. Conf. Record* 2008:4559–4561.

43. Lamare F, Turzo A, Bizais Y, Le Rest CC, Visvikis D. Validation of a Monte Carlo simulation of the Philips Allegro/GEMINI PET systems using GATE. *Phys. Med. Biol.* 2006;51:943–962.

44. Lazaro D, Buvat I, Loudos G et al. Validation of the GATE Monte Carlo simulation platform for modelling a CsI(Tl) scintillation camera dedicated to small-animal imaging. *Phys. Med. Biol.* 2004;49:271–285.

45. Buvat I, Castiglioni I, Feuardent J, Gilardi MC. Unified description and validation of Monte Carlo simulators in PET. *Phys. Med. Biol.* 2005;50:329–346.

46. Merheb C, Petegnief Y, Talbot JN. Full modelling of the MOSAIC animal PET system based on the GATE Monte Carlo simulation code. *Phys. Med. Biol.* 2007;52:563–576.

47. Netter E, Pinot L, Ménard L, Duval MA, Janvier B, Lefebvre F, Siebert R, Charon Y. The Tumor Resection Camera (TReCam), a multipixel imaging probe for radio-guided surgery. *IEEE. Nucl. Sci. Symp. Conf. Records* 2009:2573–2576.

15

Introduction to Task-Based Assessment of Image Quality for Investigators Employing Monte Carlo Simulation of Imaging Systems

Michael A. King

CONTENTS

One of the important applications of the Monte Carlo methods discussed in this book is to simulate realistic images from imaging systems which already exist or new designs for systems/collimators one is prototyping prior to construction. Once you have these images, and possibly used them to create derived images (reconstructions with perhaps some form of correction), then likely the next thing you will want to do is to evaluate their image quality in some manner. There are a number of possible metrics one can use for an evaluation (see pp. 914–922 of [1]). However, since medical images are acquired for a purpose, it has been argued that the most meaningful measure of the quality of an image would be how useful it is while performing tasks that closely approximate those for which it would be employed clinically [1,2]. The goals of this chapter are to review the elements of task-based assessment of image quality, and to provide some practical advice on performing the evaluations for investigators employing Monte Carlo simulation of imaging systems.

Mathematical Preliminaries

Patients and phantoms are not discrete. They have continuous distributions of activity and attenuation. However, the results of imaging are represented in a computer as discrete values. Thus the most appropriate model of imaging is a *continuous-to-discrete mapping* (see pp. 12–13 of [1]). This should be remembered when modeling imaging via Monte Carlo simulation which typically requires a discrete representation of the source and attenuation distributions, as in this case one is doing a *discrete-to-discrete mapping*. To partially address this difference in the mapping, it is recommended that one start with the activity and attenuation distributions sampled at least a factor of 2 finer in detail when doing Monte Carlo simulations. Also, use subsampling of the analytical geometry of the structures so that the voxel values represent the partial presence of structures within the voxels.

The general equation for an imaging system forming an image of some object using a discrete-to-discrete mapping can be written as

$$\mathbf{g} = H\,\mathbf{f} + \mathbf{n}, \tag{15.1}$$

where $\mathbf{f}$ is an $L \times 1$ column vector of voxel values that represents the object one is imaging (e.g., three-dimensional (3D) source distribution for emission imaging and attenuation distribution for transmission imaging input into one's Monte Carlo program), H is the transfer function for the imaging system which is what your Monte Carlo code is modeling (it is an $M \times L$ 2D matrix detailing how each of the L entries in the source are transformed into each of the M image pixels in the projections you are making), $\mathbf{g}$ is an $M \times 1$ column vector of pixel values that is the set of numbers collected during the imaging process (this is the output of your Monte Carlo Program arranged as projections at different projection angles about $\mathbf{f}$), and $\mathbf{n}$ is typically a zero-mean Poisson-noise $M \times 1$ column vector. In other chapters of this book, the authors discuss in detail the Monte Carlo code used to model H. In this chapter, the focus is on the measurement of the image quality of the projection images $\mathbf{g}$, or more commonly the reconstructed images derived from $\mathbf{g}$.

Elements of Task-Based Assessment

Following the principles laid out by Barrett [2] and Barret and Myers [1], there are four essential elements to the objective assessment of image quality by task performance. The first element is the *task* to be performed. Will it be a classification task such as lesion detection in some part of the body or will it be a quantification task such as the estimation of the ejection fraction of the left ventricle during cardiac contraction? A must-read paper for all investigators considering setting up task-assessment studies is the classic paper in this field by Barrett [2]. In this paper, he considers in detail three metrics which can be used for each type of task and then the relationships between them. The focus of this chapter is on lesion-detection tasks, but the reader is reminded that estimation is also of great interest in medical imaging, particularly with the growing use of emission imaging to monitor and perhaps predict the results of therapies.

The second element is the definition of the *imaging process* that creates the data as defined in Section Mathematical Preliminaries above. That is, one needs to define the source being imaged as well as the imaging chain which produces the images of interest. The source could be the tracer and attenuation distributions in a simple geometrical object such as spheres embedded in a background, a digital anthropomorphic phantom such as the MCAT, NCAT, and XCAT phantoms [3–6], actual physical phantoms, or a population of clinical patients. Actual clinical images are obviously the most realistic, but it can be a very expensive undertaking to obtain a sufficient set of images with known truth as to the presence or absence of lesions. In addition, the validity of any clinical assessment of "truth" is ultimately open to question. Since this text is on Monte Carlo methods, the digital sources are the ones most likely to be of relevance to readers of this chapter. An alternative which might be of interest is the use of "hybrid" images, which are "normal" clinical acquisitions to which simulated lesion data have been added [7]. For digital phantoms, the imaging process may be defined analytically, but again, Monte Carlo simulation will be the most likely methodology employed by the readers of this chapter. Besides the creation of the projection images one may need to also consider the methodology of reconstruction if tomographic slices are to be assessed. However, consideration of reconstruction is not always needed as it is possible to do estimation tasks directly from projection data, for example [8].

The third element is the definition of the *observer* employed to perform the task. For the detection tasks discussed in this chapter, these will either be human or mathematical (also known as "numerical") observers. With actual clinical images, physicians are de facto the preferred observers, albeit expensive to employ even if one can obtain their time. With simulated images which lack some of the subtleties of clinical images, my group has observed that scientists and physicians perform similarly. Overall, studies with human observers take too much time to allow for more than the comparison of just a few imaging strategies or parameters at one time. Thus, there is considerable interest in numerical observers as a substitute for human observers.

The fourth element for objective assessments is the definition of the *metric* by which observer performance will be judged. There are a number of possibilities [1,2]. Metrics for detection tasks typically involve some variant of receiver operating characteristics (ROC) analysis, but there are other choices.

The next section will be devoted to a review of the ROC paradigm and some of its variants. Then, in the following two sections, assessment of detection accuracy by first human and then numerical observers will be reviewed.

ROC Methodology and Metrics for Assessing Detection Accuracy

There are four possible outcomes for the binary decision task of determining whether an image contains a lesion. These outcomes can be placed into a 2×2 contingency table as illustrated in Table 15.1. Assume that the null hypothesis to be tested is that no lesion is present. If there truly is no lesion present and the observer correctly decides that no lesion is present, then this is a true-negative (TN) outcome. However, if there is no lesion present and the observer mistakenly decides there is one present, then this is a false-positive (FP) outcome. Since the observer is rejecting the null hypothesis when it should have been accepted, this is the type-I error discussed in statistics. If the lesion is present and the observer says it is present, then this a true-positive (TP) outcome. Finally, if the lesion is present and the observer says that it is not present, then the observer makes a mistake which is called a false-negative (FN) outcome. This is the type-II error of statistics.

After the observer has read a number of images, one can define the observer's *accuracy* or probability of making a correct decision as the total number of TP and TN outcomes divided by the total number of images. One might think that accuracy would be the best measure of an observer's performance, but there is a significant problem in that accuracy does not take into account the prevalence of the lesion. Suppose we are investigating a screening test which would be applied to a large number of individuals where most of the time (say 99%) there is no disease present. The observer could obtain an accuracy of 0.99 by ignoring the images and just saying there is no lesion. The observer has obtained an outstanding accuracy, but individuals with disease are never identified and the images were not employed at all. Thus, in this case the accuracy has told us nothing of the utility of the images.

Proper metrics for judging the utility of images in a decision task should take into account the chances of making both type-I and type-II errors.

TABLE 15.1

Contingency Table of Outcomes for a Binary Detection Task

	Truth as to Lesion's Presence	
Observer's Decision	Lesion Present	Lesion Absent
Lesion present	True-positive (TP)	False-positive (FP)
Lesion absent	False-negative (FN)	True-negative (TN)

Toward this end, one can define the observer's *sensitivity* [or true-positive fraction (TPF)] as the number of TP outcomes divided by the number of lesion-present images. Similarly, the observer's *specificity* [or true-negative fraction (TNF)] is the number of TN outcomes divided by the number of lesion-absent images. But now there are two metrics to be reported and the problem is that for a given set of images the observer can easily improve one of these (say increase their TPF) just by relaxing the mental threshold they employ for classifying an image as having a lesion. However, this will result in a decrease in their TNF because they will now likely call more images without lesions as having lesions. Thus, it is difficult to use these two metrics to compare the image quality for two different imaging systems or processing strategies unless the strategies are different enough that the observer improves both their TPF and TNF, something which cannot be depended upon to happen even if the two really are different.

The way around this problem has been to have the observer give up reporting an absolute decision as to lesion presence and instead provide a confidence rating as to the presence of a lesion on some discrete or continuous scale. Then, with the definition of a scalar detection threshold t (the "mental threshold" referred to in the previous paragraph), one can say that a given image was classified as lesion-present at threshold t if the rating was greater than or equal to t.

For example, on a "five-point scale" the observer might report a 1 if they believed the image was definitely or almost definitely negative, 2 if probably negative, 3 if possibly positive, 4 if probably positive, and 5 if definitely or almost definitely positive. After the observer reports on a set of images the researcher can then use these confidence ratings to calculate the observer's TPF and false-positive fraction (FPF). The FPF is 1—TNF or the number of FP cases they report divided by the total number of cases without a lesion. Calculation of the TPF and FPF for a "five-point scale" is illustrated in Table 15.2. By plotting the observer's TPF as the ordinate (*y*-value) with

TABLE 15.2

Example of How the Points for an ROC Curve Are Calculated When a "Five-Point Scale" Is Employed

Rating	Number of Cases Read as TP	Number of Cases Read as FP	TPF for All Cases with Given Rating or Higher	FPF for All Cases with Given Rating or Higher
5	TP(5)	FP(5)	TP(5)/TP(All)	FP(5)/FP(All)
4	TP(4)	FP(4)	TP(4+)/TP(All)	FP(4+)/FP(All)
3	TP(3)	FP(3)	TP(3+)/TP(All)	FP(3+)/FP(All)
2	TP(2)	FP(2)	TP(2+)/TP(All)	FP(2+)/FP(All)
1	TP(1)	FP(1)	TP(1+)/TP(All)	FP(1+)/FP(All)

Note: TP(#), FP(#), TN(#), and FN(#) indicate the number of cases (images) the observer rated with a rating of # (i.e., 1, 2, 3, 4, or 5), TP(#+) and FP(#+) mean the number of cases when those having a rating of # or greater are combined, and TP(All) and FP(All) are the total number of cases truly with and without lesions, respectively.

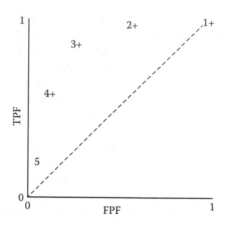

FIGURE 15.1
Example of generation of ROC curve by using the TPFs and FPFs for the different ratings of Table 15.2. Note that a number followed by a "+" sign means a lesion is called present for that rating and all numerically larger ratings.

their FPF as the abscissa (*x*-value) for each of the different rating levels as tabulated in Table 15.2, one generates an ROC curve as shown in Figure 15.1. With a "five-point scale," one will actually have six points that can be used. The starting point not shown in Figure 15.1 would be the lower-left corner of the plot where the TPF and FPF are both zero because this would be the point when the observer rated none of the images as having a lesion present. The second point would be when only the most stringent rating of 5 is used to define images as having a lesion present. Since it uses the toughest threshold for calling images as having a lesion present, it will have the lowest TPF and also the lowest FPF. Thus it will be the point closest to the lower-left corner of the plot. The point 4+ for ratings of 4 or greater (i.e., 4 and 5) will have a higher TPF but more FPs will also likely occur. Thus, it will be higher up on the plot and more to the right. The same will be true for each of the lower rating points, with the point for the rating of 1+ being at the upper-right corner of the plot as all images will be called as having a lesion. That is this will be the point for the TPF and FPF both being equal to 1.

A nonparametric ROC curve can be generated by simply connecting these points by a series of straight lines. The *area under the ROC curve* (AUC) for the nonparametric curve can then be calculated by hand as shown by Hanley and McNeil [9]. The AUC is the standard ROC metric of observer performance. It represents the observer's average sensitivity over all specificities.

The observer rating data are noisy, so some functional model with adjustable parameters is often used to fit the data and estimate the ROC curve. Typically, a binormal model for the ratings (conditional on the images being lesion-present or lesion-absent) is employed. The ROC curve and AUC are then based on maximum-likelihood estimation of the model parameters [10,11]. Details as to where to download ROC fitting software can be found in section Image Quality Resources on the Web of this chapter. With the binormal assumption, an ROC curve plots as a straight line on "normal deviate" axes (z_{TPF} vs. z_{FPF}). If the underlying distributions are not normal, then in many cases it can be assumed they can be transformed to normal by a generally unknown transform. This assumption has been shown to be valid over a broad variety of situations. However, use of this approach can yield inappropriate ROC curve shapes when cases are few and/or when the rating scale is discrete and operating points are poorly distributed [11].

It is important to note two limiting cases for AUC. The highest possible AUC is 1.0, which occurs when a TPF of 1.0 is achieved before any FP findings occur. In terms of the binormal rating model alluded to above, this limit is achieved when the two rating distributions are nonoverlapping. The lower bound on AUC is 0.5, which happens when TPF is equal to the FPF at each rating. This bound is the result of having identical rating distributions for the lesion-present and lesion-absent images, in which case the observer is essentially guessing as to the lesion status.

Relative assessments of performance are usually performed by comparing the AUCs of the competing imaging systems or processing methods. Reliance on the AUC as the criteria to judge relative performance can be problematic if the ROC curves intersect or a comparison at the specific region or point in the plot is most relevant clinically. Other metrics associated with the ROC curve are the *detectability index* (d_A) and the *signal-to-noise ratio* (SNR) [1].

A more general issue with ROC studies is that the observer can correctly identify a lesion-present image but for the wrong reason (i.e., their rating was based on an image feature unrelated to the lesion). As an example, suppose an observer provides a rating of 5 for the slice from a Monte Carlo simulated Ga-67 citrate study which is shown in Figure 15.2a. This slice in fact does contain a lesion just below the sternum at the location indicated by the cross-hair in Figure 15.2b. However, the observer might have been focusing on the noise blob located above the spine. Such cases create bias in the measure of observer performance and can decrease the statistical power of the ROC study.

One way to address the above issue is to instruct the observer where to look for evaluation of the presence of a lesion in the image by a cross-hair or some other indicator which can be toggled on or off. That is, to have the observer perform the location-known-exactly (LKE) task. An example of this approach is illustrated in Figure 15.2b where the appearance as presented to an observer with the cross-hair present is shown. The problem with LKE studies is that to obtain AUCs within the recommended range for best precision when comparing as discussed later in this chapter, one may have to decrease the lesion contrast or noise levels such that simulated images are clinically unrealistic [12]. This then brings into question the clinical relevance of the optimization or comparison being performed.

Several generalizations of the ROC paradigm avoid this problem by requiring the observer to supply ratings and also indicate (mark) the suspected

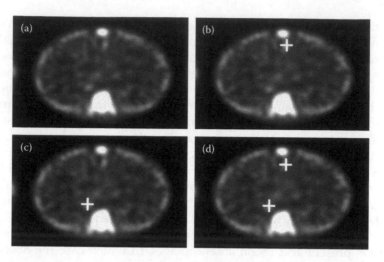

FIGURE 15.2
Transverse reconstructed slice through an MCAT-based Monte Carlo simulation of ^{67}Ga citrate imaging in the human thorax region. (a) Slice with a lesion-present which is visualized as the small bright area just below and slightly to the right of the sternum (very bright elliptical structure at the center anteriorly in the slice). (b) Same slice as (a) but with a white cross-hair centered on the lesion. (c) Same slice as (a) but with a white cross-hair centered on a noise-blob to the left and anterior to the spine. (d) Same slice as (a) but with white cross-hairs centered on a lesion and a noise-blob.

lesion locations typically by clicking with a mouse on their sites in the displayed images. Adding localization to the task often makes it more difficult and clinically relevant. The *localization ROC* (LROC) paradigm [13] is applicable for images containing no more than one lesion. The LROC curve is a plot of the fraction of lesion-present images that are TP with correct localization (*y*-axis) versus the FPF (*x*-axis). In practice, a localization is scored as correct when it falls within an acceptance radius [or *radius of correct localization* (R_C)] about the true location. How to set the R_C is discussed below. Area under the LROC (AUC_L) curve is a standard metric for observer performance with LROC. Another is the ordinate for the rightmost point on the curve, referred to as the *probability of correct localization* (P_C). This is the marginal probability that a lesion-present image will receive a correct localization. Both the AUC_L and P_C will lie in the range [0,1]. The method of Swensson [14] for maximum-likelihood fitting LROC data has become the standard methodology for LROC analysis.

Figure 15.2 can again be used to illustrate how an LROC study proceeds. The observer would be presented with the slice of Figure 15.2a and asked to mark the location where they have the greatest confidence (or most suspicion) a lesion is located. If they indicate the location shown in Figure 15.2b, then they would have correctly found the lesion. If they indicate the location illustrated in Figure 15.2c, then they would not have correctly found the lesion and should not be given credit for doing so.

Since it is unlikely that an observer will exactly determine the actual center of a lesion when indicating their perceived lesion center, a margin for error or R_C must be specified for use in determining whether the observer's location is correct or not. One way of doing this is to plot the observer's P_C versus candidate R_C values. Generally, the resulting curve will at first show a rise in P_C with R_C. In this region, one can assume that the observer has found the actual lesion and this is a correct localization. At some point the curves rate of increase in P_C will decrease to a relative constant rate from then on. These localizations are likely due to chance and thus not correct. The point at with this change in slope occurs is generally used to determine the R_C. More discussion of this and an example of use of LROC with nuclear medicine images can be found in the investigation of Wells et al. [15].

In clinical practice, the physician is not generally constrained to report only one lesion in an image. In fact they may find none or several. Bunch et al. [16] proposed *free-response ROC* (FROC) analysis to treat these multiple-report tasks. Thus, with FROC the observer would report locations and ratings for all of the sites for which they had suspicion of lesion presence. For example, when the observers are presented the slice shown in Figure 15.2a as part of an FROC study, they might be indicated the two locations shown in Figure 15.2d with a confidence rating for each. Chakraborty [17] presented a formal FROC model and algorithm (FROCFIT) for fitting the model parameters. In a subsequent paper, Chakraborty and Winters [18] developed the alternative FROC (AFROC) method of scoring FROC data which employs existing ROC analysis software. AFROC studies assessing acquisition and reconstruction protocols for whole-body PET have been carried out by Lartizien et al. [19,20]. The FROC and AFROC paradigms have been infrequently employed. This has been attributed to assumptions that both make concerning the independence of the reports on an image [21]. The recently proposed jackknife AFROC (JAFROC) paradigm [22] avoids the independence assumptions of the earlier FROC and AFROC models and has been shown to offer more statistical power (smaller confidence intervals [23] than AFROC and ROC analysis [24]). At the 2006 SPIE meeting, Volokh et al. [25] compared JAFROC analysis to standard ROC for nuclear medicine images. They observed JAFROC to provide better statistical power than ROC for simulated scintimamographic images, and that there was a reversal in rankings of reconstruction strategies between ROC and JAFROC for a limited set of clinical cardiac perfusion images.

The ROC family of paradigms for assessing image quality provides the most complete description of observer performance as both the curve portraying performance as a function of confidence rating, and metrics derived from it such as the AUC are available [26]. However, it requires a lot of effort and cooperation on the part of observers to provide a confidence rating that is assumed to be relative to a constant scale. Simpler families of testing methodologies which provide just a summary statistic are those of the *forced-choice* paradigm. In two-alternative forced-choice (2-AFC), the observer views a pair of images, one of which is with a lesion and the other without. The observer is then required to pick which of the pair is the one with a lesion. The fraction of the time they are correct in their selection ranges from 0.5 with the observer performing at chance to 1.0 when they have perfect performance. With this bounding of the possible performance, it is probably not surprising that this metric is equal to the AUC of an ROC study which would have been performed with these images, but obtained with less effort on the part of the observer and experimenter [9,26].

Assessment by Human Observers

In this section, guidelines are discussed for setting up and performing ROC-type studies with human observers which my group has adopted in our own studies. These guidelines are slanted toward the performance of studies with simulated images as this is a book on Monte Carlo methods. These are recommendations and not intended to be rules. For further guidance, I point the interested reader to an article [11] and Chapter 15 of Beutel et al. [27] by Metz, and pp. 923–952 of Chapter 14 of Barrett and Myers [1].

When performing studies with simulated images, one should strive to make the images as clinically realistic as possible. Obtain the organ/structure relative uptakes you use such that those in your reconstructed slices match those in clinical images. The same is true with count levels. Also vary relative uptakes, count levels, and the size and shape of the anatomical structures so that you are investigating a population of cases, as opposed to making the assessment with just a single configuration and noise level. Investigating just a single case leads to questions regarding the generalizability of your results to clinical application [28,29]. In the past, using a single or few backgrounds for the lesions was justifiable on the basis of the computation time required to investigate a population; however, over the last decade computation time has become significantly less of a burden.

Given that it is recommended to capture clinical realism in the images, it is not surprising that it is also recommended to have a knowledgeable physician highly involved with the research. A physician's insights can be helpful for anything from the selection of the task, through creation of the images, to the functionality of the graphical computer interface used to conduct the ROC study. The person creating the images should not be one of the observers whose results are reported for the study. They should, however, be the first person to actually perform the study, and should do so using the final set of images and software intended for the study observers. Doing this allows the experimenter to catch any bugs and unwanted features. The unpleasant alternative is to have the actual observers find these problems after the study has begun. Finally, it can be worthwhile to also have an experienced member of the research team go through the complete study beforehand in order to catch subtleties like incorrect file permissions that the experimenter might not experience.

The size and shape of the lesions should be picked based on clinical criteria. The contrast of the lesions relative to the normal backgrounds should be adjusted at a minimum such that lesions with known location can be seen when images without added noise are displayed thresholded the same way as the noisy images would be. With this lower bound then for ROC studies, the contrast should be adjusted using pilot studies such that the average AUC for the various modalities lies in the range between 0.8 and 0.95 [11]. In checking his current recommendations as part of writing this chapter, Dr. Charels Metz informed us he still recommends the above range but goes further to recommend that "the AUC of an existing imaging system against which one wants to compare a presumably better system should be set at around 0.80 or 0.85" (personal communication). This leaves room for the hopefully improved strategy to fall in the upper region of the range.

For cardiac nuclear medicine images, one should follow standard practice by displaying images such that the maximum displayed intensity is associated with the maximum value in the heart and the minimum is set to zero [30]. For tumor detection studies, my group has applied an adaptive upper threshold to the slices. The thresholding is performed as described in [15]. The maximum-count value among all the lesion pixels (denoted *MLC* for "maximum lesion count") is determined for each lesion-present image for a given modality. The maximum and the standard deviation over this set of *MLC*s (respectively, max(*MLC*) and σ(*MLC*)) is then obtained. The original pixel value $I(x,y)$ at position (x,y) in the images is then adjusted to value $I'(x,y)$ according to the formula

$$I'(x,y) = \min\left[I(x,y), \max(\text{MLC}) + n \times \sigma(\text{MLC})\right], \qquad (15.2)$$

where n is a modality-independent constant selected to prevent the lesions from always being the brightest image structure. My group's studies have used $n = 1$ or 2. The important thing is that whatever thresholding procedure is chosen it should increase lesion contrast while leaving the noise structure in the region of the lesion unaffected.

In the clinic, physicians are free to vary both the upper and lower threshold with which images are displayed. In keeping in line with matching the clinical task, allowing the observer to vary the contrast can be considered. However, I recommend some strategy such as the above ones be employed to at least provide a reasonable starting point. Also, the experimenter should be aware that if they are going to compare their human observer results to those of a numerical observer, allowing the human observer to vary the threshold makes matters more difficult for them as the same images should be employed for both human and numerical observers.

In the past, my group has conducted our observer studies in a darkened room using a monitor with a perceptually linearized grayscale [31]. To match the clinical task, one might prefer instead to make sure the monitor meets the American College of Radiology (ACR) standards for display performance [32], or whatever the equivalent standards are where the study is to be performed. My group has allowed each observer to adjust their viewing distance according to their preference, and to take as much time as they deemed necessary.

Observer training is an important part of study design. It is recommended that you train the observers by having them perform ROC studies with images from each of the modalities they will be presented with during the investigation. During training, the observer should be given feedback as to presence and location of any lesions. Use of display of the without added noise images and a marker which can be toggled off and on to indicate lesion location is a useful way of doing this. No training image should be used as a test image when conducting the actual study. They should be kept as a separate set from which images can be used in retraining the observers at the start of each session, but not used to gather data for reporting observer performance. The AUCs from the initial training sessions should be calculated however. They are useful in determining whether an observer has obtained an acceptable level of performance before they are allowed to proceed on to the study sessions. If they have not, then they should be asked to perform the training sessions again. AUCs from pilot studies with knowledgeable observers can be used to establish expectations as to the acceptable lower level of AUCs new observers should obtain.

During training, the observers learn how to operate your GUI and devise their strategy for assigning confidence ratings. Since strategies may vary with image modality, only images of a single modality should be rated in a given session. For example, an observer's rating of a cool inferior wall in the short-axis slice of a cardiac perfusion study would likely vary depending upon whether the slice had been corrected for attenuation, or not. A period of retraining at the start of each test session to allow the observer to reestablish their strategy for assigning confidence ratings should be employed. It is important that they do so as it is assumed in the ROC analysis of their ratings that the observer does not vary their criteria for specifying their ratings during their rating session. In my group's studies, we do not inform the observers what strategy they are rating. Instead, we provide them with a randomized letter coding so they can remember this is strategy A and for strategy A, I rate this way. The retraining at the start of the study helps to reinforce this.

In introducing ROC earlier in this chapter, the example of a five-point rating scale was used. However, it has been recommended that a continuous scale, or at least a discrete scale with more than five ratings, be employed [33]. This is because it is important in fitting their data to the binormal model that the observers distribute their ratings over the scale instead of reporting mostly a couple of ratings. With a continuous scale, observers have greater flexibility in ratings to report so even if they use a limited portion of the scale it is still possible to have their data spread sufficiently over the curve to allow meaningful fitting.

With one exception, the presentation order of the images shown during a given observer session should be random to avoid reading order effects becoming systematic over observers. The exception has to do with the customary usage with simulated or hybrid studies of employing 50% lesion absent and 50% lesion present images in each set. With this design scenario, it is important to ensure that the pair of images is not shown to the observer one after the other. It is amazing how this can help the observer by providing them a "background-know-exactly" (BKE) task.

The order in which different modalities are read should not be randomized but rather systematically designed by the experimenter so as to distribute the possible effects of learning the study images and loss of zeal as the sessions progress [11]. For example, if you have three observers and three modalities called A, B, and C, then the first observer might rate the modalities in the order A, then B, and finally C. The second observer could then employ a modality reading order of B, then C, and finally A. The third observer would then employ a reading order of C, then A, and finally B. In this way each modality is read first by one of the observers, each second by one of the observers, and each third by one of the observers.

Observers tire and this can alter their performance. Thus, to have enough rating data it may be necessary to break the number of images of a given modality into two or possibly three viewing sessions. When this is done, the data from an individual observer can be pooled into one set for analysis of that observer's performance under the assumption that their rating scale was unchanged between the sessions; however, the ratings from different observers should not be pooled into a single set. Instead the ratings of each observer should be used to fit an ROC curve for that observer. An average ROC curve for all observers can be obtained by averaging the individual points from the curves themselves (i.e., TPF as a function of FPF). An average curve should not be obtained by averaging the parameters of the binormal models used to create the curves.

Statistical testing of the differences between two or more modalities is based on the multireader multicase (MRMC) paradigm as first formulated by Swets and Picket [34]. This methodology allows one to generalize their analysis for variation in both readers and images, within in the ranges included in the study's experimental design. One methodology for analyzing MRMC studies is that due to Dorfman et al. [35].*

Assessment by Numerical Observers

This section is intended to be an introduction to numerical observers for ROC studies. For more detailed information, the reader is referred to pp. 952–991 of [1], Chapters 9–12 of [27], the introductory paper in [36], and papers from my group [37–39].

A numerical observer is generally applied either to the projection imaging data (the $M \times 1$ vector $\mathbf{g}$ of Equation 15.1) or the object estimates $\hat{\mathbf{f}}$ obtained through reconstruction of the data. For convenience, the vector $\mathbf{g}$ will be used in the following to represent either. The numerical observer will combine the values of $\mathbf{g}$ in some manner depending on the type of observer and produce a single number λ. This number is the observer's scalar test statistic reflecting its assessment of the presence of a lesion. If λ is larger than some threshold t, then the lesion is judged to be present, if not it is absent. By varying t and determining the TPF and FPF, an ROC analysis can be performed, and the AUC determine as previously described.

With $\mathbf{g}$ a 2D image with pixel dimensions $n \times n$ (so that $n^2 = M$), the $M \times M$ pixel covariance matrix for the images is defined as

$$\mathbf{K}_g = \left\langle [\mathbf{g} - \bar{\mathbf{g}}][\mathbf{g} - \bar{\mathbf{g}}]^{\,t} \right\rangle, \tag{15.3}$$

where the superscript t represents a transpose, the brackets denote expectation over $\mathbf{g}$ and $\bar{\mathbf{g}} = \langle \mathbf{g} \rangle$.

Let H_0 and H_1 represent the conditions that an image is tumor-absent (the background only images in simulation) and tumor-present (background plus lesion in simulation), then the conditional mean image given condition H_i will be denoted as $\langle \mathbf{g} \rangle_i$, where for a binary task i is either 0 or 1. The conditional covariance matrices are $\mathbf{K}_{g|0}$ and $\mathbf{K}_{g|1}$.

According to a categorization by Barrett and Myers [1], numerical observers are either *optimal* or *suboptimal*. To quote those authors, "By definition, optimal observers are the best possible in some sense. The Bayesian or ideal observer makes optimal use of all available information in the data ... to achieve the highest AUC attainable." The *ideal observer* has been recommended for system design work with the intent of maximizing in some way the information content of the imaging data. For the binary lesion-detection tasks in ROC studies, this observer is constructed as the log-likelihood ratio [1]

$$\lambda_{\text{ideal}} = \ln \left[\frac{p(\mathbf{g}|H_1)}{p(\mathbf{g}|H_0)} \right]. \tag{15.4}$$

In this equation, $p(\mathbf{g}|H_0)$ is the conditional probability distribution on $\mathbf{g}$ given that the image is tumor-absent (condition H_0) and $p(\mathbf{g}|H_1)$ is the conditional distribution for images with tumors (condition H_1).

Given the existence of optimal observers, why would one ever want to not do the best by using a suboptimal observer? One answer is that human observers are suboptimal. It has been shown that they do not make usage of all of the available information as do optimal observers. For example, humans have trouble dealing with correlated noise [40]; however, ideal observers decorrelate (prewhiten) noise as part of their application to the images. Thus, if your goal is to use a numerical observer as a surrogate for human observers, then you will likely want to use a "human" or "humanoid" numerical observer which will be suboptimal.

Another way to categorize numerical observers is by whether the test statistic is a *linear* or *nonlinear* function of the image. In general, both the ideal and human observers are nonlinear. Humans, for example, have a logarithmic response to variations in luminance. On the other hand, the large majority of numerical observer work to date has employed linear numerical observers, and I will exclusively do so in this section.

With the restriction of just considering linear observers, numerical observers are a template (vector) made up of a set of weights ($\mathbf{w}$). These weights are multiplied times the pixel values in the image and the result summed to calculate λ, or in equation form:

$$\lambda = \mathbf{w}^t\mathbf{g}, \tag{15.5}$$

where the superscript t indicates the transpose of the column vector $\mathbf{w}$ to a row vector with the result that the operation of the right-hand side is the *dot-product* of the row vector $\mathbf{w}^t$ with the column vector $\mathbf{g}$.

The ideal observer is linear for the basic *signal-known-exactly/background-known-exactly* (SKE–BKE) detection task of detecting a lesion (signal) $\mathbf{s}$ at a fixed location in a known background. Assume one has complete knowledge about the lesion characteristics (including shape, size, and location). The lesion $\mathbf{s}$ can then be calculated as the difference

$$\mathbf{s} = \langle \mathbf{g} \rangle_1 - \langle \mathbf{g} \rangle_0. \tag{15.6}$$

However, the task is complicated by the addition of Gaussian random noise to the image pixels. With the noise described by covariance matrix $\mathbf{K}_g$, the ideal observer template for the SKE–BKE task is

$$\mathbf{w}_{\text{ideal}} = \mathbf{K}_g^{-1}\mathbf{s}. \tag{15.7}$$

When $\mathbf{K}_g$ is the identity matrix, the template is simply $\mathbf{s}$ and Equation 15.7 describes a *non-prewhitening matched filter* (NPWMF) [1].

An example of how this observer operates for an SKE–BKE task is illustrated in Figures 15.3 and 15.4. Note that when $\mathbf{g}$ has pixel dimensions of 128×128, $\mathbf{K}_g$ is of dimensions $16{,}384 \times 16{,}384$ and calculation of its inverse can be difficult. This problem is avoided in the example by considering the case of zero-mean Gaussian noise with covariance $\sigma^2\mathbf{I}$ ($\mathbf{I}$ is the identity matrix). Then, the test statistic for the ideal observer becomes

$$\lambda = \frac{\mathbf{s}^t\mathbf{g}}{\sigma^2}. \tag{15.8}$$

At the top of Figure 15.3 is the 128×128 "lesion" image $\mathbf{s}$ consisting of a 9×9 square of pixels having magnitude 1 against a zero background. Below this image is a set of 25 noisy test images created using $\sigma = 3$. Note the images are scaled such that both positive and negative values are displayed.

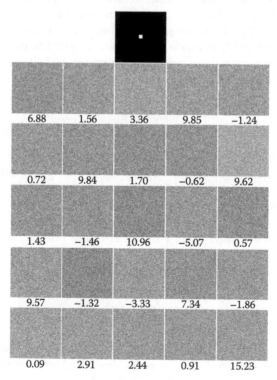

FIGURE 15.3
Shown at the top center is the signal consisting of a 9×9 pixel square with zero background. Below is a set of 25 images of this object to which random Gaussian noise has been added. λ for the Hotelling numerical observer calculated as given in Equation 15.8 is given below each image.

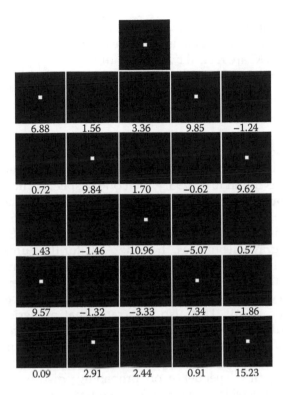

FIGURE 15.4
At the top center is the signal consisting of a 9×9 pixel square of magnitude 1 centered in a 128×128 image with a zero background. Below is the set of 25 images of Figure 15.3 before the noise was added. From these images, the truth as to in which images the signal was present is obvious.

The lesion is present in some of these images and absent in others. Can you pick out which images are which? As a hint, the value of λ as calculated from Equation 15.8 is given below each image. In making this calculation, the values in the image pixels were kept in the floating point format in which there were created. The image truths are shown in Figure 15.4.

The performance of the ideal observer in this experiment could be determined by using the test statistics to generate an ROC curve and then estimating AUC. An alternative approach for numerical observers is to compute the SNR

$$\text{SNR} = \frac{\left(\langle\lambda\rangle_1 - \langle\lambda\rangle_0\right)}{\sqrt{\left(\frac{1}{2}\sigma_1^2 + \frac{1}{2}\sigma_0^2\right)}}. \tag{15.9}$$

The numerator in this expression is the difference between the expected values of λ under conditions H_1 and H_0 while the denominator makes use of the conditional standard deviations σ_0 and σ_1 in λ. When λ is normally distributed under both conditions (as is the case for the preceding example), then SNR and AUC are related by the formula [1]

$$AUC = \frac{1}{2} + \frac{1}{2}\text{erf}\left(\frac{SNR}{2}\right), \tag{15.10}$$

where erf() denotes the error function.

Following this SNR approach, the AUC for the ideal observer of the previous example is found to be 0.98, which one suspects is much higher than any human could perform. But there is one important catch in that the numerical observer was analyzing floating-point valued images while the image format in Figure 15.3 is an eight-bit grayscale. If Equation 15.8 is used with the grayscale images, then the AUC for the numerical observer falls to 0.84 which is lower and thus closer to what a human observer might do. For this revised task, however, we are no longer working with the ideal observer because the pixel quantization introduces another random noise process [41]. This also illustrates that numerical and human observers should be presented with the same images when you are trying to validate your numerical observer predictions of human observer performance [42].

Unfortunately, the dependence on probability distributions makes the ideal observer intractable for most tasks of clinical interest. A popular alternative is the ideal linear or *Hotelling observer* [1]. Among all linear models, the Hotelling observer maximizes the SNR. The Hotelling template is given by

$$\mathbf{w}_{\text{Hot}} = \mathbf{K}^{-1}\,\mathbf{s}, \tag{15.11}$$

where $\mathbf{K}$ is the average of the pixel covariance matrices $\mathbf{K}_{g|0}$ and $\mathbf{K}_{g|1}$. In many instances, the effect of the lesion on $\mathbf{K}$ can be assumed to be negligible, so that $\mathbf{K} \approx \mathbf{K}_g$. The substitution of Equation 15.11 into Equation 15.5 yields the Hotelling test statistic

$$\lambda = \mathbf{s}^t\,\mathbf{K}^{-1}\,\mathbf{g}. \tag{15.12}$$

The *channelized Hotelling observer* (CHO) [43] is a Hotelling variant that has served as a popular model for human observers. This observer simulates the human visual system by preprocessing the 2D images through a set of C frequency-selective digital filters (or channels). Suppose $\mathbf{u}_c$ represents the cth channel response obtained by inverse Fourier transforming the filter to the spatial domain. Then $\mathbf{u}_c^t\mathbf{g}$ is the channel output for a given image and the full set of outputs is denoted as $\mathbf{U}^t\mathbf{g}$, where $\mathbf{U}$ is an $M \times C$ matrix whose columns contain the channel responses. With these channel responses as the input to the Hotelling observer, the CHO test statistic is

$$\lambda = \mathbf{s}^t \mathbf{U} \mathbf{K}_{CHO}^{-1} \mathbf{U}^t \mathbf{g}, \tag{15.13}$$

with the $C \times C$ channel covariance matrix $\mathbf{K}_{CHO} = \mathbf{U}^t \mathbf{K}_g \mathbf{U}$. Computation of $\mathbf{K}_{CHO}$ is greatly simplified by recognizing the equivalent definition

$$\mathbf{K}_{CHO} = \left\langle [\mathbf{U}^t \mathbf{g} - \mathbf{U}^t \overline{\mathbf{g}}][\mathbf{U}^t \mathbf{g} - \mathbf{U}^t \overline{\mathbf{g}}]^t \right\rangle, \tag{15.14}$$

where the channels are applied to the images prior to calculation of the covariance so that the $M \times M$ covariance matrix $\mathbf{K}_g$ does not need to be calculated. Thus, setting C to be much smaller than the number of image pixels not only furnishes a mechanism for degrading the prewhitening ability of the Hotelling observer to be on par with human abilities [40], but also dramatically simplifies the covariance calculation and inversion.

Channel design can be a key factor in whether the performance of the CHO will agree with the performance of human observers. Thus, it is important that the selection of the channels be validated through comparisons to human-observer studies [38,44]. A number of 2D channel models have been proposed, including both radially symmetric [38,43,44] and nonsymmetric ones. The latter generally require a larger C to partition frequency space but may be better suited for nonspherical lesions [45].

Figure 15.5 illustrates two symmetric models which have been employed. On the left are square (SQR) channels which follow the constant-Q form originally proposed by Myers and Barrett [43]. This is a series of nonoverlapping band-pass filters whose width increases as a multiple of 2. To the right in Figure 15.5 are difference-of-Gaussian (DOG) channels [46]. These three channels were derived from a set of Gaussians having standard deviations that varied in size by a factor of 1.6 [47]. It is important to note that in both of these models the zeroth term is 0.0. This is because the total number of counts in simulated studies provides a clue to numerical observers when lesion present studies are made by adding the lesion to (or subtracting the lesion from) the lesion absent images such that there is on average a difference in the number of counts between lesion present and lesion absent images. Note that in our previous example the average number of counts in the lesion present images was 9, whereas in the lesion absent images it was 0. This fact was used by the numerical observer (compare the λs given below images for the two cases), but human observers would be unaware of the difference when just visually rating the images.

The calculation of the channel outputs can be done in either the frequency or spatial domain. In the spatial domain, the channel responses $\mathbf{u}_c$:$c = 1,\ldots,C$ must be centered on the known lesion location before computing the dot

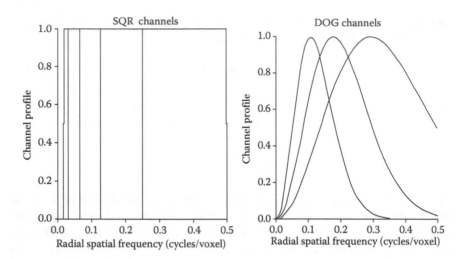

FIGURE 15.5
Frequency domain plots of sets of channel profiles for two different channel models employed with the channelized Hotelling observer. On the left is shown five nonoverlapping square (SQR) channels where the filter coefficients is either 0.0 or 1.0. On the right are shown three overlapping difference-of-Gaussians (DOG) channels where the filter coefficients vary from 0.0 to 1.0. Note in both cases, the zeroth or "DC" frequency value is 0.0.

product with $\mathbf{g}$. The initial pixel dimensions of the responses may need to be larger than $n \times n$ in order to facilitate this shift; once shifted, the portion of $\mathbf{u}_c$ that extends beyond $\mathbf{g}$ can be truncated. Examples of shifted responses are displayed in Figure 15.6. In this figure, the contrast of the rings about the lesion location has been enhanced for better illustration of the positive (brighter rings) and negative (darker rings) oscillations about zero. The response on the left corresponds to the low-frequency SQR channel from Figure 15.6 while the right-hand response is from a higher-frequency SQR channel.

Formulating numerical observers that can perform lesion detection and localization is an important step toward achieving clinical relevancy. One option is the *scanning linear observer* [39], which generalizes the Hotelling-type observers to LROC and FROC studies. For the basic LROC task of locating a known lesion $\mathbf{s}$ within some region of interest (ROI) of $\mathbf{g}$, a scanning observer first computes a statistic at every potential location. The maximum statistic indicates the rating and localization for the image. In [39], the statistic for the kth location is

$$\lambda_k = \mathbf{w}_k^t(\mathbf{g} - \mathbf{g}_b), \tag{15.15}$$

in which $\mathbf{g}_b$ is the mean lesion-free background corresponding to $\mathbf{g}$, and $\mathbf{w}_k$ is the appropriate observer template for the kth location. The template for

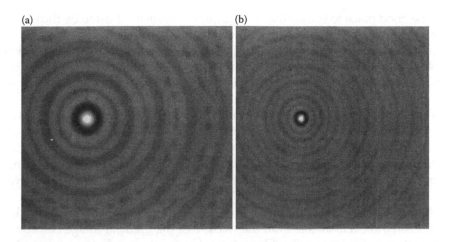

FIGURE 15.6
2D spatial domain forms of (a) a lower-frequency SQR channel and (b) a high-frequency SQR channel. The channel is centered over the site of potential lesions in the images to which it will be applied. Rings fluctuate about 0.0 and their contrast is exaggerated to illustrate better their appearance. (Reproduced with permission from Gifford HC et al. *J. Nucl. Med.* 2000;41(3):514–521.)

the NPWMF model is shift-invariant and thus very easy to apply, whereas the CHO template is shift-variant because $\mathbf{K}_{CHO}$ varies with location. The *channelized NPW* (CNPW) observer is a compromise between the CHO and NPWMF models that is both shift-invariant and which has shown good agreement with human observers [39].

From my group's point of view, it is crucial to have a numerical observer that can predict human performance at realistic detection tasks. Scanning observers based on Equation 15.15 have potential drawbacks in this regard. For one, humans usually do not have knowledge of statistical quantities like $\mathbf{g}_b$ and the scanning observer can wildly overestimate human performance as a result. Also, the experimenter may not even have access to this statistical information, as is the case with studies that use hybrid images combining normal clinical backgrounds with synthetic lesions. Current research in the group [48] is looking at ways to apply numerical observers with reduced statistical information by considering existing paradigms of how clinicians interpret images.

Image Quality Resources on the Web

There are a number of websites where the interested reader will find tutorials, bibliographies, and software.

One good place to start is the "image quality website" hosted by the Center for Gamma-Ray Imaging at the University of Arizona (http://www.radiology.arizona.edu/cgri/IQ/index.html). There one can find a discussion of "What is image quality," a list of many of the fundamental papers in the field, downloadable software, and slides from a short course on the assessment of image quality.

A site with essential information and software for performing and analyzing ROC studies is that of the Kurt Rossmann Laboratories for Radiologic Image Research at the University of Chicago (http://xray.bsd.uchicago.edu/krl/roc_soft.htm). There one can find an extensive listing of the ROC literature, and downloadable software for ROC analysis, including Dr. Metz's LABMRMC program mentioned earlier.

An extension of the LABMRMC program by Dr. Kevin Berbaum as well as other software is available from the website of the Medical Image Perception Laboratory at the University of Iowa (http://perception.radiology.uiowa.edu/Software/ReceiverOperatingCharacteristicROC/tabid/120/Default.aspx).

Another alternative for analyzing MRMC rating data using the Obuchowski and Rockette methodology [49] is available from the website of the Department of Quantitative Health Sciences at the Cleveland Clinic (http://www.bio.ri.ccf.org/html/rocanalysis.html).

One can obtain software to fit LROC data using the methodology of Swensson [14] from the website of Dr. Philip Judy (http://www.philipfjudy.com/Home/lroc-software).

A discussion of the basics of FROC and guidance on conduction FROC studies as well as downloadable software for analyzing FROC data can be found at the website of Dr. Dev Chakraborty (http://www.devchakraborty.com/).

An excellent place to look for reference lists, software, and details of their meetings which are held about every 2 years is the website of the Medical Image Perception Society (http://www.mips.ws/).

Finally, in February every year, the Society of Photo-Optical Instrumentation Engineers (SPIE) holds a meeting on Medical Imaging. This meeting is actually eight conferences in one. One of the conferences is entitled "Image Perception, Observer Performance, Technology Assessment," and each year many advances in the field of image quality assessment are presented at this conference. One can find information on past and future conferences at http://spie.org/x12166.xml.

Final Note

What I have provided in this chapter is just an introduction to task-based assessment of image quality by human and numerical observer task-performance studies. I hope this introduction provides the reader with the tools to better understand the literature and conduct task-based studies of image quality of their own using Monte Carlo simulated images.

Acknowledgments

Numerous discussions over the years with colleagues in the community of medical imaging researchers have influenced the presentation of this material. In particular, I thank Dr. Charles Metz of the University of Chicago for his assistance with my group's observer studies and for allowing access to lecture slides he presented at the 2002 IEEE Nuclear Science Symposium and Medical Imaging Conference. I thank Dr. Harrison Barrett of the University of Arizona for his guidance and insights on task-based performance and numerical observers. I also thank Dr. Howard Gifford for editing a draft of this manuscript and for providing the ideal observer example. This work was supported by the National Institute for Biomedical Imaging and Bioengineering under grant R01-EB02798. Its contents are solely the responsibility of the author and do not necessarily represent the official views of the National Institute of Biomedical Imaging and Bioengineering.

References

1. Barrett HH, Myers KJ. *Foundations of Image Science.* Hoboken: John Wiley & Sons; 2004.
2. Barrett HH. Objective assessment of image quality: Effects of quantum noise and object variability. *J. Opt. Soc. Am. A.* 1990;7(7):1266–1278.
3. Pretorius PH, King MA. A study of possible causes of artifactual decreases in the left ventricular apex with SPECT cardiac perfusion imaging. *IEEE Trans. Nucl. Sci.* 1999;46(4):1016–1023.
4. Segars WP, Mahesh M, Beck TJ, Frey EC, Tsui BMW. Realistic CT simulation using the 4D XCAT phantom. *Med. Phys.* 2008;35(8):3800–3808.
5. Segars WP, Lalush DS, Tsui BMW. Development of an interactive software application to model patient populations in the spline-based MCAT phantom. *IEEE Med. Imag. Conf.* 2000:2527–2531.
6. Tsui BMW, Frey EC, Zhao X, Lalush DS, Johnston RE, McCartney WH. The importance and implementation of accurate 3D compensation methods for quantitative SPECT. *Phys. Med. Biol.* 1994;39:509–530.
7. Pereira NF, Gifford HC, Price JC et al. An evaluation of iterative reconstruction strategies based on Mediastinal Lesion detection using hybrid Ga-67 SPECT images. *Med. Phys.* 2008;35:4808–4815.

8. Moore SC, Kijewski MF, El Fakhri G. Collimator optimization for detection and quantitation tasks: Application to Gallium-67 imaging. *IEEE Trans. Med. Imag.* 2005;24(10):1347–1356.
9. Hanley JA, McNeil BJ. The meaning and use of the area under the receiver operating characteristic (ROC) curve. *Radiology.* 1982;143:29–36.
10. Metz CE. Basis principles of ROC analysis. *Semin. Nucl. Med.* 1978;8:283–298.
11. Metz CE. Some practical issues of experimental design and data analysis in radiological ROC studies. *Invest. Radiol.* 1989;24(3):234–245.
12. de Vries DJ, King MA, Soares EJ, Tsui BMW, Metz CE. Effects of scatter substraction on detection and quantitation in hepatic SPECT. *J. Nucl. Med.* 1999;40(6):1011–1023.
13. Starr SJ, Metz CE, Lusted LB, Goodenough DJ. Visual detection and localization of radiographic images. *Radiology.* 1975;116(533):538.
14. Swensson RG. Unified measurement of observer performance in detecting and localizing target objects on images. *Med. Phys.* 1996;23(10):1709–1725.
15. Wells RG, Simkin PH, Judy PF, King MA, Pretorius H, Gifford HC. Effect of filtering on the detection and localization of small Ga-67 lesions in thoracic single photon emission computed tomography images. *Med. Phys.* 1999;26(7): 1382–1388.
16. Bunch PC, Hamilton JF, Sanderson GK, Simmons AH. A free-response approach to the measurement and characterization of radiographic-observer performance. *J. Appl. Photogr. Eng.* 1978;4:166–171.
17. Chakraborty DP. Maximum-likelihood analysis of free-response receiver operating characteristic (FROC) data. *Med. Phys.* 1989;16:561–568.
18. Chakraborty D, Winter LH. Free-response methodology: Alternative analysis and a new observer-performance experiment. *Radiology.* 1990;174:873–881.
19. Lartizien C, Kinahan PE, Comtat C. A lesion detection observer study comparing 2-dimensional versus fully 3-dimensional whole-body PET imaging protocols. *J. Nucl. Med.* 2004;45:714–723.
20. Lartizien C, Kinahan PE, Swensson RG et al. Evaluating image reconstruction methods for tumor detection in 3-dimensional whole-body PET oncology imaging. *J. Nucl. Med.* 2003;44:276–290.
21. Edwards DC, Kupinski MA, Metz CE, Nishikawa RM. Maximum likelihood fitting of FROC curves under an initial-detection-and-candidate-analysis model. *Med. Phys.* 2002;29(12):2861–2870.
22. Chakraborty DP, Berbaum KS. Observer studies involving detection and localization: Modeling, analysis, and validation. *Med. Phys.* 2004;31(8): 2313–2330.
23. Metz CE. Quantification of failure to demonstrate statistical significance: The usefulness of confidence intervals. *Invest. Radiol.* 1993;28:59–63.
24. Zheng B, Chakraborty DP, Rockette HE, Maitz GS, Gur D. A comparison of two data analyses from two observer performance studies using jackknife ROC and JAFROC. *Med. Phys.* 2005;32(4):1031–1034.
25. Volokh L, Liu C, Tsui BMW. Exploring FROC paradigm—Initial experience with clinical applications. *Proc. SPIE.* 2006;6146:61460D-61461–61460D-61412.
26. Sharp P, Barber DC, Brown DG et al. Medical imaging—The assesssment of image quality. *ICRU Report 54,* 1996.
27. Beutel J, Kundel HJ, Metter RL. *Handbook of Medical Imaging: Physics and Psychophysics.* Bellingham: SPIE, 2000.

28. He X, Frey EC, Links JM, Gilland KL, Segars WP, Tsui BMW. A mathematical observer study for the evaluation and optimization of compensation methods for myocardial SPECT using a phantom population that realistically models patient variability. *IEEE Trans. Nucl. Sci.* 2004;51(1):218–224.

29. He B, Du Y, Segars WP et al. Evaluation of quantitative imaging methods for organ activity and residence time estimation using a population of phantoms having realistic variations in anatomy and uptake. *Med. Phys.* 2009;36(2):612–619.

30. Narayanan MV, King MA, Pretorius PH et al. Human-observer ROC evaluation of attenuation, scatter, and resolution compensation strategies for Tc-99m myocardial perfusion imaging. *J. Nucl. Med.* 2003(44):1725–1734.

31. Nawfel RD, Chan KH, Wagenaar DJ, Judy PF. Evaluation of video gray-scale display. *Med. Phys.* 1992;19:561–572.

32. Badano A, Chakraborty D, Compton K et al. Assessment of display performance for medical imaging systems: Executive summary of AAPM TG18 report. *Med. Phys.* 2005;32(4):1205–1225.

33. Rockette HE, Gur D, Metz CE. The use of continuous and discrete confidence judgments in receiver operating characteristic studies of diagnostic imaging techniques. *Invest. Radiol.* 1992;27(2):169–172.

34. Swets JA, Pickett RM. *Evaluation of Diagnostic Systems: Methods from Signal Detection Theory.* New York: Academic Press; 1982.

35. Dorfman DD, Berbaum KS, Metz CE. Receiver operating characteristic rating analysis. Generalization to the population of readers and patients with the jackknife method. *Invest. Radiol.* 1992;27(9):723–731.

36. Abbey CK, Barrett HH, Eckstein MP. Practical issues and methodology in assessment of image quality using model observers. *Proc. SPIE.* 1997;3032:182 194.

37. Gifford HC, Farncombe TH, King MA. Ga-67 tumor detection using penalized-EM with nonanatomical regularizers. *Proceedings of 2002 IEEE Medical Imaging Conference.* 2003:M9-3.

38. Gifford HC, King MA, de Vries DJ, Soares EJ. Channelized hotelling and human observer correlation for lesion detection in hepatic SPECT imaging. *J. Nucl. Med.* 2000;41(3):514–521.

39. Gifford HC, Pretorius PH, King MA. Comparison of human- and model-observer LROC studies. *Proc. SPIE.* 2003;5034:112–122.

40. Myers KJ, Barrett HH, Borgstrom MC, Patton DD, Seeley GW. Effect of noise correlation on detectability of disk signals in medical imaging. *J. Opt. Soc. Am. A.* 1985;2(10):1752–1759.

41. Burgess AE. Effect of quantization noise on visual signal detection in noisy images. *J. Opt. Soc. Am. A.* 1985;2(9):1424–1428.

42. Narayanan MV, Gifford HC, King MA et al. Optimization of iterative reconstructions of Tc-99m cardiac SPECT studies using numerical observers. *IEEE Trans. Nucl. Sci.* 2002;49:2355–2360.

43. Myers KJ, Barrett HH. Addition of a channel mechanism to the ideal-observer model. *J. Opt. Soc. Am. A.* 1987;4(12):2447–2457.

44. Abbey CK, Barrett HH. Human- and model-observer performance in ramp-spectrum noise: Effects of regularization and object variability. *J. Opt. Soc. Am. A.* 2001;18:473–488.

45. Wollenweber SD, Tsui BMW, Lalush DS, Frey EC, LaCroix KJ, Gullberg GT. Comparison of Hotelling observer models and human observers in defect detection from myocardial SPECT imaging. *IEEE Trans. Nucl. Sci.* 1999;46(6):2098–2103.

46. Abbey CK, Barrett HH. Observer signal-to-noise ratios for ML-EM algorithm. *Proc. SPIE.* 1996;2712:47–58.
47. King MA, de Vries DJ, Soares EJ. Comparison of the channelized Hotelling and human observers for lesion detection in hepatic SPECT imaging. *Proc. SPIE.* 1997;3036:14–20.
48. Gifford HC, King MA, Smyczynski MS. Accounting for anatomical noise in SPECT with a visual-search human-model observer. *Proc. SPIE.* 2011;7966:79660H1–8.
49. Obuchowski NA, Rockette HE. Hypothesis testing of the diagnostic accuracy for multiple diagnostic tests: An ANOVA approach with dependent observations. *Commun. Stat. Simul. Comput.* 1995;24:285–308.

16

Monte Carlo Method Applied in Nuclear Medicine Dosimetry

Michael Ljungberg

CONTENTS

Dosimetry is the underlying concept generally used to relate imparted energy by radiation exposure in a volume of tissue to a potential biological effect. This effect can be stochastic in nature, such as the induction of cancer after a substantial time after the exposure, or deterministic where particular damages and cell death can be predicted after reaching certain levels of radiation exposure. The basic unit commonly believed to relate an amount of imparted energy to a biological effect (or the risk for an effect) is the absorbed dose expressed as the mean imparted energy in a mass element (J/kg) of the tissue. The process of calculating the absorbed dose is very complicated since it involves several types of coupled photon and charged-particle interactions in a radiation transport that has a stochastic nature. The absorbed dose cannot therefore directly be measured *in vivo* and, as a consequence of this, all results are really estimates with relatively large error bars. The Medical Internal Radiation Dose Committee of the Society of Nuclear Medicine introduced a formalism for calculating the absorbed dose for medical use of radiopharmaceuticals in the late 1960s [1] and have since then written several publications in the area. The formalism is described by the general equation

$$\bar{D}_{rr} = \tilde{A}_{rs} \cdot S_{rr \leftarrow rs} \tag{16.1}$$

where $\bar{D}$ is the mean absorbed dose to a target volume, $\tilde{A}$ is the cumulated activity (e.g., total number of disintegrations during a time interval) in the source volume, and S is a factor that defines the mean absorbed dose to the target volume per unit cumulated activity in the source volume. For internally administered radionuclides, the cumulated activity depends on the physical half-life of the radionuclide and biological half-life determined by the biokinetic behavior of the radionuclide (or radiopharmaceutical). The cumulated activity can therefore be calculated from

$$\tilde{A} = A_\mathrm{o} \int_0^\infty f(t)\,\mathrm{d}t \tag{16.2}$$

where A_o is the initial activity administered at injection time and $f(t)$ is a time-dependent function. Mostly, $f(t)$ is a combination of one or several exponential functions that, in principle, is specific for each individual and radiopharmaceutical. The S value can be described as

$$S_{r_T \leftarrow r_S} = \frac{n \cdot E \cdot \phi_{r_T \leftarrow r_S}}{m_{r_T}} \tag{16.3}$$

where n and E are the number of particles emitted per nuclear transition and the energy of the particles, respectively, and m is the mass of the target volume. The term ϕ is called the absorbed fraction and defines the fraction of the energy emitted from the source volume that will be absorbed in the target volume. When broken down into its smallest constituents, Equation 16.1 becomes

$$\bar{D}_{r_T} = \frac{\sum\limits_{r_S}\left(\tilde{A}_{r_S} \cdot \sum\limits_{i} n_i \cdot E_i \cdot \phi_i(r_T \leftarrow r_S)\right)}{m_{r_T}} \tag{16.4}$$

where the absorbed fraction $\phi(r_T \leftarrow r_S)$ is defined as the fraction of energy emitted from the source r_S and absorbed in the target r_T. The principles of the MIRD-formalism remain valid for magnitudes ranging from organ levels down to cellular levels but are most commonly used in the calculation of the mean absorbed dose to whole organs. For practical purposes in dosimetry for risk estimates, the S values are precalculated using Monte Carlo calculation on a mathematical phantom.

Dosimetry in General

Dosimetry in nuclear medicine can be categorized into two areas: (1) dosimetry used for risk estimations when regarding the induce of late stochastic biological effects, such as cancer, due to radiation when using radiopharmaceuticals for diagnostic imaging, and (2) dosimetry for planning of radionuclide therapy. The first of the two area deals with events of such a nature that is so unlikely that only large populations can be considered. In this application, the S values are therefore obtained from a mathematical representation of a reference man (or woman) with the aim of representing an average of all individuals in a population. The second application for dosimetry is aimed to determine the maximum tolerated dose to normal organs to prevent toxicity when patients are treated for cancer with radiation emitted from administered radionuclides. At these high levels of absorbed doses, the damages are relatively predictable and therefore regarded as deterministic in its nature. The estimation for the risk of stochastic cancer induction is often made prior to a release of a radiopharmaceutical by evaluating biokinetic data from patients included in clinical trials. From these evaluations, it is assumed that the biokinetics for the persons involved in the trials are likely to represent the particular population. Monte Carlo calculated S values have been tabulated [2] for several stylized reference phantoms ranging from male and female adult phantoms down to infants [3]. These phantoms, explained in more detail in Chapter 3, have been developed on the basis of reference data and on simple geometrical primitives (spheres, ellipsoids, planes, etc.). The computer codes MIRDOSE3 [4] and OLINDA [5] are available to make the calculations, described by Equation 16.4, in an easy and efficient way. When planning for radionuclide therapy, however, it is the individual that is in focus and therefore the geometry used in the dose calculation and the activity distribution should represent that individual. In either way, results obtained from Monte Carlo calculations are essential.

Organ-Based 2D Dosimetry

The activity in an organ at different time points is mostly measured by a scintillation camera either as a series of planar images or by SPECT. The planar imaging methodology has known limitations in that it is only a 2D imaging method leaving the source depth unresolved. In the attenuation compensation step, this can partly be compensated for if using the conjugate-view quantification method in which the calculation of the geometrical mean of opposite projections somewhat cancels out the source depth dependence on photon attenuation leaving the attenuation only as a function of the total patient thickness. This can then be measured by transmission measurements using an external radiation source on the opposite side of the patient.

Transmission images can today also be calculated from CT scout-view acquisitions and serve as a map of the patient thickness that is necessary in the conjugate-view method [6]. Contribution from over- and underlying activity can, however, seriously impact the accuracy in the activity quantitation and needs to be addressed [7]. Furthermore, if the activity is heterogeneously distributed within the organ, this is sometimes difficult to resolve. From planar images, data can be extracted using a region-of-interest analysis as a function of time to be used to fit a time–activity curve and calculate the area under this curve, which represents the total number of disintegrations, $\tilde{A}$, during the time interval. Since no depth information is available the organ volumes cannot be resolved and therefore 2D images-based dosimetry usually is based on S values of a reference phantom. To partly compensate for differences in geometries (phantom/patient), the masses of the patient organs (possible to estimate from measurement in CT images) can be used to rescale the standard S values for the reference man. However, the cross-dose between organs will still represent the organ distribution in the mathematical phantom.

We have developed at Lund University a dosimetry package, called LundADose [8], that performs both 2D and 3D dosimetry. Included in the 2D quantification procedure is scatter correction based on an inverse-filtering procedure where Monte Carlo simulated kernels are used, including both the distribution of scatter and events caused by septum penetration in the collimator. The kernels have been precalculated using the SIMIND code (Chapter 7) and tabulated as function of the source depth in a uniform elliptical water phantom. Figure 16.1 shows an example of the improvement for ^{111}In images (245 keV) and ^{90}Y bremsstrahlung images (105–195 keV) when using this method.

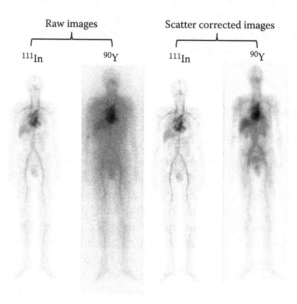

FIGURE 16.1
An example illustrating how accurate Monte Carlo-calculated filtering methods can improve image quality event for more complex imaging, such as bremsstrahlung imaging. Examples are given for ^{111}indium and ^{90}yttrium (bremsstrahlung).

Organ-Based 3D Dosimetry

The MIRD-formalism has been an accepted tool for calculating the absorbed dose also for the purpose of estimating the outcome in a radionuclide therapy. As stated above, the common way of using the tabulated S values for the reference phantoms is generally considered to be inaccurate, and patient-specific dosimetry, based on multiple SPECT measurements of the activity distribution followed by some type of 3D calculations of the absorbed dose, is promoted. The reason is that the SPECT methodology allows for the reconstruction of a set of 2D images representing specific sections (slices) of the patient. The images therefore resolve the depth and a properly defined volume-of-interest (VOI) can exclude over- and underlying activity. Figure 16.2 shows a schematic of the different steps in 3D dosimetry.

Activity Quantitation

Image reconstruction with iterative methods has increased in the recent years and with this method better compensation for scatter and attenuation is achieved. Often, energy window-based scatter corrections are used but

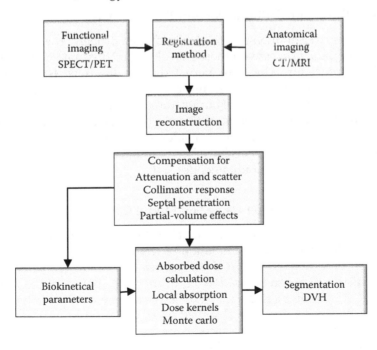

FIGURE 16.2
Flowchart describing the basic steps toward 3D-based dosimetry. Often, the registration problem is solved by a hybrid SPECT/CT system (or PET/CT system) and the activity quantification methods are included in an iterative reconstruction algorithm.

there also exists a method for compensation of scatter and collimator penetration that uses precalculated kernels obtained from Monte Carlo simulation. The ESSE method [9] is one such analytical approach. In dosimetry for radionuclide therapy, several isotopes emit photon of high energies that will not be stopped by the collimator. Also, bremsstrahlung imaging has gained increased interest due to new treatment modalities, such as ^{90}Y-labeled antibody high-dose myeloablative treatments and radio-embolization using μ-spheres. In order to be able to quantify activity, energy window-based method does not work because of the lack of a distinct photo-peak. However, analytical methods for scatter and collimator response compensation have shown to be possible if these effects can be properly modeled by Monte Carlo simulations and from this information be incorporated into a compensation method. Minarik et al. have shown from experimental measurements [10] and simulations of ^{90}Y and in a feasibility study [11] that the ESSE method in combination with proper Monte Carlo-based kernels for ^{90}Y can produce quantitative accurate bremsstrahlung images. These images can be of great importance when verifying the planning of a radionuclide treatment! In the last few years, research has focused on using Monte Carlo algorithm for scatter compensation directly in the reconstruction method [12].

Absorbed Dose Calculation with Point-Dose Kernels

If a 3D activity distribution can be obtained from properly corrected SPECT images, it is possible to convert these into absorbed dose rate images using a convolution procedure where precalculated point-dose kernels [13] described the radial distribution of energy from a charged particle in a uniform media. This method is generally fast if Fourier transform methods are used but the method has a general limitation in that the kernels are spatially invariant. Thus, it is not possible to take into consideration large dose-gradient close to boundaries. Point-dose kernels have been tabulated as a function of energy or decay scheme for many radionuclides [14].

Absorbed Dose Calculation Using Full Monte Carlo Simulation

The most accurate way of obtaining an absorbed dose distribution is to perform a full 3D Monte Carlo simulation of the radiation transport (see, e.g., [15]). From a registered SPECT/CT study, both the patient geometry and the activity distribution can be used as input for a calculation of the radiation transport. It is assumed that each voxel emits a number of particles proportional to the voxel value. The absorbed energy deposit in nearby voxels is then scored as the particle interacts along its trajectory and thereby loose energy. In principle, this calculation is equal to Equation 16.1 if one voxel is regarded as the source volume and all other voxels target volumes. In practice, this calculation is made on demand and no S values are tabulated. Only the end results, that is, the 3D absorbed energy distribution is finally stored

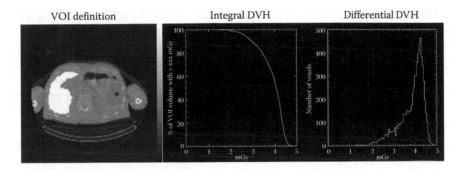

FIGURE 16.3
Two different types of dose–volume histograms (differential and integral) obtained from voxel-based absorbed doses within a volume of interest. The differential histogram shows the frequency of absorbed dose (voxel values) within a particular range. The integral dose–volume histogram describes the probability that a voxel within the volume of interest will have a value less or equal to a certain absorbed dose.

for further processing. An absorbed dose rate image can be obtained by normalizing scored energy in all voxels to each voxel mass. The main advantage with this type of calculation is that heterogeneities can be included and that the absorbed dose is now calculated on a voxel level. Thus, it is possible to obtain a distribution of regional absorbed doses within a VOI. Furthermore, the accuracy in the segmentation of the whole-organ volume is not critical because the absorbed dose is now not dependent on the definition of full-organ volume as is the case when using *S* values. Figure 16.3 shows an image where an ROI has been defined for the liver and two different types of dose–volume histograms (integral and differential) are presented.

Limitations with 3D Dosimetry

Dosimetry based on multiple 3D SPECT images in combination with explicit Monte Carlo simulation is a natural improvement over 2D planar methods using precalculated *S* values. However, there are several fundamental issues that still make absorbed dose calculation for radionuclide therapy. It must be remembered that a SPECT image does not represent the true activity distribution and will probably never do. This is due to the limited spatial resolution of the imaging modality. In many cases, the spatial resolution produces a more blurred image than the blurring effect caused by the trajectories of the secondary electrons. Therefore, a full Monte Carlo calculation might not be necessary in those cases where the pathlengths of the electrons are less than the expected spatial resolution in the images. The partial-volume effect caused by the limited resolution generally also results in an underestimation of the activity concentration in small volumes due to spill-out effect. This directly affects the absorbed dose estimation on a voxel level. Depending on the reconstruction parameters in an iterative procedure, the image noise can be quite large in the SPECT images. This will affect the distribution of the absorbed dose (DVH) within a volume. The iterative methods also tend to be nonlinear regarding the necessary number of iterations for a given result. For large activity volumes, less number of iterations is needed to reach convergence but for small volumes, such as tumors, a larger iteration number is necessary. Thus, care must be taken when evaluating and interpreting the results.

Finally, even if all problems, described above, are solved, it will not be possible to obtain information about the activity distribution within a voxel. It is generally believed that there is a heterogeneous uptake of the radiopharmaceutical on a small-scale tissue level. This means that the energy release from radioactive decay is most likely to not be evenly distributed within the voxel volume. This might therefore underestimate local absorbed doses since unexposed tissue masses are included in a 3D absorbed dose calculation based on SPECT. Therefore, in the future, it is likely that tissue-specific small-scale models, based on Monte Carlo modeling with resulting specific S values, will complement the macroscopic SPECT activity measurements.

References

1. Loevinger R, Berman M. A schema for calculating the absorbed dose from biologically distributed radionuclides: MIRD pamphlet no. 1. *J. Nucl. Med.* 1968; 9:7–14.
2. Stabin MG. RADAR—The Radiation Dose Assessment Resource. http://www.doseinfo-radar.com/.
3. Xu XG, Eckerman KF. *Handbook of Anatomical Models for Radiation Dosimetry.* Boca Raton, FL: CRC Press/Taylor & Francis Group; 2009.
4. Stabin MG. MIRDOSE: Personal computer software for internal dose assessment in nuclear medicine. *J. Nucl. Med.* 1996;37(3):538–546.
5. Stabin MG, Sparks RB, Crowe E. OLINDA/EXM: The second-generation personal computer software for internal dose assessment in nuclear medicine. *J. Nucl. Med.* 2005;46(6):1023–1027.
6. Minarik D, Sjögreen K, Ljungberg M. A new method to obtain transmission images for planar whole-body activity quantification. *Cancer Biother. Radiopharm.* 2005;20(1):72–76.
7. Sjogreen K, Ljungberg M, Strand SE. An activity quantification method based on registration of CT and whole-body scintillation camera images, with application to 131I. *J. Nucl. Med.* 2002;43(7):972–982.
8. Sjögreen K, Minarik D, Ljungberg M. The LundADose method for activity quantification and absorbed dose assessment in radionuclide therapy. *Cancer Biother. Radiopharm.* 2005;20(1):92–97.

9. Frey EC, Tsui BMW. A new method for modeling the spatially-variant, object-dependent scatter response function in SPECT. *Conference Records of the IEEE Nuclear Science and Medical Imaging Conference*, Anaheim, CA. 1996:pp. 1082–1086.

10. Minarik D, Sjogreen Gleisner K, Ljungberg M. Evaluation of quantitative (90) Y SPECT based on experimental phantom studies. *Phys. Med. Biol.* 2008;53(20): 5689–5703.

11. Minarik D, Sjogreen-Gleisner K, Linden O et al. 90Y Bremsstrahlung imaging for absorbed-dose assessment in high-dose radioimmunotherapy. *J. Nucl. Med.* 2010;51(12):1974–1978.

12. Dewaraja YK, Ljungberg M, Fessler JA. 3-D Monte Carlo-based scatter compensation in quantitative I-131 SPECT reconstruction. *IEEE Trans. Nucl. Sci.* 2006;53(1):181.

13. Giap HB, Macey DJ, Bayouth JE, Boyer AL. Validation of a dose-point kernel convolution technique for internal dosimetry. *Phys. Med. Biol.* 1995;40(3):365–381.

14. Prestwich WV, Nunes J, Kwok CS. Beta dose point kernels for radionuclides of potential use in radioimmunotherapy. *J. Nucl. Med.* 1989;30:1036–1046.

15. Dewaraja YK, Wilderman SJ, Ljungberg M, Koral KF, Zasadny K, Kaminiski MS. Accurate dosimetry in 131I radionuclide therapy using patient-specific, 3-dimensional methods for SPECT reconstruction and absorbed dose calculation. *J. Nucl. Med.* 2005;46(5):840–849.

Index